SEMICONDUCTORS AND SEMIMETALS

Edited by *R. K. WILLARDSON*
ENICHEM AMERICAS INC.
PHOENIX, ARIZONA

ALBERT C. BEER
BATTELLE COLUMBUS LABORATORIES
COLUMBUS, OHIO

VOLUME 26

III–V Compound Semiconductors and Semiconductor Properties of Superionic Materials

ACADEMIC PRESS, INC.
Harcourt Brace Jovanovich, Publishers

Boston San Diego New York
Berkeley London Sydney
Tokyo Toronto

ACADEMIC PRESS, INC.
1250 Sixth Avenue, San Diego, CA 92101

United Kingdom Edition published by
ACADEMIC PRESS, INC. (LONDON) LTD.
24-28 Oval Road, London NW1 7DX

Library of Congress Cataloging-in-Publication Data

Semiconductors and semimetals. — Vol. 1- — New York: Academic Press, 1966-
v.: ill.; 24 cm.
Irregular.
Each vol. has also a distinctive title.
Edited by R.K. Willardson and Albert C. Beer.
ISSN 0080-8784 = Semiconductors and semimetals

1. Semiconductors—Collected works. 2. Semimetals—Collected works. I. Willardson, Robert K. II. Beer, Albert C.

QC610.9.S48 621.3815'2—dc19 85-642319

ISBN 0-12-752126-7

PRINTED IN THE UNITED STATES OF AMERICA

88 89 90 91 9 8 7 6 5 4 3 2 1

To the memory of
Professor Zou Yuanxi
(1915–1987)

Contents

Chapter 4 **Semiconductor Properties of Superionic Materials**

Yu. Ya. Gurevich and A. K. Ivanov-Shits

Contributors

Numbers in parentheses indicate the pages on which the authors' contributions begin.

PALLAB K. BHATTACHARYA (143), *Department of Electrical Engineering and Computer Science, The University of Michigan, Ann Arbor, Michigan*

SUNANDA DHAR (143), *The Institute of Radiophysics and Electronics, University of Calcutta, Calcutta, India*

YU. YA. GUREVICH (229), *Institute of Electrochemistry, USSR Acad. Sci., Moscow, USSR*

ANDREW T. HUNTER (99), *Hughes Research Laboratories, 3011 Malibu Canyon Road, Malibu, California 90265*

A. K. IVANOV-SHITS (229), *Institute of Chrystallography, USSR Acad. Sci., Moscow, USSR*

HIROSHI KIMURA (99), *Hughes Research Laboratories, 3011 Malibu Canyon Road, Malibu, California 90265*

ROBERT E. LEE (99), *Hughes Research Laboratories, 3011 Malibu Canyon Road, Malibu, California 90265*

HARVEY WINSTON (99), *Hughes Research Laboratories, 3011 Malibu Canyon Road, Malibu, California 90265*

ZOU YUANXI (1), *Shanghai Institute of Metallurgy, Academy of Sciences of China, 865 Chang Ning Road, Shanghai 200050, China*

Preface

Few scientists in the western world have followed the extensive investigations of III–V compounds that have been undertaken in China during the past century or know that about one third of the gallium that they have been using to make GaAs for the past decade originated in China. In Chapter 1, Professor Zou Yuanxi reviews the last decade of these investigations, which were largely published in Chinese. You will note the investigators' excellent analytical work in the determination of trace impurities and their struggle to obtain high-purity GaAs without pBN crucibles, which they were not able to import. Also noteworthy was the need to construct their own equipment such as the MBE system and the manufacture of pBN crucibles for this system. The thermodynamics and kinetic analysis are thorough, as is the work on the "mobility killer" in GaAs and other defect levels such as the D-X centers in GaAlAs.

The late Professor Zou Yuanxi was the Honorary Director of the Shanghai Institute of Metallurgy where a staff of over one hundred have been investigating III–V compounds. There is an effort of similar magnitude at the General Research Institute of Nonferrous Metals in Beijing. Work in six other major institutes and universities is described. The Chinese are positioning their fledgling electronics industry to export equipment with all kinds of GaAs devices, ranging from LEDs and lasers to microwave devices and integrated circuits, in addition to supplying their huge, potential, internal market. This chapter emphasizes the skill they have in metallurgy and how well they are adapting it to the semiconductor industry. It is also worth noting in 1987 Chinese researchers successfully grew a GaAs crystal from the melt under microgravity conditions. The next decade or two could prove as rewarding for them as has been the past decade for the Japanese.

Variations in the activation efficiency of implanted ions across a GaAs wafer cause variations in threshold voltages of devices, and this lack of uniformity limits the complexity of integrated circuits that can be fabricated with reasonable yields. Lower dislocation densities are expected to lead to more uniform characteristics across a wafer. One useful technique for reducing or eliminating dislocations in GaAs is substituting indium for a small part of the gallium. The addition of InAs to reduce dislocation densities and improve GaAs substrates for ion-implanted ICs is the subject of Chapter 2.

Harvey Winston and co-authors review the subject of dislocations, their formation during bulk crystal growth and epi layer formation and their effects on crystal structure, electrical and optical properties, as well as device characteristics. They discuss spatial fluctuations in stoichiometry, arsenic excesses and clusters as they are influenced by dislocations. The use of both InAs and InP additions permits some degree of control of the lattice constant as well as the dislocation density of the growing crystal. By reducing threshold voltage variations to less than 15 mV, a 16-K bit SRAM circuit with 100,000 enhancement/depletion mode FETs appears possible.

In Chapter 3, Pallab Bhattacharya and Sunanda Dhar discuss other kinds of defects in III–V compounds. Deep-level traps in molecular-beam epitaxial (MBE) layers of GaAs and GaAlAs, their formation and their characteristics are described. Heterostructures and synthetic modulated structures are needed to provide high-performance electronic, opto-electronic and optical devices. Advances in MBE growth have created a new era of device technology based on precise thickness and composition control, and the ability to grow modulated structures. Thin defect-free single-crystal films are required for devices. The significance and characteristics of deep states in III–V semiconductors are emphasized. Trapping and associated effects on heterostructure performance as well as commonly observed deep levels are reviewed. The dominant electron trap in GaAlAs is treated in detail.

Superionic materials, which are the subject of Chapter 4, comprise a special class of solids that has become a subject of ever-increasing interest. One of the most important features of superionic conductors is their anomalously high ionic conductivity, which is comparable to that in concentrated solutions of strong electrolytes. An explosion of investigations of superionic materials dates back two decades to when a solid electrolyte was designed to provide an electrochemical power source to be used in space.

Most superionic materials have mixed ionic and electronic conductivity. The electronic component possesses the nature of a semiconductor. In this chapter the composition, structure, and nonstoichiometry of superionic materials are described. Thermoelectric, current-voltage, ambipolar diffusion, ionic and electronic transport, opto-electronic, photovoltaic and other phenomena are discussed.

Electrochemical capacitors (ionistors) have small geometric size, high energy storage capability, and can store the charge for long periods of time. Yu. Ya. Gurevich and A. K. Ivanov-Shits also review a new branch of electronic instrumentation called microionics, as well as electrochromic devices.

R. K. Willardson
A. C. Beer

CHAPTER 1

III–V Compounds

Zou Yuanxi (Chou Yuanhsi)

SHANGHAI INSTITUTE OF METALLURGY
ACADEMIC SINICA
SHANGHAI, CHINA

ISBN 0-12-752126-7

I. Introduction

Since the presentation of my paper on "Some Aspects of the Recent Work on III–V Compounds in China" at the Vienna Conference on Gallium Arsenide and Related Compounds in September 1980 (Zou Yuanxi, 1981), much work has been done in this field in a number of institutes, universities and factories in China. As these investigations are largely written in Chinese, they are not easily accessible to scientists of the Western World. Therefore a more complete, up-to-date review seems to be in order.

The work on GaAs in China dates from the early sixties. The first Symposium on Gallium Arsenide and Related Compounds was, however, held in Shanghai as late as 1972, due to the interruption by the Turmoil Period. The Second Symposium was held in 1977 in Liuzhou of Guangxi Autonomy. In 1979, the First Symposium on Semiconducting Compound Materials, Solid Microwave Devices and Opto-electronic Devices was held in Nanjing, and the Second Symposium in 1982 in Kunming of Yunnan Province. Both the Third and the Fourth Symposia on Gallium Arsenide and Related Compounds were held in Shanghai, in 1981 and 1983 respectively.

The present review is largely based on the papers selected from those presented at the previously mentioned Kunming and Shanghai Symposia, held in the years 1981 and 1983. These papers reflect a rapid growth of research interest in China in this important field of III–V compounds. Consequently, there has been some difficulty in the selection of materials for this review within the scope of a chapter. Despite this difficulty, we have managed to include in this review some results obtained in several institutes and universities—either recently published in the periodicals or to be published—to give a more or less complete picture of the recent advances in III–V compounds in China.

Part II begins with the growth of bulk GaAs crystals by both the LEC CZ and HB methods, with a special emphasis on undoped semi-insulating materials in view of their importance in the fabrication of GaAs integrated circuits. This is followed by the synthesis and crystal growth of InP, and then a brief discussion on GaP, InAs and InSb. Part III deals with epitaxial growth, including the recent advances in MOVPE and MBE. Part IV discusses impurity effects, space charges and deep levels. The discussion on deep levels, together with surface properties and heterogeneous equilibria among phases, is continued in Part V. In Part VI, the recent work on characterization and photoluminescence, as well as on trace impurities analysis, is presented. Finally, Parts VII, VIII and IX discuss the recent progress in technology, microwave devices and optoelectronic devices respectively, including some preliminary work on GaAs integrated circuits.

For the sake of brevity, the following abbreviations will be used throughout this chapter for the respective institutes and universities:

(1) GINM—General Research Institute of Nonferrous Metals;
(2) HSRI—Hebei Semiconductor Research Institute;
(3) IS—Institute of Semiconductors;
(4) NSR—Nanjing Solid State Devices Research Institute;
(5) SIM—Shanghai Institute of Metallurgy;
(6) BJU—Beijing University;
(7) FDU—Fudan University;
(8) JLU—Jilin University.

II. Bulk Crystals

The research interests in bulk crystals are growth of low dislocation materials, direct synthesis in the high pressure puller, semi-insulating materials, effects of different dopants and high-purity crystals, etc. The materials concerned include GaAs, InP, GaP, InAs and InSb.

1. GaAs

a. Doped GaAs

The research work on GaAs has been largely concerned with the growth of low dislocation crystals. The dopant effect of silicon is utilized by Yin Qingmin *et al.* (1981) of GINM for growing GaAs crystals weighing 400 g and with EPD $\leq 100\ \mathrm{cm}^{-2}$ by the HB 3*T*-zone method. The attainment of this low dislocation density depends on the combination of a series of factors, including a tilting HB furnace, special precautions against wetting between the melt and the quartz boat, a growth direction close to $\langle 111 \rangle_B$ in the (110) zone, a suitable thermal gradient, a strictly controlled arsenic pressure and, most important of all, a Si doping concentration of about $10^{18}\ \mathrm{cm}^{-3}$ at the seed end. Some typical distributions of dislocations along the axis of the crystals are listed in Table I.

A similar work on the growth of low dislocation Si-doped GaAs has been carried out by Mo Peigen *et al.* (1985a) of SIM. In order to obtain a slightly convex growing interface, a modified horizontal gradient-freezing furnace with a quasi-double parabolic temperature profile was used. According to these investigators, heavy doping is not strictly necessary for growing low dislocation ingots, although doping with Si offers a great advantage in increasing the yield strength of the crystals. Other precautions taken by Mo Peigen *et al.* included preventing wetting between the melt and the crucible, choosing suitable seed orientations like $\langle 111 \rangle_B$ and $\langle 211 \rangle_B$ and maintaining

TABLE I

DATA ON ZERO DISLOCATION Si-DOPED GaAs GROWN BY THE HB 3T-ZONE METHOD[a]

Ingot No.	Distance from seed end mm	Dislocation cm^{-2}	n 10^{18} cm^{-3}	μ cm^2 V^{-1} s^{-1}
81-Si-1	20	300	0.966	2430
	50	≤ 100	0.921	2480
	165	< 100	2.66	1910
	220	< 100	5.73	1220
81-Si-2	7	920	1.01	2490
	30	250	0.958	2330
	55	≤ 100	1.26	2300
	195	< 100	5.2	1490
	225	0	5.6	1200
81-Si-6	14	≤ 100	0.925	2450
	45	< 100	1.01	2370
	185	0	2.99	1830
81-Si-7	20	500	0.987	2410
	40	< 200	0.947	2410
	200	< 100	4.03	1540

[a] From Yin Qingmin *et al.* (1981). Reproduced with permission of the authors.

a strictly controlled arsenic pressure (610 ± 2°C). Further advantages of the gradient freeze technique claimed by Mo Peigen *et al.* are a lower compensation ratio and an improvement in homogeneity of the crystals. Some typical data are shown in Table II.

On the other hand, Chu Yiming *et al.* (1981) of IS have studied the influence of crystalline polarity on the facet effect in heavily doped GaAs crystals and found that the distribution ratios for Te in the two kinds of facets are significantly different from each other. The reason for this difference in behaviour is briefly discussed.

b. SI GaAs

Both Cr-doped and undoped SI GaAs crystals have been grown by the LEC CZ and the horizontal methods. For the sake of brevity, only the results on undoped SI SaAs will be discussed.

An attempt to grow undoped SI GaAs by the HB method has been made by Tan Lifang *et al.* (1982) of SIM. These investigators used a specially treated quartz (STQ) boat with a view to minimizing Si contamination. Most

TABLE II

DATA ON LOW DISLOCATION Si-DOPED GaAs GROWN BY THE GRADIENT FREEZE METHOD[a]

Ingot No.	Seed orientation	Dislocation cm^{-2}	n 10^{18} cm^{-3}	μ $cm^2\ V^{-1}\ s^{-1}$
GF81-1 H	⟨211⟩B	0	0.163	2560
M(T)		0	1.83	2150
GF81-2 H	⟨211⟩B	<500	1.05	2230
T		<500	5.17	1860
GF81-3 H	⟨111⟩B	<800	0.408	3020
T		0	5.95	1600
GF81-5 H	⟨111⟩B	930	1.18	2290
M		0	1.74	2110
T		<500	7.6	860
GF81-6 H	⟨111⟩B	620	1.29	2100
M(T)		0	3.51	1690

[a] From Mo Peigen *et al.* (1983). Reproduced with permission of Elsevier Science Publishers, The Netherlands.

of the crystals grown by this technique were, however, lightly doped with Cr, and the results will not be discussed further. It will suffice to say that Si contamination can actually be reduced without creating a corresponding increase in the oxygen content, by the use of the STQ boat, as can be seen from Table III. One crystal grown by Tan Lifang *et al.* in an STQ boat is, however, undoped, and, yet, the surface resistance after heat treatment for 90 min at 775°C remains high ($\geq 3.5 \times 10^9\ \Omega/\square$). This result is encouraging, as it is comparable to the average of 5 crystals ($7.8 \times 10^9\ \Omega/\square$) grown in a quartz boat and doped with $(5.0\text{–}6.6) \times 10^{-2}\%$ Cr.

TABLE III

RELATIVE INTENSITY OF ION CURRENTS FROM Si AND O IN BOAT-GROWN GaAs[a]

	STQ boat-grown GaAs		Quartz boat-grown GaAs	
Element	BD-7	BD-5	BD-80-27	BD-80-3
16^{O^+}	0.00133	0.00404	0.014	0.004
28^{Si^+}	0.0020	0.013	0.090	0.33

[a] From Tan Lifang *et al.* (1982). Reproduced with permission of Shiva Publishing Limited, England.

A modified low-pressure LEC method called the LEF (liquid encapsulated freeze) method adopting certain features of the LEK technique has been employed by Mo Peigen *et al.* (1985b) for growing undoped low dislocation SI GaAs crystals. By synthesizing in the furnace through the use of an arsenic injection cell and by using the LEF technique, crystals of up to 60 mm diameter have been grown from quartz crucibles, and the crystals grown under properly-controlled conditions exhibit a high resistivity ($>10^7\ \Omega$ cm) from top to bottom. The best crystals having a diameter of 30 mm grown by this novel technique have an average EPD of $<7.5 \times 10^3$ cm^{-2}, and an EPD as low as 10^2 cm^{-2} within selected areas. A typical EPD distribution along a wafer diameter in ⟨110⟩ direction for a low EPD wafer is shown in Fig. 1. In comparison with that of a wafer from an ingot grown by the conventional LEC method, the radical variation of EPD is seen to be less pronounced in the low EPD wafer.

The results of resistivity and thermal stability measurements on typical samples are given in Table IV.

From Figs. 2 and 3, it can be seen that the radical profiles of EPD and leakage current I_L are opposite to each other for the high dislocation wafer, in agreement with those reported by Miyazawa *et al.* (1982), while both are approximately W-shaped for the low disclocation wafer. The reason for this difference will be discussed in Section 9b. The distribution of sheet resistivity over a low dislocation wafer of (100) orientation is quite uniform, as can be seen from Fig. 4.

The LEC growth of low dislocation density SI GaAs continues to draw intensive attention from a number of institutes. The diameter of the crystals

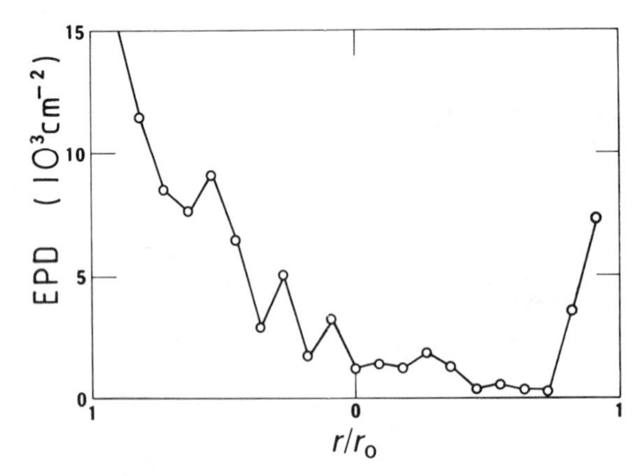

FIG. 1. Typical EPD distribution along a low EPD wafer diameter in ⟨110⟩ direction, r_0 = 13 mm. [From Mo Peigen *et al.* (1985), reproduced with permission of Shiva Publishing Limited, England.]

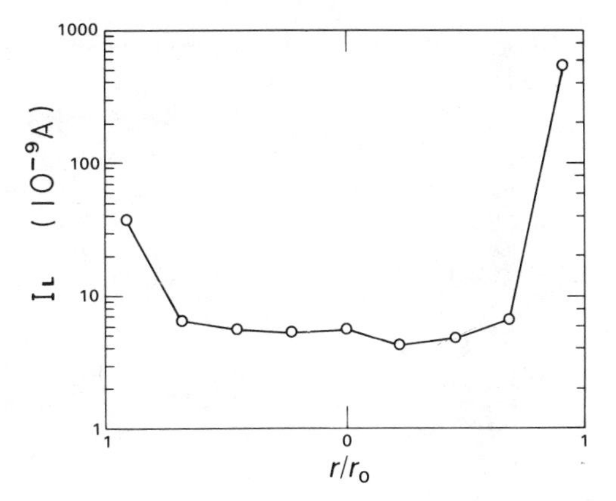

FIG. 2. Typical leakage current distribution along a low EPD wafer diameter in ⟨110⟩ direction, r_0 = 13 mm. [From Mo Peigen *et al.* (1985), reproduced with permission of Shiva Publishing Limited, England.]

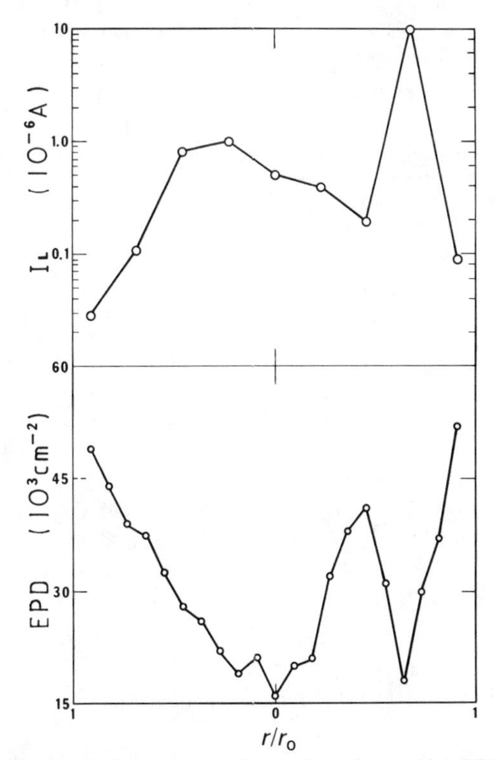

FIG. 3. Typical EPD and leakage current distribution along a high EPD wafer diameter in ⟨110⟩ direction, r_0 = 13 mm. [From Mo Peigen *et al.* (1985), reproduced with permission of Shiva Publishing Limited, England.]

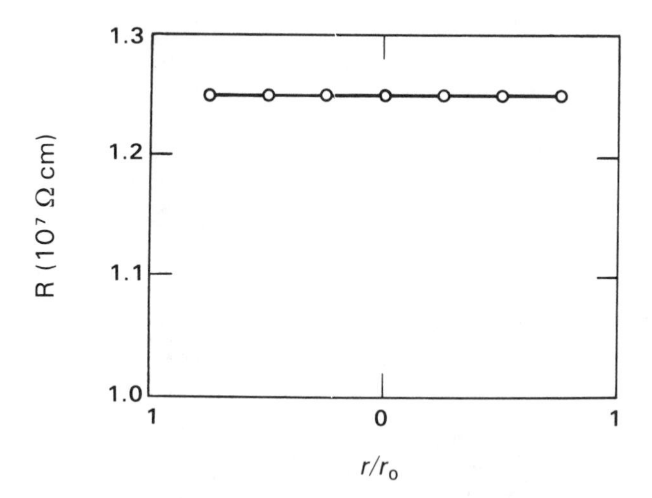

FIG. 4. Sheet resistivity along a low EPD wafer diameter in ⟨110⟩ direction, $r_0 = 13$ mm. [From Mo Peigen *et al.* (1985), reproduced with permission of Shiva Publishing Limited, England.]

TABLE IV

DATA ON RESISTIVITY AND THERMAL STABILITY FOR TYPICAL SAMPLES[a]

		ρ or ρ_s	
Sample	EPD cm^{-2}	As grown	Post annealing 800°C 20 min
LSI 49	$<10^3$	4.8×10^{10} Ω/□	3.1×10^8 Ω/□
LSI 51	$<10^3$	3×10^7 Ω cm	3×10^7 Ω cm

[a] From Mo Peigen *et al.* (1985). Reproduced with the permission of Shiva Publishing Limited, England.

is typically ~50 mm. Lin Lanying *et al.* (1985)[1] of IS and Wang Yonghong *et al* (1985)[2] of GINM have studied the beneficial effects of a low thermal gradient environment and In-doping of up to 1.6% on the reduction of dislocation density in LEC SI-GaAs. The dislocation density obtained by Wang Yonghong *et al.* in the central part of a typical wafer (amounting to ~60% of wafer area) is 10^2–10^3 cm^{-2}. Similar results have been obtained by Mo Peigen and Wu Ju (1985)[3] by using In-doping.

[1] *Abstracts, 1985 Emei Symposium on GaAs and Related Compounds,* p. 3.
[2] *Ibid.*, p. 7.
[3] *Ibid.*, p. 8.

Efforts have also been made to improve the quality of n-type GaAs by further purifying the raw materials and taking precautions to minimize contamination during synthesis and crystal growth. An outstanding result of this work is the successful growth of two ingots by the HB method, of which four wafers exhibit μ_{77K} of 3.85×10^4 to 5.86×10^4 cm^2/V s after proper annealing (Zou Yuanxi, SIM, 1976; Zou Yuanxi and Lu Naikun, Shenyang Smelter, 1977), although these investigators have not yet been able to reproduce this result on other ingots.

Shi Huiying *et al.* (1986, unpublished) of SIM have successfully developed a new purification process for arsenic. The superpure arsenic obtained has a S content <0.01 ppm, or below the detection limit of the radioactive tracer method. Another common impurity Si is found to be absent in qualitative spectroscopic analysis. As to C, the gas chromatographic analysis indicates a 75–80% reduction in concentration compared with the commercially available 6 "9" product.

As can be expected, it would be feasible now to prepare superpure GaAs crystals by the use of this arsenic in conjunction with B_2O_3 purified by a process developed by Fan Zhanguo *et al.* (1975, unpublished) of SIM. But no such effort will be made in the foreseeable future.

Besides, Zou Yuanxi *et al.* (1984, unpublished) and Zhao Shilong *et al.* (1985, patent pending), both of SIM, have succeeded in developing, respectively, a convenient process for purifying quartz wares and a simple process for obtaining 6 "9" H_2 comparable to the Pd-diffuser gas through the purifying action of a Fe–Ti alloy. These processes should be useful in the epitaxial growth of GaAs and related compounds.

It is the belief of the present writer that the combination of high purity and low dislocation will result in the production of high quality SI GaAs wafers suitable for the fabrication of large scale integrated circuits.

2. InP

The work on InP may be divided into two parts: (i) synthesis of polycrystalline InP by gradient freezing and direct synthesis in the high-pressure puller, and (ii) crystal growth by the LEC technique.

a. Synthesis of Polycrystalline Material

The synthesis of polycrystalline InP by the gradient freezing method is described by Tan Litong *et al.* (1982) of SIM. The purity of the crystals is reflected by some unpublished Hall data[4]: n_{RT} 6.1×10^{15} cm^{-3}, μ_{RT} 3800 cm^2/V s; n_{77K} 4.5×10^{15} cm^{-3}, μ_{77K} 31000 cm^2/V s.

[4] Private communication from Tan Litong of SIM.

In the 1980 review paper (Zou Yuanxi, 1981), brief mention was made of the synthesis of material from its elements inside the puller at HSRI by Sun Tongnian *et al.* (1982). A similar work has now been done by Peng Zhenfu *et al.* (1982) of NSR. Both Sun Tongnian *et al.* and Peng Zhenfu *et al.* have been able to reduce Si contamination to such an extent that SI InP ingots can be obtained by the presence of as low as 0.01–0.02 w% Fe in the melt.

The results of Sun Tongnian *et al.* on the relationship between the resistivity of the SI InP crystals and the Fe content in the melt are shown in Table V. It is seen that one sample (8-12-Fe) grown by these investigators using 7 N In and 6 N P as raw materials and an Fe content of 0.03 w% in the melt has a resistivity as high as 8.35×10^7–$1.81 \times 10^8\ \Omega$ cm. The advantage of synthesis *in situ* in the puller is apparent from a comparison of the resistivities of the samples 80-43-Fe and 81-12-Fe.

A few comments on the purity of the raw materials seem to be in order. Zhang Minquan *et al* (1983) of SIM have plotted the carrier concentration versus the Si content on the basis of the data on InP crystals reported by Cockayne *et al.* (1980), as shown in Fig. 5. It can be seen that practically all the points lie within a narrow band, with the lower bound intersecting the ordinate at $n = 3.4 \times 10^{15}$ cm^{-3}, which corresponds to about 0.08 ppma. The most probable impurity responsible for this donor concentrations seems to be S.

TABLE V

RELATIONSHIP BETWEEN RESISTIVITY OF SI InP AND Fe CONTENT IN THE MELT[a]

Sample No.	Fe W%	Purity of source materials		Resistivity Ω cm	Remarks
		In	P		
79-41-Fe	0.05	6 N	5 N	10^6–10^7	Synthesis
79-51-Fe	0.025	6 N	5 N	10^6–10^7	Synthesis
79-43-Fe	0.01	6 N	5 N	Top <1 middle and tail $>10^6$	Synthesis
80-43-Fe	0.03	6 N	6 N	3–6×10^7	Polycrystalline InP
81-12-Fe	0.03	7 N	6 N	8.35×10^7– 1.81×10^8	Synthesis

[a] After Sun Tongnian *et al.* (1982). Reproduced with permission of Shiva Publishing Limited, England.

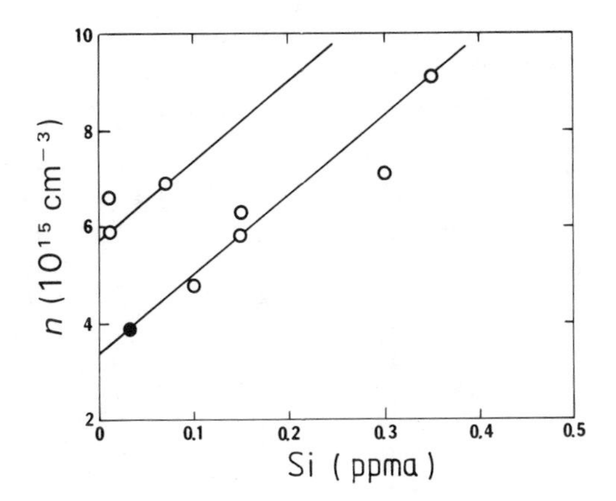

FIG. 5. Relationship between carrier concentration and Si content for bulk InP crystals based on data of Cockayne *et al.* (○), 1980, and Sun Tongnian *et al.* (●), 1982. [From Zhang Minguan *et al.* (1983), reproduced with permission of the authors.]

One experimental point based on data of Sun Tongnian *et al.* is also shown in Fig. 5. It is gratifying to note that this point lies nicely on the lower bounds of the band, reflecting the high purity of the sample. Using mass spectrographic analysis, Sun Tongnian *et al.* found the S content of this sample to be 0.06 ppma, which is sufficiently close to the 0.08 ppma estimated above. This comparison provides, therefore, adequate support to our suggestion that S is the residual donor. As regards the source of S, it seems pertinent to quote the S content of the 6 N/7 N In currently available in China, which has been reported by Zhang Minquan *et al.* to be 0.03–0.04 ppmw, based on fluorescent colorimetric analysis (see Table VII). It is interesting to find that this amount of S correspond excellently to 0.08–0.11 ppma in InP, the value estimated above. It is our opinion, therefore, that the purity of 6 N P used would be sufficient for the synthesis of high-purity InP, so far as S contamination is concerned. The comparison made in Fig. 5 also suggests that the In used by Cockayne *et al.* and that the In used by the Chinese investigators would be similar in S content, and, therefore, might have been purified by a similar process.

b. Growth of Single Crystal

The growth of InP single crystals by the LEC method with various dopants has been studied by Fang Dunfu *et al.* (1983, 1984) of SIM. The dopants used include S, Sn, Zn and Fe. The electrical properties and etch pit densities of typical InP single crystals of about 30 mm in diameter are summarized in Table VI.

TABLE VI

ELECTRICAL PROPERTIES AT 300 K AND EPD OF TYPICAL InP SINGLE CRYSTALS[a]

Sample No.	Dopant	n or p cm^{-3}	μ $cm^2/V\ s$	ρ $\Omega\ cm$	EPD cm^{-2}
IP-10-10	—	1.06×10^{16}	4290	1.38×10^{-1}	1.28×10^{4}
IP-T-44	Sn	6.37×10^{17}	2390	3.88×10^{-3}	4.1×10^{4}
IP-T-37	Sn	7.08×10^{18}	1210	7.3×10^{-4}	3.6×10^{4}
IP-S-18	In_2S_3	7.5×10^{17}	2370	3.12×10^{-3}	1.74×10^{4}
IP-S-9	In_2S_3	2.7×10^{18}	1470	1.58×10^{-3}	1.42×10^{3}
IP-S-17	In_2S_3	5.3×10^{18}	1500	7.47×10^{-4}	2.58×10^{2}
IP-S-6	In_2S_3	1.0×10^{19}	1150	5.9×10^{-4}	~0
IP-Z-3	ZnP_2	4.25×10^{17}	105	1.4×10^{-1}	3.8×10^{3}
IP-Z-26	ZnP_2	1.25×10^{18}	79.3	6.7×10^{-2}	2.4×10^{2}
IP-Z-21	ZnP_2	2.66×10^{18}	67.3	3.7×10^{-2}	~0
IP-Z-8	ZnP_2	4.34×10^{18}	54.3	2.26×10^{-2}	~0
IP-F-8	Fe_2P	3.0×10^{7}	2400	8.7×10^{7}	2.4×10^{4}
IP-F-5	Fe_2P	1.4×10^{7}	2100	2.2×10^{8}	6.0×10^{4}

[a] From Fang Dunfu *et al.* (1984). Reproduced with permission of Elsevier Science Publishers, The Netherlands.

The conclusions made by Fan Dunfu *et al.*[5] in summarizing their work on the growth of InP single crystals are as follows:

(1) By maintaining a carefully-adjusted thermal field distribution and further improving the quality of both the polycrystals and dehydrated B_2O_3 encapsulant, twin-free InP single crystals can be obtained, even with a shoulder angle of up to 54°, contrary to the experience of Bonner (1980).

(2) A comparison of S- and Sn-doped InP crystals shows that the crystal perfection and homogeneity of the former are superior to those of the latter.

(3) Dislocation-free, Zn-doped p-type InP single crystals of up to 30 mm in diameter, free from precipitates can be grown without difficulty if the carrier concentration exceeds $2 \times 10^{18}\ cm^{-3}$.

As regards the quality of SI InP crystals grown in different laboratories, it will suffice to point out that the samples IP-F-5 (Table VI) and 81-12-Fe (Table V) are similar in resistivity. It may also be pointed out that the SI InP crystals grown by Peng Zhenfu *et al.* as well as by Fang Dunfu *et al.* have sufficiently good thermal stability to make them promising for device fabrication. Besides, it is pointed out by Peng Zhenfu *et al.* that Gunn diodes and PIN photodiodes of good performance have been successfully fabricated from epilayers grown on their Sn-doped InP substrates.

[5] From Fang Dunfu *et al.* (1984). Adapted with permission of Elsevier Science Publishers, The Netherlands.

Liu Sitian *et al.* (1985)[6] of GINM have reported on the LEC growth of ⟨100⟩ Sn- and Zn-doped InP single crystals of 30–40 mm in diameter. It is shown that the thermal environment in the puller, the shouldering angle and the appearance of floating scum on the surface of the melt all play an important role in the growth of twin-free crystals.

Both Liu Xunlang *et al.* (1985)[7] of IS and Xu Yongquan *et al.* (1985)[8] of SIM have studied the effect of isoelectronic doping on the reduction of dislocation density in LEC *n*- or *p*-type InP. It is shown that Ga is a very effective isoelectronic dopant in the growth of low dislocation InP crystals and that the amount of S or Zn needed may be significantly reduced in this way.

Recent news from SIM (1986, to be published) concerns the successful growth of a particularly twin-free LEC ⟨100⟩ InP crystal, ~50 mm in diameter, although much work remains to be done to improve the reproducibility.

3. GaP, InAs AND InSb

a. GaP

The GINM remains the sole supplier of Te- and S-doped GaP crystals in China. In a recent paper, Liu Xitian *et al.* (1983) of GINM have compared the effects of Te and S on the crystal perfection of GaP ingots grown in a high-pressure puller by the direct synthesis LEC technique. It is found that the EPD for Te-doped GaP crystals of 45 mm diameter is about 5×10^5 to 10^6 cm^{-2}, while that for S-doped crystals of a similar size is generally 7×10^4 to 3×10^5 cm^{-2}, or about one order of magnitude lower, indicating the better doping effect of the latter. It is also found that the radial distribution of dislocation density is usually W-shaped, as in GaAs crystals.

This could probably lead to a similar profile for a EL2-like deep level in GaP, if it exists, as speculated by Zou Yuanxi (1983a).

According to Liu Xitian *et al.*, another difference between Te- and S-doped GaP crystals lies in the fact that "anomalous black spots" are found in the central portion of all the Te-doped GaP crystals, but have not been found in undoped or S-doped GaP. The diameter of the "anomalous black spots" can be as large as 3–4 mm. These spots are thought to have originated from the segregation of Te in the form of some complex of Te and native defects. This is considered another advantage of S as a dopant for GaP.

[6] *Ibid.*, p. 10.
[7] *Ibid.*, p. 9.
[8] *Ibid.*, p. 12.

b. *InAs*

The Emei Institute of Semiconductor Materials is currently producing InAs by the LEC method. The InAs crystals are of adequate purity, as reflected by the 77 K mobility ($>4.5 \times 10^4\ cm^2/V\ s$) reported by Wang Dingguo (1977).

Wu Jisen *et al.* (1984a) of SIM have made a serious attempt to explore the nature of the residual donor concentration in bulk InAs, which is usually around $10^{16}\ cm^{-3}$. There are two schools of thought on this problem, viz. the chemical impurity and the natural defect hypotheses. The former is advocated by Zou Yuanxi (1983b) on the grounds that in a plot of N_D calculated from literature data for bulk, VPE and LPE InAs—the shallow donor concentration versus N_A, the total acceptor concentration—the experimental points lie nicely on the same curve. Such a relationship is not found with GaAs, where separate curves are drawn for bulk, VPE and LPE materials.

In Fig. 6 (Wu Jisen *et al.*, 1984a), a plot of N_D versus R is shown, which is the ratio of the strength of secondary ionic currents for four samples with one arbitrary chosen sample as the standard, calculated by Wu Jisen *et al.* from their SIMS data. Four impurity elements, S, Na, Si and Cu, are selected because they are believed to be the principal donors in InAs. The curves in Fig. 6 indicate that Si and Na are probably the principal donors in InAs, followed by S and Cu, although the experimental points for the latter two elements are somewhat scattered. Besides, the donor nature of Cr in InAs has been suggested by Balagurov *et al.* (1976). Therefore, Wu Jisen *et al.* have concluded that the residual donor concentration of about $10^{16}\ cm^{-3}$ in bulk InAs could be accounted for by the sum of the impurities Si, Na, S, Cu and Cr.

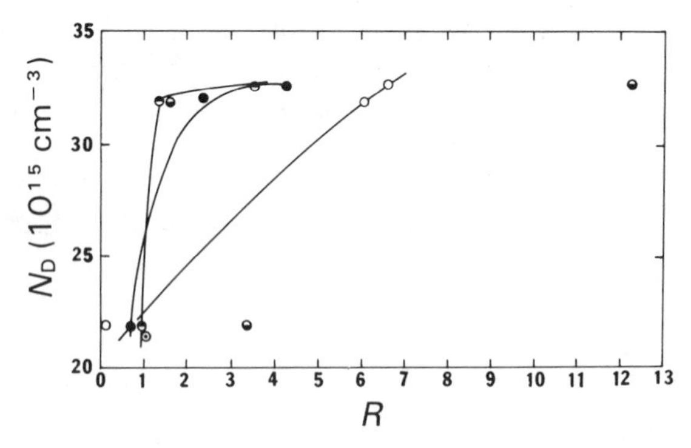

FIG. 6. Relationship between N_D and relative ionic current strength for several InAs crystals. ○ S, ● Na, ○ Si, ○ Cu, ⊙ standard sample. [From Wu Jisen *et al.* (1984), reproduced with permission of the authors.]

TABLE VII

ANALYSIS OF IMPURITIES IN In AND As[a]

Impurity	Source and concentration ppmw		Contribution to n cm^{-3}
Si	In	0.06–0.08	$4.5\text{–}6 \times 10^{15}$
S	In	0.03–0.04	6×10^{15}
	As	0.07–0.09	
Na	As	0.01	6×10^{15}
Cu	In	0.04	1.5×10^{15}
Cr	In	0.02–0.06	$0.8\text{–}2.4 \times 10^{15}$
Total			$1.88\text{–}2.19 \times 10^{16}$

[a] From Wu Jisen *et al.* (1984). Reproduced with permission of the authors.

In order to find the chief source of the above-mentioned impurities, their concentrations in the source In and As are listed in Table VII. The data indicate that the raw materials could contribute a major part of the donor impurities, although contamination during synthesis and crystal pulling might also be responsible to a certain extent.

c. *InSb*

The preparation of InSb as a semiconductor material usually consists of (i) vacuum treatment of In and Sb, (ii) synthesis and zone refining of the compound, and (iii) crystal pulling. It is the opinion of Wang Dingguo *et al.* (1981) of the Emei Institute of Semiconductor Materials, however, that the above procedure, involving the transfer of materials from one container to another several times, would inevitably result in contamination by some impurities to a certain extent. Therefore, these investigators adopt the procedure of synthesizing, removing the surface film and zone refining in the same container, and then transferring to a crucible in an induction furnace for crystal pulling. It is claimed by these investigators that high-purity InSb can be reproducibly grown by this modified technique.

In a recent paper, Wang Dingguo and Wang Quansheng (1984) reported on the growth of low-dislocation InSb crystals in an induction furnace by using an improved thermal field in conjunction with a proper necking procedure. Low dislocation InSb single crystals of 20–30 mm in diameter with an EPD $\sim 10^2\ \text{cm}^{-2}$ have been reproducibly grown in this way.

Yu Zhenzhong *et al.* (1980a, b) and Jin Gang *et al.* (1980, 1981) of the Shanghai Institute of Technical Physics have performed some more fundamental work on such topics as (i) facets and twin formation in the growing of InSb single crystals, (ii) anomalous segregation of impurities, and

(iii) instantaneous segregation of impurities. Their experimental results indicate that the external forms of the crystals, as well as the formation of growth twins, are closely related to the development and properties of {111} facets on the solid growth surface. It is their opinion that all their experimental results can be interpreted successfully on the basis of a {111} plane octahedron or two-octahedron model of the zinc-blende structure. The impurity distributions of the cross section of Te-doped and undoped InSb single crystals determined by Hall measurements and mass spectroscopic analysis confirm the anomalous impurity segregation on the facet. An expression based on thermodynamic considerations have been derived by these investigators to explain the observed anomalous segregation phenomena. By experimental measurement of temperature variations on the solid–liquid interface during crystal pulling by means of an immersed thermocouple, Jin Gang *et al.* (1981) have been able to explain the morphology of the impurity striation reported in their previous paper (Jin Gang *et al.*, 1980).

III. Epitaxial Growth

The work on epitaxial growth, including work on LPE, VPE and MOVPE, has continued to expand in recent years, especially in the cases of LPE and VPE. Some preliminary work on MBE has also been carried out at a few institutes. The materials grown include GaAs, InP and some other III–V compounds, with emphasis on GaAs.

4. Liquid Phase Epitaxy

a. GaAs and $Al_xGa_{1-x}As$

Liquid phase epitaxy is used to grow GaAs layers for the fabrication of Hall devices, Gunn diodes, IMPATT diodes and optoelectronic devices, etc. It is found by Ding Moyuan *et al.* (1984) of GINM that the use of a double-doped substrate can significantly improve the longitudinal carrier concentration distribution in the epilayer necessary for the fabrication of Gunn diodes, while Sn-doped LPE material is satisfactory for Hall devices (Ding Moyuan *et al.*, 1982). By using a sliding quartz boat and back etching, Gong Xiuying and Zhao Jianshe (1981) of IS have been able to grow high quality GaAs epilayers of strictly controlled thickness, suitable for Hi–Lo junction IMPATT diodes, without the use of a buffer layer.

Lin Lanying *et al.* (1982a) of IS have achieved a significant success on the growth of high-purity LPE GaAs. By using a special quartz sliding boat, a low growth temperature (700°C) and an improved epitaxial process, these investigators have been able to grow reproducible epilayers with $n = 10^{13}$–$10^{14}\ \mathrm{cm}^{-3}$, $\mu_{\mathrm{RT}} = 9000\ \mathrm{cm}^2/\mathrm{V\,s}$ and $\mu_{77\mathrm{K}} = (1.5\text{–}1.9) \times 10^5\ \mathrm{cm}^2/\mathrm{V\,s}$. The values for the best sample are $n = 3.97 \times 10^{13}\ \mathrm{cm}^{-3}$, $\mu_{\mathrm{RT}} = 11500\ \mathrm{cm}^2/\mathrm{V\,s}$ and $\mu_{77\mathrm{K}} = 1.95 \times 10^5\ \mathrm{cm}^2/\mathrm{V\,s}$. The 77 K mobility is comparable with that reported by Greene (1973) for quartz boat epitaxy.

A model for the growth of current-controlled LPE GaAs layers based on two growth mechanisms involving an As^- ion and an As atom respectively has been derived by Wang Zhongchang *et al.* (1983) of JLU. The growth rate is found to be inversely proportional to the square root of time, and a relationship between growth rate and current density is obtained.

By using a low temperature (590°C) LPE and rapid cooling, Tu Xiangzheng *et al.* (1982) of IS have been able to eliminate the dislocation nets and, at the same time, lower the vacancy concentration in $\mathrm{Al}_x\mathrm{Ga}_{1-x}\mathrm{As}$ LPE layers, as inferred from the shape of the rocking curves obtained from double-crystal X-ray diffraction analysis. A low temperature (760°C) LPE has also been used by Xu Guohua *et al.* (1982) of SIM for growing single and multiple channel structures for planar stripe $\mathrm{GaAs/As}_x\mathrm{Ga}_{1-x}\mathrm{As}$ lasers.

As a basis for the LPE of $\mathrm{Al}_x\mathrm{Ga}_{1-x}\mathrm{As}$, Shi Zhiyi and Zhou Qiumin (1984) of the Central-South Institute of Mining and Metallurgy have calculated the liquid–solid equilibrium in the Ga–Al–As ternary system using both the quasi-regular solution (QRS) and the quasi-chemical approximation (QCA) models.

It is shown that the results of calculation based on the QRS model are in better agreement with the experimental data reported in the literature than with the calculations of other investigators. Similar work has been done by Jiang Xuezhao (1985) of the Beijing Nonferrous Metal and Rare Earths Research Institute. The calculated phase diagram is in reasonable agreement with experimental data in the temperature range 1112–1273°C. The Al–Ga–As ternary diagram at 800°C has also been experimentally determined by Yu Jinzhong *et al.* (1982) of IS. From their experimental results, Yu Jinzhong *et al.* (1985) have also derived empirical expressions for calculating the weight of GaAs and Al to be used per gram of Ga in the LPE of $\mathrm{Al}_x\mathrm{Ga}_{1-x}\mathrm{As}$ at 780–850°C with $x = 0$–0.85. Besides, the effects of the x value on such parameters as growth rate, homogeneity, mobility and doping concentration, for a given amount of Mg, Ge, Te and Sn as the dopants, have also been investigated. As pointed out by these investigators, their experimental results should find practical applications in the fabrication of visible heterostructure lasers (see Section 22a).

It is worthwhile to mention the work on measurement of the P/N transition temperature T_{tr} by the induced current dynamic method in the LPE of Si-doped $\mathrm{Al}_x\mathrm{Ga}_{1-x}\mathrm{As}$ layers performed by Yao Ganzhao *et al.* (1983)

of the South China Institute of Technology. The transition temperatures corresponding to different Si concentrations in the melt for the LPE of $Al_xGa_{1-x}As:Si$ have been determined.

The surface morphology of $Al_xGa_{1-x}As$ epilayers and the defects therein have been investigated by Zhou Min (1982) of the Beijing Nonferrous Metal and Rare Earths Research Institute, resulting in successful fabrication of laser diodes. The $Ga_{1-x}Al_xAs/Ga_{1-y}Al_yAs$ laser diodes in the visible range have been grown by Shi Zhiwen *et al.* (1982) of IS on grooved GaAs substrates with reasonable success.

b. InP, $In_{0.53}Ga_{0.47}As$ and $In_{1-x}Ga_xAs_yP_{1-y}$

The work on LPE InP includes the suppression of thermal degradation of the substrate with the use of a specially-designed graphite boat (Zhang Yunqin and Tang Daiwei of IS, 1985, determination of the solubility of P in dilute solutions in In and of the diffusion coefficient for P in In as a function of temperature (Wang Yiping *et al.* of NSR, 1983) and the LPE growth of satisfactory Sn-doped InP layers (Wang Yunsheng and Zhuang Enyou of IS, 1981).

Much progress has been made on the LPE growth of $In_{0.53}Ga_{0.47}As$ on a InP substrate. Mao Yuguo *et al.* (1984) of NSR have studied the various factors for compositional control and derived a simple formula for melt source preparation. While the purity of the epilayers obtained is reasonably good, it has to be improved further in order to meet the requirements for long wavelength photodetectors. The highest room temperature mobility reported so far is $9450\,cm^2/V\,s$[9], obtained by Wang Shutang *et al.* (1984) of IS. Another work is that of Wu Xiangsheng *et al.* (1983) of SIM on the growth of Zn-doped p-$In_{0.53}Ga_{0.47}As$ contact layer on a ⟨100⟩ InP substrate by the single epitaxy process. The lattice mismatch $\Delta a/a$ is $<3 \times 10^{-4}$, and a carrier concentration $p > 3 \times 10^{18}\,cm^{-3}$ is obtained.

The LPE $In_{1-x}Ga_xAs_yP_{1-y}$/InP DH multilayer structure with respect to doping, lattice mismatch, under-cooling, rate of growth, surface morphology, etc., has been studied by a number of investigators (Wu Xiangsheng *et al.* of SIM, 1982a,b; Wu Fuming *et al.* of HSRI, 1982; He Jianhua *et al.* of HSRI, 1982; Wang Xianren of JLU, 1985), resulting in the successful fabrication of long wavelengths LEDs and LDs. The lattice mismatch problem has been solved successfully by Yang Yi *et al.* (1983) of SIM by optimizing such parameters as cooling rate, growth temperature, melt composition, etc. A study by Lu Jiangcai and Chen Jun (1983) of The Yongchuan Institute of Photoelectric Technology has also been made on the effect of the last two parameters on $\Delta a/a$ in the growth of $1.55\,\mu m$ $In_{1-x}Ga_xAs_yP_{1-y}$/InP heterogeneous structure.

[9] A recent result of Shi Huiying *et al.* of SIM has raised this value to $>10{,}000\,cm^2/V\,s$.

In order to ensure the location of the *p–n* junction within the active region, Zhang Xueshu *et al.* (1983) of HSRI have compared the *p*-type dopants Zn and Cd with respect to volatilization and contamination. The latter dopant is favored. The transition layer at the interface of the $In_{1-x}Ga_xAs_yP_{1-y}/InP$ heterojunction has been studied by Li Mingdi *et al.* (1983) of Jiao Tong University by Auger spectroscopy. The P profile in the transition layer is shown by these investigators to be approximated by a tangent hyperbolic curve.

c. *GaP, $Ga_xIn_{1-x}P$, InSb and $InSb_{1-x}Bi_x$*

Fang Zhilie and Li Chunji (1983) of FDU have successfully adopted the closed rotating boat method, while Ding Zuchang *et al.* (1983) of Zhejiang University have chosen to use the half-closed rotating boat method in order to make the LPE of GaP compatible with the large-scale production of LEDs.

Yu Lisheng *et al.* (1984) of BJU have successfully grown $Ga_xIn_{1-x}P$ epilayers on GaAs substrates by LPE. On the basis of Stringfellow's data (Stringfellow, 1970) on the Ga–In–P ternary phase diagram, it is found that the x value can be maintained at around 0.51 so as to match the epilayer well with the substrate.

Yu Zhenzhong *et al.* (1983) of the Shanghai Institute of Technical Physics have reported the presence of two kinds of stacking faults in InSb LPE layers—namely, the dumbell- and pole-shaped faults. The mechanism for generating the stacking faults is discussed.

Sun Qing and Wu Wenhai of Shanghai University of Science and Technology and Zou Yuanxi and Si Huiying of SIM (1984) have studied the growth of $InSb_{1-x}Bi_x$ with the current-controlled LPE process. These investigators have been able to grow $InSb_{1-x}Bi_x$ epilayers with x as high as 0.0124 using a growth temperature of 335°C and an average growth rate of 0.5–1 μm/min. This composition corresponds to a room temperature absorption limit of 9.5 μm, which lies within the atmosphere window of 8–14 μm.

5. Vapor Phase Epitaxy

a. *GaAs*

VPE growth of undoped GaAs with the $AsCl_3$–Ga–N system has been studied by Lin Lanying *et al.* (1982b) of IS. By using as low a growth temperature as possible, these investigators have succeeded in growing high-purity GaAs with $n \sim 10^{13}$–10^{14} cm^{-3} and $\mu_{77K} \sim (1.6$–$2.11) \times 10^5$ cm^2/V s. They have also studied the effect of vacuum treatment of the Ga source on the purity of the epilayer, and found that this treatment lowers N_A and raises N_D. The only detectable impurity is found to be C if only 1.2% H_2 is added to the N_2 system. The N_2:VPE process has also been studied by Lu Fengzhen *et al.* (1981) of SIM. The residual impurities are found to be S and C, the source of the former being $AsCl_3$. In addition, Lu Fengzhen *et al.* have studied in detail the various parameters for N_2:VPE in order to make this process applicable to the fabrication of Gunn diodes as well as FETs.

Lin Yaowang *et al.* (1984) of IS have also studied VPE growth of high-purity GaAs in an argon atmosphere. A 77 K mobility of 2.05×10^5 cm/V s has been obtained, and Hall measurements on comparable samples have shown that the $AsCl_3/Ga/Ar$ (1–2% H_2) system is more suitable for the preparation of high-purity GaAs than is the system using N_2 as the carrier gas.

The usual $AsCl_3/Ga/H_2$ system has been used by Peng Ruiwu *et al.* (1982) of SIM for growing multilayer structures suitable for fabricating FETs. The appearance of a knee in some epilayers is attributed to certain complex defects, such as $V_{Ga}O_i$. Another work is that of Mo Jinji *et al.* (1984) of SIM on the growth of undoped or lightly doped GaAs epilayers. The concentration of EL2 in the buffer layer and the growth rate are found to increase as temperature increases within the range 680–740°C. The temperature dependence of the growth rate is explained on the basis of the mixed absorption model suggested by Cadoret *et al.* (1975) and Laporet *et al.* (1980).

A novel double-chamber technique for growing multilayer structures for the fabrication of FETs and IMPATT diodes has been adopted by Wang Feng *et al.* (1983) of the Tienjin Electronic Materials Research Institute. According to Fan Maolan (1983) of the Emei Institute of Semiconductor Materials, the use of a three-temperature-zone furnace in conjunction with a suitable thermal field significantly improves the longitudinal carrier distribution, resulting in a higher yield of satisfactory Gunn diodes. Multilayer structures of VPE GaAs have also been grown in a vertical furnace by Cheng Qixiang (1981) of NSR for the fabrication of varactor and mixer diodes.

The use of a pure solid GaAs source is claimed by Zhang Wenjun *et al.* (1981) of NSRI to have certain advantages over Ga as the source. Low temperature, small gradient VPE of GaAs has been studied by Wang Tiancheng *et al.* (1981, 1983) of the Beijing Electronic Tube Factory, resulting in reduced Si contamination and improved steepness of the concentration profile of double-layer structures. A back liquid encapsulation technique for getting impurities from the substrate has been developed with a view to growing VPE GaAs with a low compensation ratio by Cheng Qixiang (1982) of NSR.

b. Dopants in GaAs

The doping behavior of the dopants S, Sn, Cr, Fe and Cd has been studied by a number of investigators. Peng Ruiwu (1980) and Sun Shangzhu *et al.* (1981), both of SIM, have studied S doping by using elemental S as the dopant.

The influence of growth parameters on the physical and electrical properties of the epilayers is investigated and discussed. Yang Yinghui *et al.* (1981) of GINM have studied S doping by the use of S_2Cl_2 as the dopant. The expression

$$K_S = \frac{[S^+] \cdot P_{AS_4}}{2.24 \times 10^{22} \cdot P_{H_2S}} \tag{1}$$

is suggested as the distribution coefficient for S with H_2S as the dopant. The K_S value for elemental S as the dopant is found to be 0.18, while that for S_2Cl_2 is 0.06. This difference is attributed to the effect of S_2Cl_2 on the growth rate. In another paper, the distribution coefficient for Sn expressed as

$$K_{Sn} = \frac{\text{Sn (in GaAs)}}{\text{Sn (in liquid Ga)}} \tag{2}$$

is found by Yang Yinhui *et al.* (1985) to be 3.8×10^{-3}. The distribution of Sn and S with $SnCl_4$ and S_2Cl_2 as the dopants respectively has been studied by Wang Feng and Zhang Ruihua (1981) of the Tienjin Electronic Materials Research Institute, and their results indicate a linear relationship between n and the partial pressures of the dopants within a wide concentration range. The distribution of S has also been studied by Wu Saijuan *et al.* (1982) of IS. The results obtained indicate the dependence of the incorporation of S on substrate orientation and misorientation, as well as on growth temperature at constant partial pressures of $AsCl_3$ and S_2Cl_2.

The use of solid Cr as the dopant has been studied by Peng Ruiwu *et al.* (1983) of SIM, and found to be suitable for growing the buffer layer for FETs.

Xu Chenmei *et al.* (1985) of SIM have studied the distribution of Cd, and found the distribution coefficient to be 10^{-2} to 10^{-3}. The doping of Fe with $Fe(CO)_5$ as the dopant has been studied by Tang Houshun *et al.* (1982) of FDU. As a liquid, $Fe(CO)_5$ is found to have certain advantages over Fe powder.

Peng Ruiwu (1984) has derived the following equation for calculating the distribution coefficients of various dopants in the VPE of GaAs on the basis of the experimental measurements made at SIM:

$$k_i = \frac{X_i}{P_i} = K_i \frac{N_c}{n(\text{or } p)}, \tag{3}$$

where

$$K_i = \frac{k^3 Z_i}{(2\pi m_i kT)^{3/2} kT} \exp\left\{\frac{-\frac{1}{2}n_s(2\xi_{\text{dop-Ga}} - \xi_{\text{Ga-As}}) - E^{\text{strain}}}{kT}\right\}. \tag{4}$$

N_c is the state density in the conduction or valence band; n(or p) is the surface free electron (or hole) concentration at the Fermi level pinning; n_s is the bond number for each atom in the crystal; Z_i is the statistical quantity of impurity atom vibration; ξ_{Ga-As} is the bond energy of Ga–As atom pair; E^s is the strain energy in the crystal.

The calculated distribution coefficients at 1000 K for Zn and Cd are 27.12 and 2.3×10^{-2} respectively, in agreement with the corresponding experimental values. The discrepancy between the calculated and experimental values for S is, however, as large as 5 orders of magnitude, and, therefore, a correction should be applied to the ξ_{S-Ga} value to make them agree with each other. A corrected value of -28.55 kcal/mol for ξ_{S-Ga} has been obtained by Peng Ruiwu from the experimentally-determined $\partial \ln K_i/\partial(1/RT)$. The corresponding k_S value is found to be 9.3, which is sufficiently close to the experimental value.

c. *InP, $GaAs_{1-x}P_x$:N and GaN*

Huang Shangxiang (1983) of NSR has reported on VPE growth of undoped and S-doped InPs by the $In/PCl_3/H_2$ process. The electrical properties of the undoped epilayers grown on vapor phase etched InP substrate are $n_{300K} = 3.6 \times 10^{14}$–$9 \times 10^{16}$ cm^3, $\mu_{300K} = 3580$–4280cm/V s. The source of contamination in VPE InP has been discussed by Shen Songhua *et al.* (1983) of SIM. The S-doped epilayers obtained by a novel multilayer epitaxy technique are found to be satisfactory for the fabrication of Gunn diodes with improved outputs.

For VPE growth of InP multilayers with the $In/PCl_3/H_2$ system, three new-type reactors have been designed by Huang Shangxiang (1981). Compared with the double-chamber reactor developed in other countries, the new design is claimed to be both simpler and easier to operate. Besides, it needs only one In source and one PCl_3 flow.

Fang Zhilie *et al.* (1981) of FDU have used the $H_2/PCl_3/As/NH_3$ system instead of the hydride process for growing the $GaAs_{1-x}P_x$:N_2 epilayer on the GaP substrate with $x > 0.5$ for the successful fabrication of yellow LEDs.

The VPE of GaN has been studied by Fu Shuqing *et al.* (1981) of the Chang Chun Institute of Physics, using the HCl–Ga–NH_3–Ar system. Fan Guanghan *et al.* (1981) of the same Institute have studied the kinetics of the process resulting in homogeneous deposition over an area as large as 20 cm^2.

6. MOVPE and MBE

a. *MOVPE*

In the previous paper (Zou Yuanxi, 1981), a brief mention was made of the MOVPE process carried out at the Sichuan Solid-State Circuits Research Institute, using TMC as the Ga source. This work was started in 1972 by a group led by Sheng Linkang (1975), and the Sichuan Institute remained the only institute working on this process until 1983. In that year, a group led by Lu Fengzhen[10] (1984) of SIM began to work on a similar process, using TEG as the raw material, in collaboration with Tu Jiahao of the Shanghai Chemical Engineering Research Institute.

Sheng Linkang *et al.* (1982) have reported on the growth of device quality GaAs using the MOVPE process with TMG as the Ga source. It is found that the carrier concentration n decreases with increasing As/Ga ratio and growth temperature, in accordance with the following equation:

$$n = \frac{P^{\circ}_{H_2S}}{P^{\circ}_{TMG}} \cdot N \cdot \left\{ \frac{1}{A + B[(As/Ga) - 1]} \right\} \tag{5}$$

where A and B are constants under given growth conditions. This work has been extended, by Sheng Linkang and Jiang Jinyi (1985) of the same Institute, to grow $Ga_{1-x}Al_xAs$ layers using TMG and TMA as the source of the III elements. It is found that the growth rate decreases while the x value increases with increasing Al/III ratio in the gas phase. Both the growth rate and the x value are independent of the As/III ratio, and the former is also independent of the growth temperature. These experimental observations are explained on the basis of a model that assumes that the epitaxial growth is controlled by the mass transport of the complexes $(CH_3)_3Ga(AsH_3)_3$, $[(CH_3)_3Al\ AsH_3]_n$ $(n > 2)$ and $AsH_3 \cdot Ga(CH_3)_3 \cdot (CH_3)_3Al\ AsH_3$. Sheng Linkang *et al.* (1983), in a subsequent paper, discussed the factors affecting the epitaxial growth of $Al_xGa_{1-x}As$ by the MOVPE process and found that the most important factor is the oxygen content of the gas phase, as well as of TMA.

The preliminary work of Lu Fengzhen *et al.* has not been published elsewhere. It will suffice to say that the purity of the GaAs epilayers, reflected by the room temperature carrier concentration of (9–10) $\times 10^{14}$ cm^{-3} and the 77 K mobility of 6.4×10^4 cm^2/V s, is adequate for certain practical applications. It will be noted in this connection that both the Sichuan and the Shanghai groups synthesize their own organometallic compounds, the purity of which will certainly be improved with the accumulation of more experience. Additionally, Lu Fengzhen *et al.* have successfully grown polycrystalline GaAs layers on a variety of substrates, including molybdenum, sapphire, silicon and quartz, with a view to providing GaAs material for fabricating low-cost GaAs solar cells after proper ion-implantation and laser annealing.

SIM remains the only institute in China actively involved in the research on MOVPE of GaAs, although preparatory work is being done at several other institutes.

[10] Private communication from Lu Fengzhen of SIM.

Lu Fengzhen *et al.* (1985)[11] of SIM have reported on the purification of TEG and TEA by synthesis and rectification at a reduced pressure, and have successfully obtained a purity of >99.99%, comparable to the 1981 Japanese products. These investigators have also succeeded in growing multilayer GaAs structures of reasonable quality by using TEG as the source material.

From a thermodynamic point of view, Liang Bingwen *et al.* (1986, to be published in *Rare Metals* in Chinese) of SIM have been able to derive an equation for the rate of epitaxial growth by using TMG and TEG as the Ga source respectively. The growth rate calculated from this equation agrees fairly well in the case of TMG growth with the data reported by Reep and Ghandhi (1983).[12] The use of this equation has resulted in a significant improvement in the evenness of the epitaxial layer. The applicability of the equation to the case of TEG growth remains, however, to be explored. Some experimental results on the rate of growth using TMG and TEG as the Ga source have already been obtained by Ding Yongqing *et al.* (1986, to be published) of SIM.

b. MBE

The first MBE System with loading and growth chambers was established in China in 1972 by a cooperative group[13] consisting of the Institute of Physics, Beijing, the Institute of Physics, Lanzhou and the Shenyang Scientific Instruments Factory. The second system was established in 1980 by another cooperative group consisting of IS, the Institute of Physics, Lanzhou and the Shenyang Scientific Instruments Factory (Cooperative Research and Development Group for MBE System, 1981). The second system includes an ultra-high vacuum system with an ultimate vacuum limit of 4×10^{-10} torr and the necessary analysis equipment. Another important work is the manufacture of a PBN crucible suitable for use in MBE, by the Institute for Metals Research.

The undoped GaAs epilayers grown by both Kong Meiyin *et al.* (1984) and Sun Dianzhao *et al.* (1984) of IS are *n*-type. The electrical properties of high-purity GaAs films are $n_{300K} = (1\text{–}10) \times 10^{14}\ cm^{-3}$, $\mu_{300K} \sim 8000\ \text{cm}^2/\text{V s}$ and $\mu_{77K} \sim 80000\ \text{cm}^2/\text{V s}$. The selectively doped GaAs/n-$Al_xGa_{1-x}As$ has a peak mobility of $2.23 \times 10^5\ \text{cm}^2/\text{V s}$ and a carrier concentration of $5.7 \times 10^{11}\ \text{cm}^{-2}$ at 22 K.

A similar work has been carried out by Zhou Junming *et al.* (1983) of the

[11] "*Abstracts, 1985 Symposium of GaAs and Related Compounds*," p.17.

[12] *Journal Electrochem. Soc.* **130**, p. 675.

[13] Private communication from Zhou Junming of the Institute of Physics.

Institute of Physics. a 77 K mobility of (5–6) × 10^4 $cm^2/V\,s$ has been obtained, according to the latest information.[14]

The presence of two dimensional electron gas has been observed by the above-mentioned investigators at IS and the Institute of Physics respectively. Additionally, SdH oscillators and quantized Hall resistance of 2 DEG have been observed under a high magnetic field and at low temperatures. During the growth of the $Al_xGa_{1-x}As$ layer, Zhou Junming *et al.*[15] have observed that the island growth of undoped or lightly Si-doped epilayer changes into layer growth when the Si concentration is higher than 8 × $10^{17\ cm^{-3}}$. Further work on this peculiar behavior of Si is in progress.

Li Aizhen *et al.* (1984a) of SIM, in collaboration with engineers of the Shenyang Scientific Instruments Factory, have made an attempt to grow n^{++}-n^{+}-p (buffer layer) GaAs multilayer structures on (100) SI GaAs substrates to meet the requirements for fabricating MESFETs and coplanar diodes. Some encouraging results (transconductance 119 mS/mm for a 1 × 316 μm device and 125 mS/mm for a 1 × 40 μm device at zero bias, pinch off voltage 1.8 V) have already been obtained. These investigators have also obtained undoped p-type MBE GaAS with p = (1–10) × 10^{14} cm^{-3} and Be-doped material with hole concentration in a range of 10^{15}–5 × 10^{19} cm^{-3} and excellent hole mobility. The selective area MBE technique for GaAs coplanar diodes and GaAs SOI applications has also been studied by Li Aizhen *et al.* (1984b).

In other papers, Li Aizhen *et al.* (1983a, 1985b) and Li Aizhen and Milnes (1983) have reported on the use of Ge as an n-type dopant in MBE GaAs with particular attention paid to compensation in the range 6.7 × 10^{15} to 1.5 × 10^{20} cm^{-3}, the residual doping effect and the amphoteric n to p transition at doping levels above 5 × 10^{18} cm^{+3}. These investigators have also examined etching and growth conditions that allow selected area epitaxial growth through SiO_2 window regions (1983b).

A persistent effort is being made with a view to improving the design of the MBE system. With the cooperation of SIM and the Shenyang Scientific Instruments Factory, this effort has resulted in improved equipment with liquid-nitrogen-shrouding in both the growth chamber and the source cells, an enlarged substrate size to 2.5 cm square and a substrate heater in the loading chamber. A horizontal system with three chambers will be put into operation in 1986.

Sun Dianzhao *et al.* (1985)[16] of IS have made further efforts to improve the quality of MBE GaAs, and have obtained undoped p-type MBE GaAs

[14] Private communication from Zhou Junming of Institute of Physics.

[15] *Ibid.*

[16] *Abstracts, 1985 Emei Symposium on GaAs and related Compounds*, p. 14.

with a hole concentration of 2×10^{14} cm^{-3} and $\mu_{300K} \sim 400$ cm^2/V s. Li Aizhen *et al.* (1986, to be published) of SIM have been able to grow Si-doped MBE GaAs of improved electrical properties, $n = 1.50 \times 10^{16}$ cm^{-3} and $\mu_{300K} = 6880$ cm^2/V s, by using specially purified As as the source, compared with those, $n_{300K} = 1.97 \times 10^{15}$ cm^{-3} and $\mu_{300K} = 4510$ cm^2/V s, in epilayers grown under similar conditions but with the use of commercially available 7 "9" As.

Li Aizhen (1985)[17] of SIM has studied the growth of GaAs/Ge polar-nonpolar heterojunctions and $GaAs_{1-y}Sb_y$/GaAs lattice-mismatched structures, the latter structures by means of electron diffraction and electron probe analysis. Zhou Junming *et al.* (1985)[18] of the Institute of Physics have studied the interface of modulation doped GaAs/n-$Al_xGa_{1-x}As$ heterostructures by means of photoluminescence spectropic analysis and high-resolution electronmicroscopic observation, resulting in a significant improvement in mobility and PL characteristics.

Multiple quantum well GaAs/n-$Al_xGa_{1-x}As$ structures with well widths ranging from 30 Å to 180 Å and x varied between 0.25 and 0.4 have been successfully grown by Chen Zhonggui *et al.* (1985)[19] of IS. These structures prove to have good crystal quality, steep interface and well-controlled layer thickness, and should provide satisfactory material, in the opinion of these investigators, for the fabrication of quantum well lasers.

IV. Impurity Effects, Space Charges and Deep Levels

7. Impurity Effects

a. Na in GaAs and InAs

As pointed out by Milnes (1983), Na and K have apparently received no further study since the tentative report of Na as an acceptor in GaAs, a long time ago (Hilsum and Rose-Innes, 1961). It has been suggested by Zou Yuanxi (1972) that Na could be an unknown acceptor in GaAs, and the probable sources of Na contamination have been discussed.

Wu Jisen *et al.* (1984b) of SIM have made an investigation of the electrical properties of the Na-implanted layer in GaAs. It is found that shallow p-type layers on undoped SI GaAs substrates or p–n junctions on n-type substrates can be reproducibly obtained by implantation of Na^+. Thus, the suspected acceptor behavior of Na in GaAs is proved. It is the opinion of Wu Jisen *et al.* that Na, as an acceptor in GaAs, should contribute in some way to

[17] *Ibid.*, p. 18.
[18] *Ibid.*, p. 16.
[19] *Ibid.*, p. 16.

donor compensation in SI GaAs, and, therefore, further investigation of the behavior of Na should be warranted.

As mentioned in Section 3b, Na is expected to be one of the principal donors in InAs. Some preliminary experiments have been carried out by Wu Jisen *et al.* (1984a) in favor of this prediction, although further experimental work is needed to clarify this problem.

b. O, Ti, C and Cr in GaAs

Although oxygen on the As site has long been known to be a donor in LPE GaAs, such behavior was not observed in bulk or VPE GaAs. Kaminska *et al.* (1981) found it necessary to assign shallow donor behavior to oxygen in HB GaAs, but they modified their opinion in a later paper (1982). Wang Guangyu *et al.* (1982) of SIM have, however, suggested that the observations of Kaminska *et al.*, of a high shallow donor concentration in the interfacial region of VPE GaAs, could be explained by assuming the presence of a complex of oxygen, probably $V_{Ga}O_i$, as a shallow donor.

As will be discussed in Section 8, the SiO_x complex with $x = 0.25$–0.75 should behave like a "mobility killer" in *n*-type GaAs. By analogy with Si, it is our opinion that Ti and C may replace Si to form the corresponding complexes, which could be as effective as SiO_x so far as the adverse effect on room temperature mobility is concerned.

Shi Huying *et al.* (1983) of SIM have studied the effect of Cr on the room temperature mobility in LPE GaAs. Their results are shown in Fig. 7 as a

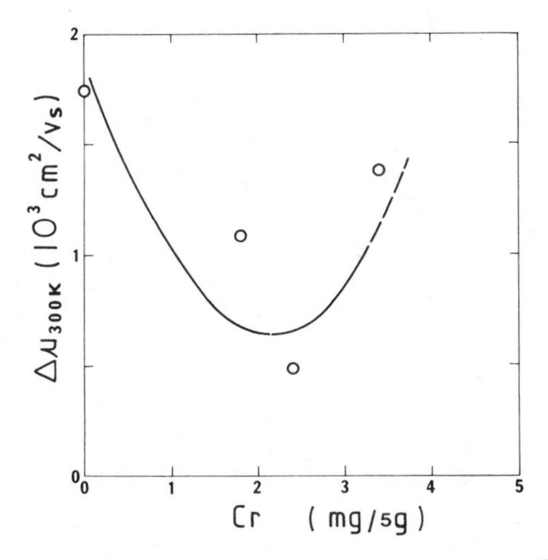

FIG. 7. Dependence of mobility lowering on Cr added to the melt for LPE GaAs. [From Shi Huiying *et al.* (1983), reproduced with permission of the authors.]

plot of $\Delta\mu_{300K}$ versus the amount of Cr added to the melt, where $\Delta\mu_{300K}$ is a measure of the depression in μ_{300K} caused by the presence of the "mobility killer." It can be easily seen that the addition of Cr, up to a certain extent, will nullify the adverse effect of the "mobility killer," probably due to the competition between Si and Cr for oxygen, resulting in a decrease of SiO_x complex. The increase in $\Delta\mu_{300K}$ with still larger addition of Cr, although not very decisive due to scarcity of data, might be caused by Cr itself being a moderate "mobility killer." It may be pointed out in passing that Sadana *et al.* (1982) have explained their data on Cr redistribution in ion-implanted and furnace-annealed GaAs in terms of Cr–O complexing combined with segregation at dislocations.

It is interesting to note that the beneficial effect of Cr on the room temperature mobility in LPE GaAs was reported by Eisen *et al.* as early as 1975, and that Nanishi *et al.* (1981) have also reported a similar effect of the presence of 0.06–0.01 ppm Cr in LEC GaAs.

c. Residual Donors in InSb

There are two schools of thought on the nature of the residual donors in high-purity InSb, viz. the chemical impurities and the natural defects hypotheses. Among the chemical impurities, Te was suggested to be the most probable candidate by Hulme and Mullin (1962) in their review paper, published a long time ago. By using a novel analytical method—to be described in Section 12c—Li Aizhen *et al.* (1965) of SIM have reported the Te content for several InSb samples. Their results are compared with the values calculated from the carrier concentration of a 1 : 1 basis in Table VIII. This comparison indicates that the analytical values amount to only about one-tenth of the calculated ones for two samples, the carrier concentrations of which are within the range 10^{14}–10^{15} cm^{-3}. Clearly, this comparison denies that Te is the chemical impurity responsible for the residual donor concentration in high-purity InSb.

TABLE VIII

TYPICAL Te ANALYSIS OF HIGH-PURITY InSb[a]

Sample No.	Te, analysed ppb	n_{77K} cm^{-3}	μ_{77K} $cm^2/V\ s$	Te, calculated ppb
Z7-3	~5	5×10^{14}	5.2 10^5	18
Z7-4	~5	1×10^{15}	—	37
Z9-2	<2	1×10^{14}	5.4 10^5	3.7
Z9-4	8	2.4×10^{15}	2.6 10^5	89

[a] From Li Aizhen *et al.* (1965). Reproduced with permission of the authors.

In an unpublished paper, Zou Yuanxi (1973) has argued that Si should be the most probable impurity responsible for the residual donor concentration in zone-refined InSb, given thermodynamic considerations, in spite of its acceptor behavior at high concentrations. To this suggestion, we can now add O, which is analogous to the proposed oxygen-related shallow donor in GaAs discussed in Section 7b. It is interesting to note that Si and O are the two major constituents of SiO_2, and, therefore, cannot be excluded in the consideration of the possible candidates for the residual donor concentration in high-purity InSb, according to Hulme and Mullin.

It should be pointed out in this connection that the probable donor behavior of oxygen in InSb was suggested a long time ago by Xu Honda and Lin Lanying (1966) of IS.

8. Space Charges

A brief discussion has been made of this problem in the previous paper (Zou Yuanxi, 1981). Stringfellow (1979) has proposed to interpret his mobility data, as well as those of Chandra and Eastman (1979), for LPE GaAs and $Al_xGa_{1-x}As$ respectively, by the presence of central-cell scattering due to C_{As}, instead of by space-charge scattering. However, Walukiewicz *et al.* (1981) have suggested that the above-mentioned mobility data could be explained by the adoption of the variation method, instead of Mathiessen's rule, in the calculations, without invoking the contribution of central-cell scattering. Therefore, the assumption of the existence of the "mobility killer" made by Weisberg and Blanc (1960) would be unnecessary. However, the experimental observation of the crossing of the μ-T curves for some bulk GaAs samples by Weisberg and Blanc, and a similar observation on some undoped LPE GaAs samples by Shi Huiying *et al.* (1984) of SIM, could hardly be explained by the calculations of Walukiewicz *et al.* using the variation method. Moreover, Shi Huiying *et al.* have studied the effect of light illumination on μ_{77K} in four undoped LPE *n*-GaAs samples. Their results are shown in Fig. 8, which plots $(\Delta\mu/\mu)_{77K}$ versus $N_{SC}A$, the product of the concentration of scattering centers and the cross section, calculated after Conwell and Vassell (1968), using the following relations:

$$\mu_{SC} = 2.4 \times 10^9 \left[N_{SC} A \left(\frac{Tm^*}{m^0} \right)^{1/2} \right]^{-1}, \tag{6}$$

$$\frac{1}{\mu} = \frac{1}{\mu_L} + \frac{1}{\mu_I} + \frac{1}{\mu_{SC}}, \tag{7}$$

where μ_{SC} represents space-charge scattering mobility and the other symbols have the usual meanings.

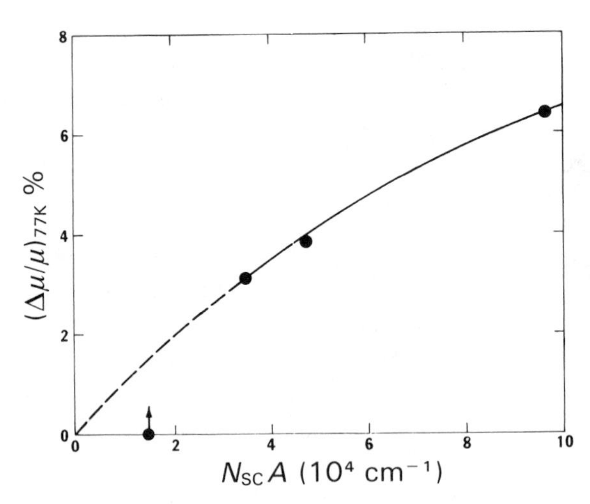

FIG. 8. Relationship between $(\Delta\mu/\mu)_{77K}$ caused by illumination and $N_{SC}A$ for undoped LPE GaAs. [From Shi Huiying *et al.* (1984), reproduced with permission of the authors.]

It can be seen that a fairly good correlation exists between $(\Delta\mu/\mu)_{77K}$ and $N_{SC}A$, in agreement with the anticipated effect of illumination, drastically shrinking the space-charge regions. Therefore, the experiments of Shi Huiying *et al.* are in accord with the concept of the 'mobility killer'' that originated from space charge scattering and that was proposed by Weisberg and Blanc a long time ago.

Shao Jiuan *et al.* (1986) of SIM have studied the same problem in undoped VPE GaAs. The shallow donor concentration N_D and the total acceptor concentration N_A are calculated and plotted with a curve calculated by Zou Yuanxi (1974) on the basis of his structural defect model. The dominant shallow donor can be taken as S for those points lying to the left of the curve and as Si for those situated on the curve. For those two groups of experimental runs, $N_{SC}A$ is calculated and plotted in Figs. 9 and 10. It can be seen that the band for the Si group has a slope of unity, and that the band for the S group has a zero slope. It seems, therefore, logical to assume that the space charge centers as Si-related complexes. A reasonable configuration of the complexes would be $Si_{Ga}(O_i)_x(V_i)_{1-x}$, similar to the acceptor $Si_{Ga}O_i$ suggested by Weiner and Jordon (1972).

For a period of about 10 years, Shi Weiying *et al.* have worked on the ''mobility killer'' problem in undoped LPE GaAs. In the previous paper (Zou Yuanxi, 1981), the ''mobility killer'' was suggested to express the configuration of the space charge center as $(Ga_{Ga})_{0.5}(Si_{Ga})_{0.5}O_{As}$ on the basis of the slope of the band, obtained by plotting $N_{SC}A$ versus N_D. In light of the results of Shao Jiuan *et al.* on VPE GaAs, it seems that the configuration

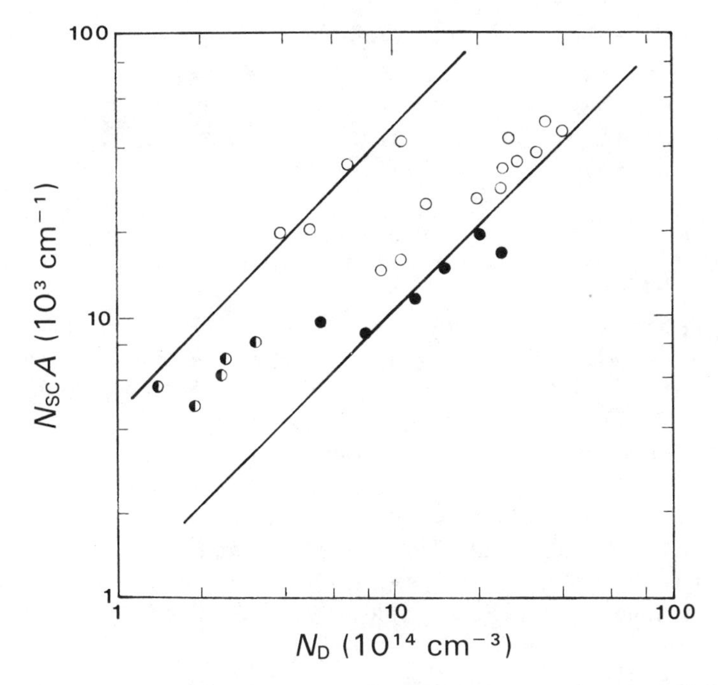

FIG. 9. Relationship between $N_{SC}A$ and N_D for VPE GaAs with Si as the principal donor. ○ authors (1985); ● DiLorenzo (1972); ◐ Lin Lanying (1982b). [From Shao Jiuan *et al.* (1985), reproduced with permission of the authors.]

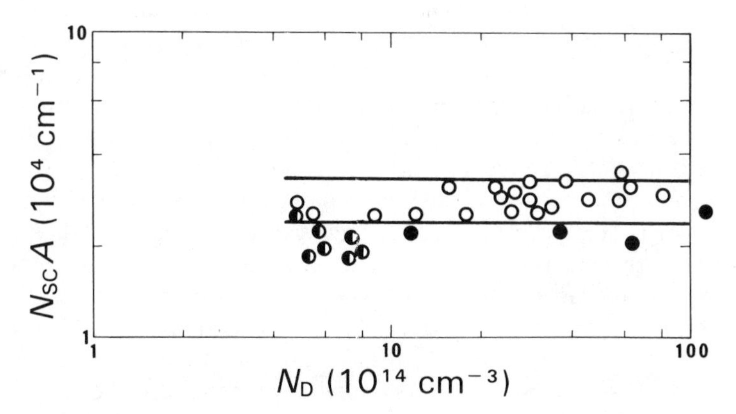

FIG. 10. Relationship between $N_{SC}A$ and N_D for VPE GaAs with S as the principal donor. ○ authors (1985); ● DiLorenzo (1972); ◐ Lu Fengzhen *et al.* (1981). [After Shao Jiuan *et al.* (1985), reproduced with permission of the authors.]

$Si_{Ga}(O_{As})_{0.5}(As_{As})_{0.5}$ should be preferred. By selecting those experimental data pertaining to undoped LPE GaAs samples grown under similar conditions, Shi Huiying *et al.* (1986) found, however, that the slope of the band was no longer 1.5, as reported in the previous paper, but 1.2 instead. Now, N_D in undoped LPE GaAs may be regarded as the sum of Si_{Ga} and O_{As}, and the ratio

$$K = \frac{O_{As}}{Si_{Ga}} \tag{8}$$

may be assumed to be greater than unity under the growth conditions used by these investigators. Since K is found experimentally to increase as N_D increases, the slope of the band in a plot of $N_{SC}A$ versus N_D should be higher when K is constant. Therefore, their data indicate an effective $x > 0.2$. A more general configuration could probably be represented by $x = 0.25$–0.75, corresponding to one to three oxygen atoms bound to one central Si atom. It seems logical to assume that the same x values would be applicable to the case of VPE GaAs. In bulk GaAs, the Si–O complex of a similar configuration may also be expected to play the role of the "mobility killer," although an experimental verification would be very desirable.

9. Deep Levels

a. A and B Hole Traps in GaAs

In the previous paper (Zou Yuanxi, 1981), we introduced a model, postulated by Zou Yuanxi in 1974, to account for two unknown acceptors in GaAs, which are predicted to be equal in concentration but different in nature, just like the A and B hole traps observed by Lang and Logan (1975) in LPE GaAs one year later. This model will be discussed in Section 15a. It will suffice to point out here that the complex defects $Ga_{As}V_{Ga}$ and $As_{Ga}V_{Ga}$ have been tentatively assigned to the A and B traps respectively. Some experimental evidences for this assignment have already been given in a previous paper (Zou Yuanxi *et al.*, 1983), and will be discussed in Section 15a, together with some recent data reported by Zou Yuanxi *et al.* (1984) of SIM.

Wang Zhanguo *et al.* (1984) of IS have, however, concluded that the A and B levels in their LPE GaAs samples are coupled and, thus, due to one single defect, as demonstrated by their photocapacitance measurements. They have further suggested that the defect may be the Ga_{As} anti-site. On the other hand, Barnes and Samara (1983) have concluded from their experimental observation of the different pressure dependence of the capture cross sections, as well as from the emission energies of these two traps that A and B levels are not associated with the same center. Therefore, the elucidation of this problem had better wait for future research.

b. EL2 and Related Electron Traps in GaAs

In the model mentioned above, the presence of $(V_{Ga})_2$ in GaAs is predicted. It will be remembered that the stability of the double Ga vacancy was first discussed by Logan and Hurle (1971). For reasons to be given in Section 15a, Zou Yuanxi (1982) has tentatively assigned $(V_{Ga})_2$ to the C electron trap observed by Ozeki *et al.* (1979) in N_2 : VPE GaAs.

Our understanding of the EL2 mid-gap electron trap has been a long time in the making. It is now believed that this trap is not related to oxygen but, rather, to a native defect. It will be recalled that the first assignment of a native defect to this trap was made by Hasegawa and Majerfeld (1976), although it has only recently been agreed by many investigators that this trap is most probably As_{Ga} or its complexes, including $As_{Ga}V_{Ga}$, $As_{Ga}V_{As}$ and $V_{Ga}As_{Ga}V_{As}$. The last-mentioned complex has been proposed by Zou Yuanxi (1982) on account of the relationship between [EL2] + [C] in the VPE GaAs and As/Ga ratio in the gas phase derived from literature data. Although this relationship holds also for the simple arsenic antisite, the complex $V_{Ga}As_{Ga}V_{As}$ is preferred on the grounds that the formation of a complex is expected to result in a reduction of strain energy and, therefore, should be promoted by the presence of dislocations, in agreement with the findings of Martin *et al.* (1980). The above-mentioned complex hypothesis is also in agreement with the experimental observation of Xin Shangheng of SIM in collaboration with Eastman *et al.* (1982) that EL2 in MBE *n*-GaAs is greatly enhanced after Si_3N_4 encapsulated annealing at 700°C, which accompanies a similar increase in the carbon concentration due to contamination. As carbon occupies the As site in GaAs, the introduction of carbon is expected to result in an increase of $(V_{Ga})_2$. Moreover, as pointed out by Xin Shangheng *et al.*, there should be strong thermal stress and possible radiation damage in the capped annealed sample. This significant increase in the EL2 concentration in capped annealed samples can be explained fairly well in this way.

Although oxygen is generally believed to have no relation with EL2, several investigators (Taniguchi and Ikoma, 1983a; Yu and Walters, 1982; Lagowski *et al.*, 1984) have nevertheless concluded that there is another deep level related to oxygen in the neighborhood of EL2. This conclusion is in agreement with the observation of Ikoma *et al.* (1982) that there exists a complex of oxygen with a native point defect in oxygen-implanted and annealed GaAs layers. Similarly, Zou Yuanxi *et al.* (1984) have found an EL2-like level in LPE GaAs grown on a high dislocation substrate. The concentration of this level is found to be proportional to the carrier concentration in the epilayer, and therefore, the configuration $V_{Ga}O_{As}V_{Ga}$ has tentatively been assigned to this carrier concentration by Zou Yuanxi *et al.* Analogous to this oxygen-related complex defect in LPE GaAs, Mo Peigen *et al.* (1985b) of SIM have observed a deep level in low dislocation LEC SI GaAs grown in a quartz crucible under wet B_2O_3. The activation energy of this GaAs is found to be 0.65 eV from temperature-dependent Hall measurements. The concentration of this deep level appears to be independent of EPD, but lower in regions near the periphery than in the central portion of the ingot, as inferred from the variation of leakage current across the wafer. According to Mo Peigen *et al.*, this level could tentatively be identified as $V_{Ga}O_iV_{Ga}$, and, therefore, its concentration is expected to depend heavily on the concentration of V_{Ga}, which should be lower in regions near the periphery, due to volatilization of As. The parallel relationship between EPD and I_L for the low dislocation wafers mentioned in Section 1b could probably be explained in this way. Besides, this complex is believed to play an important role in the compensation mechanism of undoped, low dislocation LEC SI GaAs, as does EL2 in the undoped, high dislocation material.

It is gratifying to note that the observation of an EL2-like level in LPE GaAs grown on a high dislocation substrate by Zou Yuanxi *et al.* (1984) has recently been confirmed by Xiang Xianbi *et al.* (1985) of IS in a study of the behavior of oxygen in LPE GaAs. The latter investigators, however, choose to tentatively identify their "EL2" level as an Ag_{Ga}-donor-acceptor complex.

Wang Guangyu (1986, Doctoral Thesis) of SIM has attacked the problem of identifying the atomic configuration of EL2 in SI LEC GaAs with reasonable success, on the basis of the suggestions of Makram-Ebeid (1984)[20] and Weber (1984)[21] on the use of EL2 to denominate the defect showing the characteristic photoquencing effect in addition to the dependence on dislocation density and stoichiometry.

From a prudent thermodynamic analysis of their experimental results and some related literature data, Lu Fengzhen *et al.* (1986)[22] of SIM have been able to show that EL2 could originate from $(V_{Ga})_2$, and should, therefore, be equivalent to the latter as far as deviation from stoichiometry is concerned. It is the opinion of Wang Guangyu and Zou Yuanxi [1986, to be published in *Rare Metals* (in Chinese)] of SIM that configurations such as $As_{Ga}As_i$, $As_{Ga}V_{As}$ and $As_{Ga}V_{Ga}$, postulated in the recent literature, would be unlikely to meet the stoichiometric requirement quantitatively, as these defects are stoichiometrically equivalent to $(V_{Ga})_3$, V_{Ga} and $(V_{Ga})_3$ respectively. Being equivalent to $(V_{Ga})_2$, the ternary complex $As_{Ga}V_{As}V_{Ga}$ would, then, seem to be a reasonable choice, as it may also explain the dependence on dislocation density, and possibly also account for the photoquenching phenomenon. Clearly, further investigations are needed in order to clarify this perplexing problem.

[20] *Proc. 3rd Conf. on Semi-Insulating III–V Materials, Kah-nee-ta, Oregon*, p. 184.
[21] *Ibid.*, p. 296.
[22] *Rare Metals* (in Chinese), **5**, no. 1, p. 9.

Wang Jianqiang *et al.* (1985)[23] of IS have observed a decrease in EL2 concentration in VPE GaAs grown under an Ar ambient compared with those grown in a H_2 or N_2 atmosphere. This finding is parallel to the effect of ambient gas on [EL2] in LEC GaAs found by Emori *et al.* (1984)[24] and might, in our opinion, result from the relatively low dislocation density in the Ar-grown material. An experimental confirmation of this point would be desirable.

In addition to EL2, some other deep levels, including $V_{Ga}Si_{Ga}V_{Ga}$, $Sn_{Ga}V_{Ga}$, $Ge_{As}V_{Ga}$ and $V_{III}Zn_{III}V_{III}$, have been suggested by Zou Yuanxi *et al.* (1985)[25] of SIM to be the controlling recombination level in certain critical concentration ranges from an analysis of the related diffusion length and optoelectronic data.

On the basis of the above-mentioned discussions, we wish to assume further that the deep levels As_{Ga}, $V_{Ga}As_{Ga}V_{Ga}$, $V_{Ga}O_iV_{Ga}$, $V_{Ga}O_{Ga}V_{Ga}$ and $(V_{Ga})_2$ (see Section 15a) may all belong to the EL2 family, as suggested by Taniguchi and Ikoma (1983b). Needless to say, further experimental work will have to be done in order to clarify this problem.

c. *Other Deep Levels in GaAs*

Zhou Binglin and Chen Zhenxiu (1983) of SIM have made Hall and resistivity measurements on undoped SI GaAs in the temperature range 180–500 K. For those samples having a moderately high resistivity, the activation energy measured is 0.43 eV, which is believed to be due to EL5.

Xu Shouding (1983) of IS has found seven deep levels in bulk and VPE SI GaAs by the OCTS method. These levels are 0.109, 0.14, 0.32, 0.42, 0.64, 0.8 and 0.91 eV, of which three (0.109, 0.14, 0.91 eV) are hole traps, believed to have originated from the impurities Mn, Cu and Cr, etc.

d. *Deep levels in Other III–V Compounds*

(1) Deep levels in GaP. Shi Yijing of IS in collaboration with Nishizawa *et al.* (1982) have been able to identify tentatively several deep levels in GaP LEDs. According to these investigators, the 1.8 eV level may be identified as Fe, the 2.0 eV level as P_i and the 2.1 and 2.2 eV levels as donor-V_{Ga} complexes.

Yu Suping and Huang Qisheng (1983) of Xiamen University have studied deep levels in 1 MeV electron-irradiated GaP by the DLTS method. Their results are shown in Table IX.

The trap A is found to exist before and after irradiation and is believed to be related to nitrogen and some shallow donor. The nature of the deep levels is discussed on the basis of the results of annealing experiments.

[23] *CJS.* **6**, p. 556.
[24] *Proc. 3rd. Conf. on Semi-Insulating III–V Materials, Kah-nee-ta, Oregon*, p. 111.
[25] *Proc. Matter. Res. Soc., San Francisco, California.* **48**, p. 385.

TABLE IX

DEEP LEVELS IN 1 MeV ELECTRON-IRRADIATED GaP (AFTER T^2 CORRECTION)[a]

Trap level	Label	Activation energy, eV	
		$E_C - E_T$	$E_T - E_V$
Electron traps	E_1	0.14	
	E_2	0.24	
	E_3	0.33	
	A	0.43	
	E_4	0.46	
	E_5	0.63	
	E_6	0.76	
Hole traps	H_1		0.27
	H_2		0.50
	H_3		1.0

[a] From Yu Suping and Huang Qisheng (1983). Reproduced with permission of the authors.

The electronic properties of the E_4 trap has been investigated further by Huang Qisheng *et al.* (1985), using a combination of thermal and optical measurements. A simple configuration coordinate diagram is presented, which explains several features of recombination-enhanced interactions and thermal annealing phenomena.

Shi Yijing (1982) of IS has measured the concentration of the 1.45 eV deep level in GaAlAs LEDs by the photocapacitance method. Its effect on the luminescence intensity has been measured.

(2) *Deep levels in InP.* Feng Yougang *et al.* (1983) of FDU have made DLTS measurements on LEC InP, and found that there might exist defect clusters, as manifested by the $E_c - 0.54$ eV broad band in the DLTS spectrum. Besides, two deep levels ($E_c - 0.15$ eV, $E_c - 0.40$ eV) have been found to exist at relatively low concentrations, in general agreement with the observation of Wada *et al.* (1980).

(3) *Deep levels in GaAlAs.* In addition to the DX center observed by Lang and Logan (1979) in *n*-type Te-doped GaAlAs, Zhang Yunsan and Gao Jilin (1983) of IS have reported the existence of another deep center in the vicinity of DX but entirely different from DX in physical properties. This newly-discovered deep center is considered to be also related to the *n*-type dopant (Sn, $n \sim 3 \times 10^{16}$ cm^{-3}) incorporated into the waveguide layer.

It is noteworthy that Gao Jilin *et al.* (1985) have observed, in *n*-GaAlAs : Sn by the DLTS method, the dependence of both LE_1, which corresponds to DX, and LE2 ($E_C - E_T = 0.37$ eV) on oxygen introduced into the LPE layer. This matter will be discussed again in Section 24a, in connection with slow degradation of GaAlAs/GaAs DH laser diodes. These investigators have also reported the creation of a deep level at 0.89 eV, below the conduction band of proton irradiation, in agreement with Lang *et al.* (1976).

Zhou Binglin of SIM *et al.* (1982) have studied the electrical properties of the persistent-photoconductivity centers in MBE *n*-GaAlAs : Si by capacitance spectroscopy techniques. Two electron traps at electron emission activation energies of 0.44 and 0.57 eV have been found to be the origin of the persistent-photoconductivity phenomenon in this material. The 0.44 eV trap is suggested to be the same trap center responsible for said phenomenon in LPE GaAlAs, i.e., the DX center as found by Lang and Logan, on the basis of data on activation energy and electron capture cross-section.

Furthermore, Ge Weikun and Wu Ronghan (1986)[26] of IS have measured the DX centers in LPE *n*-$Al_{0.3}Ga_{0.7}As$: Te samples by both DLAS (Deep Level Admittance Spectroscopy) and DLTS techniques. Two energy levels, which might be associated with the DX center, are resolved with ΔE_{em} ~0.35 eV and ~0.20 eV, and ΔE_{eq} ~0.10 eV and ~36 MeV respectively. Majority carrier traps have also been observed by Zhang Guicheng *et al.* (1986, private communication) of SIM in the active region of Si- and Ge-doped $Al_xGa_{1-x}As$ DH LEDs, with activation energies ~0.34 eV and ~0.42 eV respectively.

For additional information on DX centers turn to Chapter 3 of this volume, "Deep Levels in III–V Compound Semiconductors Grown by MBE," by P. K. Bhattacharya and S. Dhar.

(4) *Deep levels in* $GaAs_{0.6}P_{0.4}$. Qiu Sichou *et al.* (1984) of the Central China Institute of Technology have studied with TBC and DLTS methods, the electron traps in $GaAs_{0.6}P_{0.4}$ LEDs with the *n*-layer grown by two different VPE processes using PH_3 and P as the source of phosphoros respectively. Two electron traps located at 0.07 and 0.37 eV below the conduction band are found to exist in the VPE layers grown by both processes.

(5) *Deep levels in InGaAsP.* Bao Qingcheng *et al.* (1982) of IS have studied deep levels in InGaAsP with the DLTS method. Two electron traps, A and B, are found to exist. Some typical data are given in Table X.

The concentration of the A trap is much higher than the heavy metals contamination usually found in semiconductor materials. It is the opinion

[26] *CJS.* **7**, p. 254.

TABLE X

DLTS MEASUREMENTS ON InGaAsP[a]

Sample No.	Lattice mismatch $\Delta a/a$ %	Oxide film	Energy level eV	Capture cross section σ_∞ cm^2	Concentration N_T/N_D %
LMS-I-1	<0.21	Single (Al_2O_3)	A 0.20	1.4×10^{-16}	~2
LMS-I-2	<0.21	Composite (Al_2O_3)/anodic oxide	A 0.19 B 0.48	1.4×10^{-16} 3.8×10^{-12}	~1 ~1
LMS-II-1	<0.21	Composite	A 0.20 B 0.48	1.4×10^{-16} 3.8×10^{-12}	~1 ~1
XS-I	>0.3	Composite	B 0.46	3.0×10^{-10}	~1

[a] From Bao Qingcheng *et al.* (1982). Reproduced with permission of the authors.

of these investigators that the A trap could be due to V_P-related defects. In view of the suppression of the A trap in the B-trap region and other observations, the B trap is believed to be due to lattice mismatch as well as complex defects.

On the other hand, Gao Jilin *et al.* (1985) of IS have failed to find any detectable deep level, using both the DLTS and Admittance Spectrometry methods, in the p-$In_{1-x}Ga_xAs_yP_{1-y}$ ($x = 0.26, y = 0.6$) layer of the InGaAsP/InP four-layer structure, indicating a good lattice match, in agreement with the observation of Levinson and Temkin (1983). A series of electron traps as well as hole traps have, however, been found by Gao Jilin *et al.* with DLTS measurements on the same layer after proton irradiation at 150–300 KeV.

V. Characterization and Analysis

10. Photoluminescence and Electrochemical Studies

a. Photoluminescence Studies

Zhang Lizhu *et al.* (1983) of BJU and Yin Qingmin of GINM have studied the 77 K PL spectra of annealed n-GaAs doped with Sn, Te and Si. Three peaks located at 1.36, 1.30 and 1.26 eV are observed and discussed. A similar study has been made of the 77 K PL spectra of thermal neutron-irradiated, Si-doped GaAs by Lin Zhaohui *et al.* (1985) of BJU in collaboration with Yin Qingmin of GINM. Four peaks, at 1.49, 1.36, 1.30 and 1.26 eV, are observed and discussed. The 4.2 K PL spectra of high-purity LPE and VPE GaAs have been investigated by Chen Tingjie *et al.* (1982) of IS in collaboration with Meng Qinghui *et al.* of the Institute of Physics. Several peaks in the range 1.44–1.46 eV are observed and attributed to the first phonon replicas of BA and DA bands associated with residual acceptors. The *S* values for C_{As}, Si_{As} and Ge_{As} are obtained from the respective ratios of the intensity of the first phonon peaks, to those of the zero phonon peaks for these impurities.

The characteristics of the PL spectra (77–300 K) of $Al_xGa_{1-x}As$ as a function of the Al content have been studied by Jiao Pengfei (1983) of BJU. The PL properties of the isoelectronic impurity nitrogen in LPE $Al_xGa_{1-x}As$ ($x = 0.2$–0.8) have been studied at 4.2 K by Gong Jishu *et al.* (1982) of IS. The efficient PL emissions observed are identified as bound exciton recombinations of isolated N centers associated with X conduction band minimums and their phonon side bands. This identification is confirmed by the annealing temperature dependence, as well as by the N concentration dependence of the emission.

Yu Kun *et al.* (1981) of the Institute of Physics in collaboration with Wu Lingxi *et al.* of IS have studied the low temperature (77 K, 4.2 K, 1.8 K) PL spectra of high-purity LPE VPE and MBE GaAs, VPE InP and bulk InP. Their results indicate that the residual acceptor impurities are Si, C and Ge in LPE GaAs, and Si, C, Ge and Zn in VPE GaAs respectively. For VPE InP, the B peak at 1.377 eV is attributed to Zn, after Williams *et al.* (1973). In bulk InPs, there appears a broad peak at 1.122 eV, which is attributed to some complex of V_P. For MBE materials, the low temperature PL spectra indicates peaks at 1.4923 eV in *n*-type GaAs and at 1.3733 eV in SI GaAs, which are believed to be C and Cu respectively.

Xin Shangheng of SIM in collaboration with Eastman *et al.* (1982) have used low temperature (5 K) PL technique to detect and characterize the ~1.40 eV emission band in an Mn-doped MBE GaAs layer and heat-treated Si Cr-doped GaAs specimen. By comparing the PL spectra of these two kinds of samples, these investigators conclude that the emission at ~1.40 eV should be attributed to Mn_{Ga}, which is also the impurity responsible for the type conversion of SI GaAs.

Wang Shaobo *et al.* (1981, 1984) of SIM have succeeded in identifying certain impurities as well as native defects in *n*-type GaAs and InP by the low temperature (4.2, 77 K) photoluminescence method. The residual acceptor impurities in undoped LPE GaAs are found to be predominantly C_{As} and Si_{As}, characterized by the peaks 1.481 and 1.497 eV respectively. By ion implantation with Si followed by annealing, it has been proved that the 1.403 eV peak in the spectrum of HB, VPE and LPE GaAs is related to a complex involving Si_{As} and V_{As}. An investigation of the spectra of undoped

LEC InP crystals shows that the ~1.38 eV peak of the spectrum is caused by the C acceptor, while the 1.08 and 1.20 eV peaks are connected with V_p and V_{In} respectively. The absence of the 1.20 eV peak in the PL spectrum of Sn-doped InP and its disappearance with increasing S doping are considered evidence for interaction between S_p and V_{In}.

The low temperature (1.8, 4.2, 10 K) PL method has been applied to the study of undoped LEC InP by Wu Lingxi *et al.* of IS in collaboration with Meng Qinghui and Li Yongkang of the Institute of Physics (1984). Four groups of A and B peaks, including the 1st, 2nd and 3rd phonon replicas, are listed in Table XI. It is found by these investigators that a 1.37 eV peak, which is close to the A peak, is present in a Zn-doped sample. While the 1.377 eV peak is observed in the spectrum of a relatively pure undoped sample, the B peak is not found.

Lin Zhaochui and Zhang Lizhu (1983) of BJU have studied the effect of Cu on the 77 K PL spectrum of S-doped InP. According to these investigators, the obvious effect of Cu can probably be attributed to the formation of complexes among Cu, S and other defects. The 1.36 eV peak in the 77 K PL spectrum of Sn-doped LEC InP has been studied by Lin Zhaochui *et al.* (1984) of BJU in collaboration with Cui Yucheng of GINM. As this peak appears only in Sn-doped InPs after being annealed at 500°C, and not in similarly annealed undoped and S-doped InPs, it is concluded that the center responsible for this peak should be a complex of Sn and V_P.

The emission band at 0.95 μm has been observed in 1.3 μm and 1.55 μm InGaAsP/InP DH lasers by Zuang Weihua *et al.* (1984a,b,c) of IS. The

TABLE XI

PL SPECTRA OF UNDOPED LEC InP[a]

	Emission peak		
Symbol	Peak position ν cm^{-1}	Peak energy eV	Δ peak energy meV
A	11102	1.377	9
B	11062	1.368	
A′	10762	1.334	8
B′	10690	1.326	
A″	10410	1.291	9
B″	10339	1.282	
A‴	10062	1.248	9
B‴	9990	1.239	

[a] From Wu Lingxi *et al.* (1984). Reproduced with permission of the authors.

presence of this band is attributed to band-to-band radiative transition in the InP limiting layers. As pointed out by these investigators, such a non-lasing transition should result in carrier loss and weakened emission in the 1.3 and 1.55 μm LDs.

The temperature dependence of the PL spectra of LPE $In_{1-x}Ga_xP/GaAs$ has been reported by Yu Lisheng *et al.* (1983) of BJU in the temperature range 77–300 K. An anomalous temperature dependence of +2.5 MeV/°C has been observed for some of the samples.

Wang Shouwu *et al.* (1981) of IS have studied the luminescence spectra at different spatial positions perpendicular to the junction plane of a DH laser. Their experimental results show that the deep level radiation peaks at 1.03 and 1.09 μm (called D_1 and D_2) appear throughout the region of the *n*-GaAs substrate, *n*-GaAs buffer layer and *p*-GaAs active layer. It is the opinion of Wang Shouwu *et al.* that D_1 *and* D_2 have originated from hole traps, which could be related to the antisite complexes proposed by Zou Yuanxi (1977).

Chen Peili *et al.* (1986, to be published) of IS have confirmed, by means of photoluminescence microspectroscopy studies, that the InP limiting layers of an InGaAsP/InP DH LD are the sources of the 0.95 μm band observed by Zhuang Weihua *et al.* (1984a,b; 1985) of IS. This conclusion provides a convincing mechanism for the strong dependence of the threshold current of such LDs on the operating temperature.

Teng Da *et al.* (1986, to be published in *CJS*) of IS have observed both intrinsic and extrinsic transitions in photoluminescence experiments on GaAs/GaAlAs quantum well structures with well widths ranging from 40 to 145 Å. A PL line, named I line, has been observed for the first time and is believed to be related to the interface of the GaAs/AlAs heterojunction.

Xin Shangheng of SIM and Eastman (1985)[27] have found the occupancy of Sn atoms in MBE GaAs to be dependent on the substrate temperature and the As_4/Ga ratio. The Pl peaks at 1.507 eV and 1.35 eV are given as evidence for the existence of Sn as an acceptor.

A method for measuring the C content in GaAs at room temperature has been developed by Jiang Desheng *et al.* (1986)[28] of IS: studying the temperature dependence of the C-induced local vibration mode IR absorption and calibrating against the known C concentrations in a series of GaAs samples obtained by IR absorption at 77 K.

In GaP : N, Qian Jiayu and Lu Jinyuan (1984) of GINM have observed radiative recombination of exciton bound to isolated N atom (A line) and of excitons bound to coupled N atom (NN_i line, i = 1, 2...10).

[27] *CJS*. **6**, p. 552.
[28] *CJS*. **7**, p. 59.

The photoluminescene behavior of (N^+, Zn^+)-implanted $GaAs_{1-x}P_x$ has been studied at 1.8–4.2 K by Xu Jungying *et al.* (1984) of IS. It is found from their experimental results that the alloy composition at which N-Zn transition changes over to N bound excitonic recombinations depends on the nitrogen and zinc impurity concentrations.

b. Characterization by Electrochemical Methods

The electrochemical method is especially suitable for determining the concentration profile in *p–n* junctions and heterojunctions. This method used in conjunction with ion implantation and ion milling techniques has been applied to determine the carrier concentration and mobility profiles in submillimeter doped InP and GaAs layers by Shao Yongfu *et al.* (1981). The factors affecting the accuracy and reproducibility of the determination of semiconductor carrier concentration by the electrochemical C–V method have been investigated by Shao Yongfu *et al.* (1982) of SIM. This method has been applied to GaAs *p–n* junctions, GaInAsP/InP DH structures and Si-implanted, Cr-doped SI GaAs by these investigators. Wang Fuxian and Zuo Qipei (1983) of GINM have applied this method to a LPE p–n^+ GaP. The heterojunctions GaInAs/InP and GaInAsP/InP have been studied by Yu Guangao *et al.* (1983) of NSR, using the same technique.

Carrier concentration profiles in InP epilayers have been determined by Yang Dejia (1981) of NSR, using the newly-developed electrochemical barrier C–V method, with precautions taken to eliminate various sources of error. The results are in agreement with those obtained using the mercury probe method in conjunction with chemical stripping.

A photochemical method has been applied to the determination of the Al profile in the $Al_xGa_{1-x}As$ layer of an $Al_xGa_{1-x}As/GaAs$ solar cell in the range $x = 0$–0.84 by Chen Ziyao *et al.* (1985) of SIM. The profile obtained agrees well with that obtained by Auger analysis, but the photoelectrochemical method is less complicated.

11. Minority Carrier Diffusion Length and Lifetime

a. In HB Bulk GaAs

Mo Peigen (1984) of SIM has reported experimental data on the hole diffusion length L_p in HB Si-doped GaAs determined by the SEM-EBIC method. Measurements on some samples are checked by the surface photovoltaic method, and the results obtained by these two methods are found to agree with each other within 5%. The relationship between L_p and N is shown in Fig. 11. It is seen that a maximum L_p value occurs at a carrier concentration slightly over 10^{17} cm^{-3}. The L_p value drops rapidly with decreasing carrier

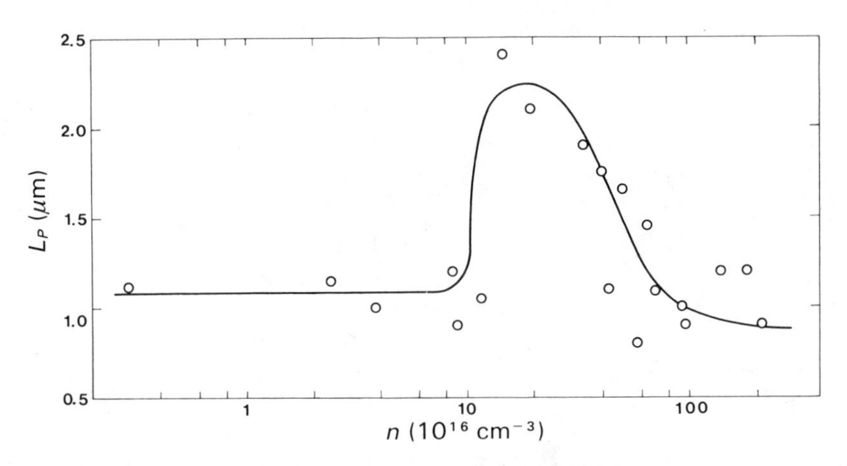

FIG. 11. L_p in Si-doped HB GaAs. [From Mo Peigen (1984), reproduced with permission of the author.]

concentrations. According to Mo Peigen, these results may be interpreted by assuming that the recombination center controlling L_p is EL2 up to a carrier concentration of 10^{17} cm^{-3}, beyond which the concentration of EL2 is expected to drop suddenly to give way to another center, assumed to be $Si_{Ga}V_{Ga}$. Mo Peigen suggests further that SiO_2 microprecipitates would form at high Si concentrations, resulting in an increase in V_{Ga} concentration.

b. In GaAs Epilayers

For LPE GaAs, it has been suggested by Partin *et al.* (1979) that the minority carrier lifetime τ_p or diffusion length L_p is determined by the A hole trap. As will be discussed in Section 15a, the A and B traps should be proportional to $n_{300K}^{1/2}$, and, therefore, L_p should be proportional to $n_{300K}^{1/4}$. This relationship has been found to hold true by Zou Yuanxi (1985) in a plot of L_p versus n_{300K} for Sn-doped LPE GaAs based on the data of Min Huifang *et al.* of SIM.

For VPE GaAs grown by the hydride process, Miller *et al.* (1977a) have found that there is a parallel relationship between the As/Ga ratio in the gas phase on the one hand, and between the τ_p^{-1} and EL2 concentration on the other. It is tempting, therefore, to consider EL2 as the recombination level. This cannot be true, however, in view of the small hole-capture cross section of that level, as pointed out by Mitonneau *et al.* (1979). Therefore, it is our opinion that some doubt may also be cast upon Mo Peigen's suggestion of the effect of EL2 on MCDL in bulk GaAs (see Zou Yuanxi *et al.*, 1985)[29].

[29] The recent result of Liang Binwen of SIM indicates that the controlling recombination level in HB GaAs could be EL5 in the range $n = 10^{16}$–10^{17} cm^{-3}.

c. *In Other III–V Compounds*

(1) *In $Al_xGa_{1-x}As$.* Liu Hongxun *et al.* (1983) of BJU have measured L_n and L_p in $Al_xGa_{1-x}As$ alloys by the EBIC method. Their results indicate that L_n decreases with increasing Al content, while L_p remains constant. The experimental results are discussed from the viewpoint of band structure.

(2) *In GaP with a discussion of the SPV method.* Chen Chao and Shen Qihua (1983) of Xiamen University have used three modifications of the SPV method for measuring L_p in *n*-type bulk GaAs, bulk and LPE GaP and bulk InP. The constant light intensity method is suggested and developed for the first time by these investigators. In a comparison of the three modifications of the SPV method, the constant light intensity method is considered to have certain advantages over the other two modifications—namely, the half-wave and the constant photovoltage methods. Some L_p values measured by the constant photovoltage method in LEC and LPE GaP are listed in Table XII.

As to the effect of sample thickness on measurements of MCDL by the SPV method, Yang Hengqing *et al.* (1984) of FDU have recently suggested a method for finding the real diffusion length from the apparent value on the basis of theroretical considerations and experimental results on Si.

Zhou Bizhong (1984) of Xiamen University has measured the effect of 1 MeV electron irradiation on the MCDL in GaP by the SPV method. It is found that both MCDL and luminescence efficiency decrease sharply with the creation of a high density of point defects, as well as of micro-defects, as a result of electron irradiation. The defects responsible for the irradiation effect observed are assumed to be vacancies, antisite defects and their complexes. A similar study on GaP : N has been carried out by Zhou Bizhong *et al.* (1985).

TABLE XII

Lp VALUES IN GaP MEASURED BY THE CONSTANT PHOTOVOLTAGE METHOD[a]

Sample	n cm^{-3}	μ_p $cm^2/V\,s$	L_p μm	τ_p 10^{-8} s
S-doped LEC	1.6×10^{17}	107	1.37	0.68
Te-doped LPE	$\sim 5 \times 10^{17}$	82		
80-1			0.32	0.05
80-3			4.51	9.55
80-5			6.05	17.2
80-11			3.15	4.66
80-17			2.90	3.95

[a] From Chen Chao and Shen Qihua (1983). Reproduced with permission of the authors.

By using the GaP LED as an example, Chen Chao and Zhang Qi (1983) of Xiamen University have modified the equations derived by Stupp and Milch (1977) so as to make them more useful for calculating MCDL from spectral response in both sides of the *p–n* junction of a GaP : N diode grown by double liquid phase epitaxy. It is claimed that this method may be applicable to diodes fabricated from other III–V compounds grown by the LPE process.

(3) *In* $In_{1-x}Ga_xAs_yP_{1-y}$. Yang Linbao *et al.* (1984) of SIM have measured the L_n and τ_n values in the *p*-$In_{1-x}Ga_xAs_yP_{1-y}$ ($x = 0.26, y = 0.61$) epilayer on an InP substrate with the laser scan and the electroluminescence decay methods. It is gratifying to note that the results obtained by these two methods agree fairly well with each other. The surface recombination velocity S and diffusion length L_n are evaluated as (6–7) $\times 10^4$ cm/s and 1.6–2.3 μm respectively, from an analysis of the results of both methods.

12. Microdefects and Trace Impurities Analysis

a. Microprecipitates and Microinclusions

In view of the important role played by C in the compensation mechanism for undoped SI GaAs, a study of C microinclusions in GaAs should merit our serious attention. The pioneer work of Wan Qun *et al.* (1984) of GINM on this subject is, therefore, especially recommendable.

Wan Qun *et al.* have studied the C microinclusions in bulk and epitaxial GaAs materials by means of the JCXA-50A electron microprobe and the JEM-1000 ultrahigh-voltage electron microscope (HVEM), with painstaking precautions taken to prevent contamination by C during preparation of the samples. It is shown that C microinclusions exist in all of the HB and LEC crystals, as well as in the VPE and LPE films. It is also shown that the microinclusions are randomly distributed in the samples. The size of the microinclusions is 2000 Å to 100 μm. Electron diffraction analysis shows that the C microinclusions have a primary cubic structure with a lattice constant of 5.55 Å. In several samples analyzed by the electron microprobe, Si and O are observed along with C. It is the opinion of these investigators that the C microinclusion is probably the nucleation center for the other precipitates. This supposition is supported by the coexistence of C, α-cristobalite and free As, as shown in Fig. 12. The main source of C microinclusions is thought to be the C in high-purity arsenic and the C compounds in $AsCl_3$.

The HVEM has also been used by Sun Guiru and Liu Ansheng (1983) of GINM for observing the form of Si–O microinclusions in various III–V

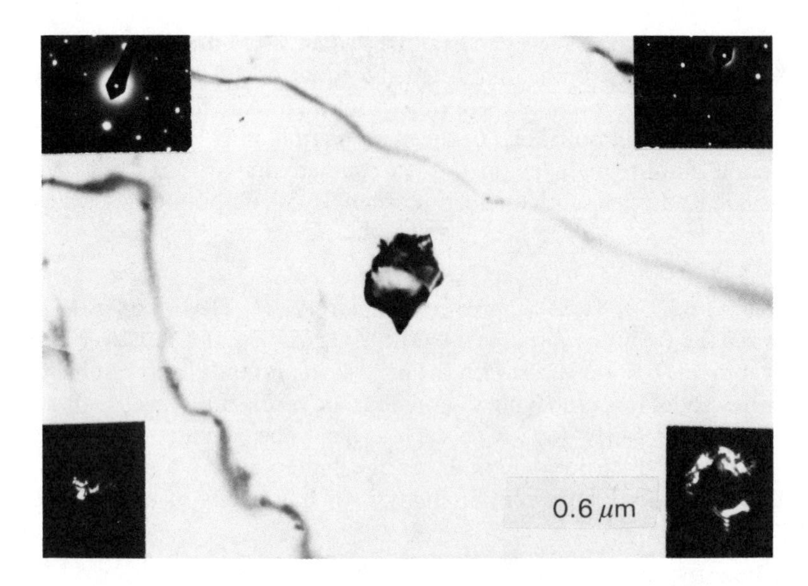

FIG. 12. Micrograph of C microinclusion and As microprecipitate in HB GaAs. Left: As (orth.). Up—$[\overline{251}]$ diffraction pattern; down—dark field image by $(1\bar{1}3)$ reflection. Right: C (cub.). Up—[123) diffraction pattern; down—dark field image by $(\bar{2}10)$ reflection. [From Wan Qun *et al.* (1984), reproduced with permission of the authors.]

materials, including HB Si-, Cr-doped and undoped GaAs, LEC Cr-doped GaAs, Sn-doped InP and Te-doped GaP. It is found that there are more microinclusions in regrown crystals than in those grown from directly synthesized materials. The electron diffraction analyses show that the microinclusions are tetragonal α-cristobalites. Observations with HVEM also reveal that the α-cristobalite microinclusions act as nuclei for the precipitation of free As or Cr_2As. The presence of microprecipitates of CrAs and Cr_2As in Cr-doped SI HB GaAs has been reported by Sun Guiru and Tan Lifang (1983).

Related to the above observations is the work on As microprecipitates by Sun Guiru (1983). Microprecipitates of free As are observed in Si-doped, Cr-doped and undoped HB GaAs, indicating that As precipitation is independent of the dopant used. Nevertheless, α-cristobalite microinclusions or microprecipitates, C microinclusions and Cr_2As microprecipitates can provide heterogeneous nuclei for the precipitation of As. The crystal structure of the As microprecipitates is confirmed by electron diffraction analysis.

Microprecipitates in Fe-doped LEC SI InP have been examined by Chai Xiyun *et al.* (1983) of IS, while those in Fe- and Zn-doped LEC InP by

Xu Yongquan *et al.* (1983) and Fang Dunfu *et al.* (1984) of SIM. It is concluded by the latter investigators that precipitate-free Fe- and Zn-doped InP can be produced under certain controlled conditions (see Section 2b).

b. Microdefects and Dislocations

It is well-known that microdefects and dislocations play an important role in the degradation of DH laser diodes. The research work done on microdefects and dislocations is, therefore, mostly related to $Al_xGa_{1-x}As/GaAs$ and $In_xGa_{1-x}As_yT_{1-y}/InP$ heterostructures, especially the former.

Chen Gaoting *et al.* (1981) of the Shanghai Institute of Optics and Fine Mechanics have observed the presence of short line defects on the surface of the GaAs epilayer grown on AlGaAs. These short lines are parallel to each other and lie along the $[1\dot{1}0]$ direction. They are 0.5–10 μm long and 0.5 μm in width, with a depth of several hundred Å. The density of these short lines depends on the purity of hydrogen and the Al content. It is suggested that these short lines are related to lattice-mismatch, oxygen contamination and preferred growth along the [110] direction.

The TEM work of Shen Houyuan *et al.* (1984) of IS on interfacial defect in the LPE $Al_xGa_{1-x}As/GaAs$ structure reveals that dislocation nets do not exist in the interfacial layer, where x is small enough, but will be generated with higher values of x. These dislocation nets are found to be able to suppress the extension of substrate dislocations to a certain extent. The dependence of the formation of dislocation nets in the interfacial layer on x is attributed to lattice-mismatch.

The configurations of dislocations in LPE $Al_x Ga_{1-x}As$ layers have been observed with HVEM by Liang Jingguo *et al.* (1981) of BJU. TEM observations show that dislocation nets are distributed inhomogeneously with the LPE layer. There are also many dislocation loops of different sizes scattered in the dislocation nets and dislocation clusters generated by inclusions. Isolated stacking faults are also observed. Models for the generation of the configurations of dislocations are discussed.

c. Trace Impurities Analysis

So much work has been done on trace impurities analysis under the impact of the development of semiconductor industry in China that only a very limited amount will be covered here. For the sake of brevity, the results of several selected investigations are listed in Table XIII. Readers especially interested in this topic may consult a recent review written by Wang Houchi (1982) of SIM.

TABLE XIII

DETERMINATION OF TRACE IMPURITIES IN III–V COMPOUNDS AND RELATED MATERIALS

Author	Method	Material	Detective limit	
Lin Zhongpeng[a] (1977)	Polarography-cathodic stripping voltammetric method	Ga	S	5 ppb
Zhang Taishao[a] (1979)	Flameless atomic absorption spectrophotometry	Ga	Na	2 ppb
Wang Rongjuan and Xiong Qinghua[c] (1980)	Catalytic polarography	In, Sb InSb	Te	0.5 ppb
Liu Qiongen *et al.* (1980)	Flameless atomic absorption spectrophotometry	As	Na Cu	0.2 ppb 0.5
		$AsCl_3$	Na Cu	0.05 0.1
		GaAs	Na Cu Fe Cu	0.2 0.5 4 2
Wu Zhuoqing and Zhou Meizhen[c] (1981)	Flameless atomic absorption spectrophotometry	Quartz	Na K Cu Fe	3 ppb 4 5 20
Lin Zhongpeng[a]	Polarography-cathodic stripping voltammetric method	As $AsCl_3$	S S	0.05 μg/g 0.005
Wu Jinying[b] *et al.* (1984/85)	Flameless atomic absorption spectrophotometry	Ga	Zn	ppb order
Wu Zhuoqing *et al.*[c] (1982)	Flameless atomic absorption spectrophotometry	In	Cu Fe	0.04 ppb 0.04
Li Changshi *et al.*[b] (1982)	Polarography using a reducing mixture	In	S	0.08 μm/g absolute
Xiong Qinghua and Wang Rongjuan[c] (1983)	Catalytic polarography	In, B_2O_3	Si	7×10^{-10} M

[a] GINM
[b] Nonferrous Alloy and Semiconductor Materials Institute
[c] SIM

VI. Surface Properties and Physical Chemistry

13. Surface and Interface Properties

a. Surface Properties

The research work on surface properties of semiconductors has been carried out exclusively by scientists of FDU.

The preferential sputtering of the (110) surface of GaAs has been studied by Dai Daoxuan and Tang Houshun (1985), using XPS and CLS, and by Dai Daoxuan and Zou Weiliang (1985), by means of AES. By using XPS, preferential As sputtering is observed on the GaAs (110) surface, sputtered with 1–8 keV Ar^+ with the Ga/As atomic ratio within the range of 1.3–1.8, which decreases somewhat after annealing at 300°C, but still remains >1.0. The CLS of the GaAS (110) surface indicates a loss peak in the neighborhood of 11.2 eV after sputtering, which is considerably reduced by annealing at 300°C. From the AES studies, it is concluded that the energy of Ar^+ should be within 1–2 keV in order to avoid the adverse effect of preferential sputtering.

The CLS spectra of InP (100) and (111) surfaces have been measured by Hou Xiaoyuan *et al.* (1984), using a primary electron beam of 300 eV energy. Seven loss peaks from 3.8 to 20.2 eV are observed. The appearance of In bulk and surface plasma loss peaks indicates the presence of In islands on the surface, introduced by Ar^+ etching. It seems to be more difficult to eliminate the In islands on the InP (100) surface than on the (111) surface by annealing.

The Ar^+-etched clean InP (100) surface has also been studied by Yu Mingren *et al.* (1983). It is found by these investigators that preferential P etching takes place within a layer about 20 Å thick beneath the surface, through Ar^+ etching. By annealing at 300°C for half an hour, the surface In/P atomic ratio drops to 1.2, which is close to unity, according to these investigators, taking the estimated 20% experimental error into consideration.

The adsorption of oxygen on the GaAs (111) surface has been studied by Ding Xunmin *et al.* (1984) of FDU by XPS. It is found that oxygen atoms are first bonded to surface Ga atoms at low oxygen exposure, and that the As atoms are also involved in the substrate-absorbate bond formation at higher exposures. After annealing the sample at temperatures below 400°C, the As–O bonds disappear, while the Ga–O bonds can be eliminated only at an elevated annealing temperature (550°C). The Ga islands, which might exist on the ion-bombarded and annealed surface, are believed to have been partly eliminated by the oxygen exposure and annealing treatment.

The interactions between the InP clean surface under electron beam radiation and the oxygen-containing species in the high vacuum chamber have been studied by Yu Mingren *et al.* (1984) by means of AES. It is found that water vapor plays a more significant role in response to electron-stimulated adsorption of oxygen (ESO) than oxygen itself. At the early stage, the oxygen atoms bond with surface In atoms only, not with the P atoms. Then, they penetrate into the bulk and combine with both In and P atoms, converting the surface layer into an oxide with a practically constant oxidation rate. By comparing the Auger depth profiles of various InP native oxides, the ESO process is found to be similar to the ordinary oxidation processes. It is also found that the P concentration in the ESO oxide is higher than in the thermal oxide but lower than in the anodic oxide.

b. Interface Properties

(1) Interface studies pertaining to oxidation. The anodic oxidation of GaAs in an oxygen plasma has been studied by a number of research groups (Zhang Guansheng *et al.* of FDU, 1981a,b, 1983; Chen Zhihao *et al.*, 1982; Dong Zhiwu of JLU, 1983). Their experimental results on the AES and XPS of the oxide are in general agreement with those reported in the literature. The presence of elementary As at the oxide-substrate interface is a common observation of all these investigators, and is explained by the interaction between As_2O_3 and GaAs, as proposed by Schwartz (1980). The density of interface state observed by Chen Zhihao *et al.* is comparable with the results of some recent publications (Zeisse, 1977; Yokoyama *et al.*, 1978; Koshiga and Sugano, 1979; Weimann, 1979).

Chen Keming *et al.* (1982) of IS have studied the In_2O_3–InP and SiO_2–InP interfaces by means of AES and ESCA. They conclude that the binding energy for In4d in In_2O_3 (InP) is 20.4 eV, while that for P2*p* is 13.4 eV. Their experimental results also provide some evidence for the occurrence of the interfacial reaction between SiO_2 and InP resulting in the formation of In_2O_3 and Si. The Al_2O_3–InP interface has been studied by Yuan Renkuan and Xu Junmin (1984) of Nanjing University.

Anodic oxidation of InGaAs and InSb has been studied by Chen Keming *et al.* (1985) of IS and Xie Boxing *et al.* (1983) of the Kunming Institute of Physics. The density of interface states is found to be 2×10^{12} cm^{-2} eV^{-1} in the case of InSb.

Some modifications have been made by Zhou Mian and Wang Weiyuan (1983) on the classical theory of M–S contacts, taking into consideration the interface layer and states between metal and semiconductor in order to explain the I–V characteristics of thin insulator MIS SB diodes. The I–V and I–1/T expressions derived from the modified theory agree well with the experimental results. The carrier transport mechanism and the relationship between the electron transmission coefficient and the effective barrier height are discussed.

(2) *Metal-semiconductor interface.* The interfacial reactions between Pt and GaAs in the temperature range of 250–450°C have been studied by Xie Baozhen and Liu Shijia (1984) of the Institute of High Energy Physics in collaboration with Zhang Jingping *et al.* of IS by means of BRS and AES. It is shown that after annealing at 300°C for 20 min., a solid state reaction takes place, resulting in the formation of $PtAs_2$ at the interface. The final equilibrium states $PtAs_2$ and PtGa appear at an annealing temperature higher than 400°C.

Compound formation at the Pd/GaAs interface has been studied by Zeng Xianfu of the South China Institute of Technology and by Chung (1982) by means of X-ray diffraction *in situ* at temperatures from 25 to 500°C. The compounds Pd_2Ga, PdGa and $PdAs_2$ are observed after 15–20 min. of heating at 350°C ion an argon atmosphere, while isothermal heating at 500°C in argon results in the appearance of PdGa alone in the form of globular protrusions ($\geq 0.1\ \mu$m in diameter), as revealed by SEM.

An investigation of the Pt/InP interfacial reaction has been made by Dai Daoxuan *et al.* (1985) of FDU by means of XPS. A mixing of Pt, In and P atoms is shown to occur at the Pt/InP interface with a high In/P atomic ratio apparently resulting from the Pt/InP interfacial reaction.

(3) *Semiconductor-semiconductor interface.* The GaAs/AlGaAs DH diode has been found to exhibit switching and memory characteristics by Lin Shiming *et al.* (1982) of IS. A model consisting of the interface state emptied by impact ionization and refilled with injected electrons is proposed. The deep level E_2 at the interface is found to have a density as high as $2.6 \times 10^{17}\ \mathrm{cm}^{-3}$ with $E = E_c - 0.34$ eV and $\sigma = 1.3 \times 10^{17}\ \mathrm{cm}^2$, and is assumed to be related to lattice-mismatch as well as to oxygen.

It is well known that the so-called inverted structure or an undoped GaAs epilayer grown on a heavily doped Si : AlGaAs substrate by MBE usually exhibits a low mobility. A preliminary study has been made of this effect by Xin Shangheng of SIM and Eastman (1983). It is suggested by these investigators that one of the factors responsible for the low mobility of the epilayer could be the presence of excessive interface states.

The photovoltage response of LPE GaAlAs/GaAs hetrojunction has been measured by Yu Lisheng and Wang Gunda (1982) of BJU. It is shown that many interface energy states exist at the hetrojunction interface, including an acceptor-like interface state in the forbidden energy gap of GaAlAs, at about 0.7–0.8 eV below E_c.

Ma Shimang and Tan Jingzi (1984) of NSR have studied the deep levels in GaAs MESFETs by DLTS measurement. Several hole traps (0.4, 0.53, 0.68 and 0.91 eV) and electron traps (0.30, 0.44 and 0.84 eV) are detected in the vicinity of the interface between the active and buffer layers. Two levels

(E_v + 0.91 eV and E_c − 0.84 eV) can be identified as Cr and EL2 respectively. A similar study has been made by Shen Chi and Zhan Qianbao (1985) of FDU and SIM respectively. Three hole traps (0.53, 0.71 and 0.91 eV) are found to exhibit a large noise with the 0.71 eV level being present in MESFETs. Besides, a wide band (E_v + 0.2 eV to E_v + 0.5 eV) is observed in those MESFETs having a large noise figure and degenerated performance.

14. Heterogeneous Reactions among Phases

a. Equilibrium among GaAs Melt, B_2O_3 Encapsulant and SiO_2 Crucible

During the growth of LEC GaAs crystals, Si contamination from the SiO_2 crucible is of significance in view of its effect on the electrical properties of undoped SI material. The reaction among the GaAs melt, B_2O_3 encapsulant and SiO_2 crucible was written as

$$Si + B_2O_3 = B + SiO_2 \tag{9}$$

by Thompson and Newman (1972) on the basis of the linear relationship between the Si and B contents of the GaAs crystals obtained from their localized vibrational modes analyses. No attempt has, however, been made by these investigators to balance the equation. Following the method of these investigators, Zhang Minquan *et al.* (1985) of SIM have obtained similar relationships between the Si and B concentrations reported in the literature (Laithwaite *et al.*, 1977; Oliver *et al.*, 1981; Fairman *et al.*, 1981; Hobgood *et al.*, 1981), as shown by the bands in Figs. 13 and 14, which have the slopes 2/3 and 1/2 respectively. On the other hand, the line obtained by Thompson and Newman has a slope of 4/3, as found by Zhang Minquan *et al.*, after interchanging the coordinates so as to conform to the bands in Figs. 13 and 14. According to the suggestion of Zhang Minquan *et al.*, this difference in the slopes may be explained by assuming the following reactions take place among the melt, encapsulant and crucible:

$$3\underline{Si} + 2B_2O_3 = 4\underline{B} + 3SiO_2, \tag{10}$$

$$3\underline{Si} + B_2O_3 + 3H_2O = 2\underline{B} + 3SiO_2 + 3H_2, \tag{11}$$

and

$$4\underline{Si} + B_2O_3 + 5H_2O = 2\underline{B} + 4SiO_2 + 5H_2. \tag{12}$$

It is also suggested by Zhang Minquan *et al.* that the amount of H_2O participating in the reaction will depend upon the crystal-pulling conditions, especially the H_2O content of the encapsulant. This suggestion is in agreement with the experimental findings that Si contamination can be reduced significantly by using a wet B_2O_3 encapsulant. It follows, therefore, that the

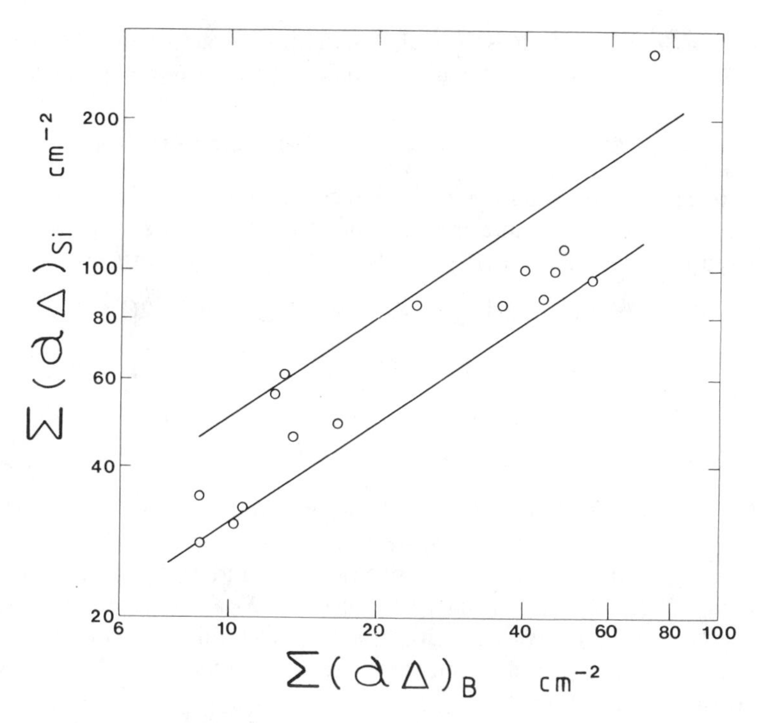

FIG. 13. Si–B equilibrium between melt and encapsulant based on data of Laithwaite *et al.* (1977). [From Zhang Minquan *et al.* (1985), reproduced with permission of the authors.]

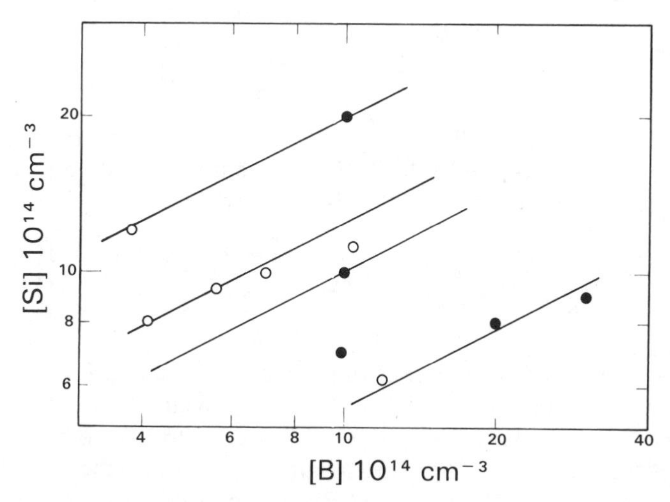

FIG. 14. Si–B equilibrium between melt and encapsulant based on data of Fairman *et al.* (○), 1981), and Hobgood *et al.* (●), 1981. [From Zhang Minquan *et al.* (1985), reproduced with permission of the authors.]

slope of an Si–B plot, such as those shown in Figs. 13 and 14, may be taken as a measure of the H_2O content, which is difficult to determine experimentally.

b. Removal of Si from In or Ga Melt by Preferential Oxidation

The thermodynamics and kinetics of Si contamination and purification of molten In and Ga melts in a quartz container have been studied by Wang Guangyu and Peng Ruiwu (1985) of SIM. For the reaction

$$SiO_2(\text{vitreous}) + 2H_2(g) = \underline{Si}(\text{in liq. In or Ga}) + 2H_2O(g), \qquad (13)$$

the differential equation

$$\frac{dX_{Si}}{dt} = \frac{FM_{In}}{2RTW_{In}}\left\{\left(\frac{K_p}{\nu_{Si}X_{Si}}\right)^{1/2} - P^{\circ}_{H_2O}\right\} \qquad (14)$$

can be derived after Greene (1973), where X_{Si} is the Si atomic fraction in the melt, F is the H_2 flow rate, ν_{Si} is the activity coefficient of Si in the liquid metal, $P^{\circ}_{H_2O}$ is the initial H_2O partial pressure and K_p is the equilibrium constant of reaction (13). It can be seen that dX_{Si}/dt becomes zero when $P^{\circ}_{H_2O} = (K_p/\nu_{Si}X_{Si})^{1/2}$, or when X_{Si} reaches the equilibrium concentration. On the basis of thermodynamic data from the literature, the following relationships are obtained:

$$\lg X^{eq}_{Si}(\text{In}) = -\frac{26600}{T} + 5.68 - 2\lg P^{\circ}_{H_2O}, \qquad (15)$$

and

$$\lg X^{eq}_{Si}(\text{Ga}) = -\frac{24800}{T} + 5.12 - 2\lg P^{\circ}_{H_2O}. \qquad (16)$$

By integrating Eq. (14) and substituting in the initial condition ($X_{Si} = X^{\circ}_{Si}$ when $t = 0$), the following equation is obtained:

$$T = \frac{4X^{eq}_{Si}RTW_{In}}{FP^{\circ}_{H_2O}M_{In}}\left[(\sqrt{p} - \sqrt{q}) + \frac{(p - q)}{2} + \lg\frac{(\sqrt{p} - 1)}{(\sqrt{q} - 1)}\right], \qquad (17)$$

where $X^{eq}_{Si} = K_P/\nu_{Si}(P^{\circ}_{H_2O})$, $p = X^{\circ}_{Si}/X^{eq}_{Si}$ and $q = X_{Si}/X^{eq}_{Si}$.

It is the opinion of Wang Guangyu and Peng Ruiwu that Eq. (17) is more convenient and intuitive than those derived by Weiner (1972) and Greene (1973), because it gives t, the shortest time needed for X°_{Si} to be lowered to X^{eq}_{Si}, explicitly in terms of the other parameters. It is also their opinion that this equation is especially useful in LPE and/or VPE of InP, where the Si content of the In melt needed for the growth of high-quality epilayers would be too low to be detected by the current methods for trace impurities analysis.

c. Distribution of Te between Liquid Al and In

A study has been made of the distribution of Te between liquid Al and In by Wang Caihao *et al.* of Shanghai University of Science and Technology and Wang Rongjuan *et al.* of SIM (1982). The distribution ratio

$$D = N_{\mathrm{Te}}^{\mathrm{In}}/N_{\mathrm{Te}}^{\mathrm{Al}} \tag{18}$$

is found to increase with dilution and tends to exceed ten when $N_{\mathrm{Te}}^{\mathrm{In}}$ is decreased to 0.01 ppma.

d. Thermodynamic and Kinetic Analyses for VPE of GaAs and InP

The thermodynamic analysis for VPE of GaAs has been made by Lu Dacheng (1982a) of IS using the $Ga/AsCl_3/H_2IG$ system (where IG stands for inert gas), taking into consideration the incompleteness of the reaction between the gallium source and the gas phase. This incompleteness is taken care of by a flow efficiency η and a reaction efficiency ν, after Morizane and Mori (1978). Three sets of simultaneous equations have been derived for the various zones of the VPE reactor, and these have been solved on a computer. The results of calculation show the general features of inert gas-hydrogen mixed-carrier systems, and should prove useful for certain practical purposes.

Lu Dacheng (1982b) has made a similar analysis of the thermodynamics of VPE of InP, with the $In/PCl_3/H_2$ system resulting in certain useful relationships among the various parameters involved in the process.

Wang Guangyu and Ding Yongqing of SIM (1985) have, on the other hand, evaluated the effect of thickness fluctuation of the solid crust on the melt surface on the composition of the downstream gas phase from a kinetic point of view. From the results of this study, it is recommended by these investigators to use an In boat of the sealed type in conjunction with a small flow rate, with a view to improving the source-melt interaction, thereby obtaining a VPE InP layer of better quality.

15. Thermochemical Models for Certain Defects

a. A Thermochemical Model for Certain Deep Levels in GaAs

A thermochemical model developed by Zou Yuanxi *et al.* in the past ten years (Zou Yuanxi, 1974, 1977, 1981, 1982; Zou Yuanxi *et al.*, 1983a,b, 1984) for certain deep levels in GaAs may be described in terms of three equilibria, viz, the A–B–C equilibrium, the C–EL2 equilibrium and the C–D (I) equilibrium.

(1) *The A–B–C equilibrium.* As mentioned above (Section 9a), an apparent equilibrium among certain deep levels in n-GaAs has been proposed by Zou Yuanxi (1974):

$$\mathrm{Ga_{As}As_{Ga}} + (\mathrm{V_{Ga}})_2^- + e^- = \mathrm{Ga_{As}V_{Ga}^-} + \mathrm{As_{Ga}V_{Ga}^-}, \tag{19}$$

where $\mathrm{Ga_{As}V_{Ga}^-}$ and $\mathrm{As_{Ga}V_{Ga}^-}$ are expected to be equal in amount, like the A and B hole traps observed by Lang and Logan (1975) in LPE GaAs. It is gratifying to note that the presence of a large amount of $\mathrm{Ga_{As}As_{Ga}}$ in GaAs can be predicted on the basis of the theoretical estimates of the enthalpy of antistructure pair formation in zinc-blende type semiconductors (Van Vechten, 1975). The presence of the double gallium vacancy has been discussed by Logan and Hurle (1971), who considered the strong binding energy between two arsenic atoms on neighboring arsenic sites. It will be shown later that the double Ga vacancy may tentatively be assigned to an electron trap C observed by Ozeki *et al.* (1979) in $\mathrm{N_2}$: VPE GaAs.

Assuming $\mathrm{Ga_{As}As_{Ga}}$ and $(\mathrm{V_{Ga}})_2^-$ to be constant under given crystal growth conditions, we have [30]

$$K_{\mathrm{ABC}} = \frac{[\mathrm{Ga_{As}V_{Ga}^-}][\mathrm{As_{Ga}V_{Ga}^-}]}{n}, \tag{20}$$

or

$$[\mathrm{Ga_{As}V_{Ga}^-}] = [\mathrm{As_{Ga}V_{Ga}^-}] = K_{\mathrm{ABC}}\, n^{1/2}. \tag{21}$$

The inequality of [A] and [B] in certain cases is explained by the instability of A in the presence of dislocations, where $\mathrm{V_{Ga}V_{As}}$ is expected to precipitate easily:

$$2\mathrm{Ga_{As}V_{Ga}^-} + (\mathrm{V_{Ga}})_2^- = 2\mathrm{Ga_{Ga}} + 2\mathrm{V_{Ga}V_{As}} + 3e^-. \tag{22}$$

The source of the double Ga vacancy is assumed to be the substrate.

(2) *The C–EL2 equilibrium.* The C–EL2 equilibrium may be written as follows:

$$\mathrm{As_{As}} + (\mathrm{V_{Ga}})_2^- = \mathrm{V_{Ga}As_{Ga}V_{As}} + e^-, \tag{23}$$

$$K_{\mathrm{CE}} = \frac{[\mathrm{EL2}]n}{[\mathrm{C}]}. \tag{24}$$

The above relation has been found to hold true by Zou Yuanxi (1982) from an examination of the data of Ozeki *et al.* (1979), thus lending support to the postulate for the existence of the C–EL2 equilibrium, as well as to the assignment of $(\mathrm{V_{Ga}})_2^-$ to the C trap.

[30] A bracket is used to represent concentration throughout this chapter.

(3) *The C–D(I) equilibrium.* The equilibrium between the C-trap and donors may be represented by the following equation:

$$2D^+ + (V_{Ga})_2^- + 3e^- = 2DV_{Ga}^-, \tag{25}$$

$$K_{CD} = \frac{[DV_{Ga}^-]^2}{[C]n^5}. \tag{26}$$

The deviation from stoichiometry Δ may be represented by

$$\Delta = [EL2] + [C] + \tfrac{1}{2}[DV_{Ga}^-], \tag{27}$$

which should be constant under given crystal growth conditions.

For bulk GaAs, the constant K_{CE} has been found by Zou Yuanxi *et al.* (1982) to be $5.1 \times 10^{16}\ cm^{-3}$ from the data of Lagowski *et al.* (1982), which employs a curve-fitting method on the assumption of constant deviation from stoichiometry, found to be $4.15 \times 10^{16}\ cm^{-3}$ by trial and error. Assuming different values for K_{CD}, Zou Yuanxi *et al.* have been able to obtain a set of [EL2]-n curves by solving Eqs. (24), (26) and (27) simultaneously on a computer. The curves are seen to be in good agreement with the DLTS data of Lagowski *et al.* and of Zou Yuanxi *et al.* themselves. This agreement provides, in our opinion, adequate support to the proposed C–D equilibrium represented by Eq. (25).

Another C–D equilibrium, represented by

$$D^+ + (V_{Ga})_2^- = V_{Ga}DV_{Ga}, \tag{28}$$

will result in the formation of the deep donor $V_{Ga}DV_{Ga}$. Examples for reactions (25) and (28) are given in Section 9b.

Needless to say, the C–D equilibrium may likewise be extended to acceptor impurities, and then it will be called the C–I equilibrium instead.

b. *Some Experimental Evidences for the A–B–C Equilibrium*

Some experiments have been carried out by Zou Yuanxi *et al.* (1984) with a view to providing support to their model.

(1) *Effects of carrier concentration and growth rate on A and B traps in LPE GaAs.* The effect of growth rate on [B] is shown in Fig. 15. It is seen that a maximum in [B] is reached at a growth rate of $\sim 0.08\ \mu m/min$ under their growth conditions. The relationship between [B] and n_{300K} for a constant growth rate is shown in Fig. 16 as a straight line with a slope of $\frac{1}{2}$, in agreement with the model.

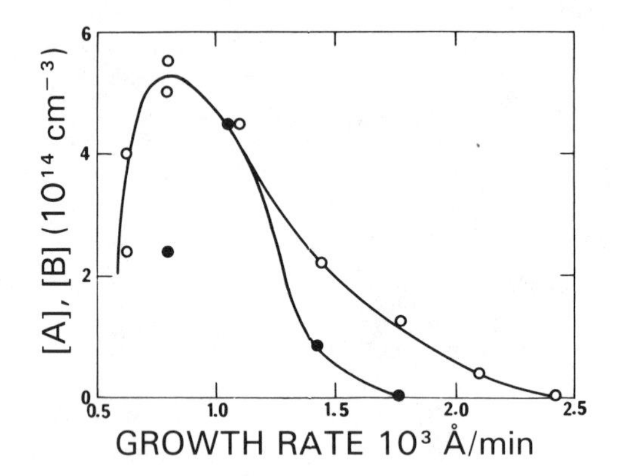

FIG. 15. Growthrate dependence of A and B traps in LPE n-GaAs. [From Zou Yuanxi *et al.* (1984), reproduced with permission from *Thirteenth International Conference on Defects in Semi-Conductors*, edited by L. C. Kimerling and J. M. Parsey, The Metallurgical Society of AIME, 420 Commonwealth Drive, Warrendale, PA 15086, U.S.A.]

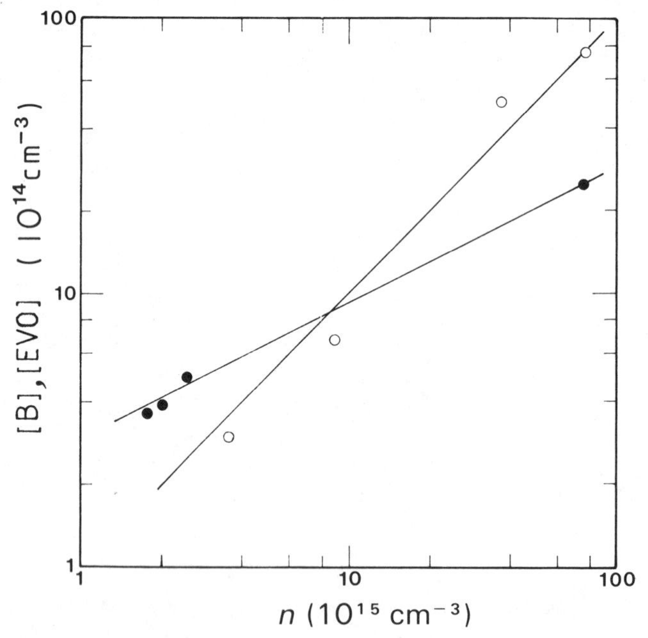

FIG. 16. Carrier concentration dependence of B (●) and EVO (○) in LPE n-GaAs. [From Zou Yuanxi *et al.* (1984), reproduced with permission from *Thirteenth International Conference on Defects in Semi-Conductors*, edited by L. C. Kimerling and J. H. Parsey, The Metallurgical Society of AIME, 420 Commonwealth Drive, Warrendale, PA 15086, U.S.A.]

(2) *Diffusion of Ga vacancy from the substrate into the epilayer.* At a growth rate, it is conceivable that substrate dopants, along with V_{Ga}, can diffuse into the epilayer and react with the latter, resulting in a decrease in A and B concentrations. As the growth rate increases, this effect will be less pronounced. With still increasing growth rates, the diffusion of V_{Ga} will lag behind the growth of the epilayer. The experimental observation of a maximum [B] at a certain growth rate can, therefore, be regarded as evidence that the substrate is the source of Ga vacancy.

Zou Yuanxi *et al.* (1983) have also studied the effect of temperature on the formation of A and B traps. A high growth temperature is found to promote the formation of these traps. This can be taken as more evidence of the diffusion mechanism in supplying Ga vacancies to the epilayer.

It may be mentioned in passing that the possibility of outdiffusion from the substrate as the origin of EL2 created in *n*-type MBE GaAs after capped heat treatment has also been pointed out by Xin Shangheng *et al.* (1982). Their suggestion is in accord with the experimental observation of Zou Yuanxi *et al.* mentioned above, as it is the belief of the latter investigators that EL2 and the A and B traps are of the same origin—the double Ga Vacancy.

c. *Equilibria Involving the Two Recombination Centers in InSb*

In order to explore the nature of the two well-known donor-like recombination centers in InSb ($E_1 = E_v + 0.071$ eV, $E_2 = E_v + 0.11$ eV), an analysis has been made of the data of Hollis *et al.* (1967) by Zou Yuanxi (1973). The results are listed in Table XIV.

Sun Qing of Shanghai University of Science and Technology, and Zou Yuanxi and Shi Huiying of SIM (1983) have made a tentative identification of E_1 and E_2 on the basis of the constancy of pN_1/N_2 and $N_1 + N_2$, shown

TABLE XIV

RELATIONSHIP BETWEEN N_1 AND N_2 IN InSb[a]

Sample	p (cm^{-3})	$E_v + 0.071$ eV N_1 (cm^{-3})	$E_v + 0.11$ eV N_2 (cm^{-3})	pN_1/N_2 (cm^{-3})	$N_1 + N_2$ (cm^{-3})
P1	4.52×10^{14}	3.0×10^{13}	7.7×10^{13}	1.8×10^{14}	1.1×10^{14}
P2	1.85×10^{14}	5.5×10^{13}	8.0×10^{13}	1.3×10^{14}	1.4×10^{14}
P3	6.6×10^{14}	2.0×10^{13}	8.8×10^{13}	$1,5 \times 10^{14}$	1.1×10^{14}
P4	41.0×10^{14}	0.6×10^{13}	13.0×10^{13}	1.9×10^{14}	1.4×10^{14}

[a] From Zou Yuanxi (1973).

in Table XIV. The following reactions are assumed to approach apparent equilibrium at a certain temperature during post-growth cooling:

$$Sb_{Sb} + (V_{In})_2^- + 3e^+ = Sb_{In}^{2+} + V_{In}V_{Sb} \tag{29}$$

$$O_i^x + (V_{In})_2^- + 2e^+ = (V_{In}O_iV_{In})^+ \tag{30}$$

$$K_{SV} = \frac{[Sb_{In}^{2+}][V_{In}V_{Sb}]}{[(V_{In})_2^-]p_T^3}, \tag{31}$$

$$K_{OV} = \frac{[(V_{In}O_iV_{In})^+]}{[(V_{In})_2^-][O_i^x]p_T^2}. \tag{32}$$

By combining Eqs. (31) and (32) with the equations

$$K_{VV} = \frac{[(V_{In}V_{Sb})]}{[V_{In}][V_{Sb}]} \tag{33}$$

and

$$K'_{VV} = [V_{In}][V_{Sb}], \tag{34}$$

the following relation is obtained:

$$\frac{p[(V_{In}O_iV_{In})^+]}{[Sb_{In}^{2+}]} = K'[O_i^x] = K_{VOS}, \tag{35}$$

where $[O_i^x]$ is assumed to be constant.

As both reactions (29) and (30) involve $(V_{In})_2^-$ as a reactant, the sum $N_1 + N_2$ may be taken as constant, on the assumption that $[V_{In}^x]$ and the retained $(V_{In})_2^-$ are low enough to be negligible, and that the deviation from stoichiometry in bulk InSb is approximately constant. Thus, the assignment of $(V_{In}O_iV_{In})^+$ and Sb_{In}^{2+} to the centers E_1 and E_2 respectively is seen to be consistent with the above-mentioned relationships between the concentrations of these two centers listed in Table XIV.

VII. Technology

16. Ion Implantation, Laser Annealing and Neutron Transmutation

a. Ion Implantation

A summary has been made by Wang Weiyuan *et al.* (1983) of SIM of their work on ion implantation in GaAs. These investigators have applied direct implantation in SI GaAs substrates for the fabrication of coplanar switching diodes and dual-gate MESFETs, with satisfactory results. The work of

Wang Weiyuan *et al.* (1982) on Si implantation in Cr-doped SI GaAs substrate shows that proper selection of the substrate material and purity control of the $\sim Si^+$ beam are essential to reproducible, high activation implantation. A study has also been made of proton implantation in Sn-doped GaAs and the subsequent annealing by Wang Weiyuan *et al.* (1979). It is shown from their experimental results that the critical dose for the amorphous layer formation is $\sim 10^{16}$ cm^{-2}, and the recovery of strain occurs in the temperature range 150–450°C. Besides, the range for various ions in amorphous targets of GaAs and related compounds has been calculated for the first time (Wang Dening *et al.*, 1980).

Ion implantation has also been used by Han Jihong *et al.* (1982) of NSR for the fabrication of low noise MESFETs. Both direct implantation in the SI GaAs substrate and implantation in a buffer layer are used. The advantages of implantation over epitaxy are discussed.

MIS SB GaAs dual-gate FETs with a heavily doped channel have been fabricated by Zhou Mian and Wang Weiyuan (1984) of SIM, using direct SI implantation in the SI substrates.

The oxygen implantation technique has been successfully adopted by Deng Xiancan (1983a) of HSRI in the fabrication of isoplanar type GaAs low noise microwave FETs and power SB FETs. These investigators have also applied this technique to the fabrication of several GaAs MICs that are claimed to have significant advantages over the conventional mesa-type structure used in China and other countries.

It may be remarked that Deng Xiancan (1977) began to work on oxygen implantation in GaAs in 1976, in the same year that Favennec (1976) reported his work on oxygen compensation. By using oxygen implantation at 600 KeV and annealing at 500–850°C, Deng Xiancan (1983a) has been able to explore the mechanism of compensation in the SI layer.

Silicon implantation in Fe-doped Si InP at room temperature, followed by capless annealing at 700°C, has been studied by Qiao Yong *et al.* (1983) of SIM. It is shown by these investigators that Si-implanted *n*-InP layers with good electrical properties can be obtained in this way. Qiao Yong *et al.* have also presented their experimental results on capless annealing. Yan Shulan and Zhou Shuxing (1982) of NSRI have studied H^+ implantation in *p*-InP and *p*-InGaAsP to yield disordered SI layers.

It can be seen from the above that research on implantation in InP and related compounds lags considerably behind that on implantation in GaAs. As to other III–V compounds, Zhang Fujia and Sun Da (1985) of Lanzhou University have successfully investigated Zn^+ implantation in the GaP bulk crystal and the (N, Zn)-doped LPE layer and subsequent annealing in a N_2 atmosphere, resulting in a significant improvement in light emitting efficiency of diodes.

b. *Laser and Electron Beam Annealing*

Ma Jinglin *et al.* (1984) of Beijing Polytechnic University have investigated laser annealing of (N, Zn)-implanted $GaAs_{0.6}P_{0.4}$ with a view of improving the light emitting efficiency of the LEDs. It is pointed out by these investigators that stability of the power output of the cw CO_2 laser, homogeneity of the laser beam spot and a suitable annealing procedure are essential to the success of the process. Compared with thermally annealed samples (850°C, 35 min), the laser annealed ones (laser power output 100, 200 W/cm^2, 1 s–1 min) show a greater light intensity.

Electron beam annealing of S-implanted GaAs, including Cr-doped SI GaAs and Zn-doped *p*-GaAs, has been investigated by Sun Huiling *et al.* (1983) of IS, in collaboration with the Electron Beam Group of Institute of Electrical Engineering. As expected, the carrier concentration profile in the EB annealed layer is better than that in the thermally annealed layer because of the absence of diffusion in the former. Additionally, the mobilities in these two differently annealed layers are comparable to each other, in contrast with the case of Si-implanted samples. It is the opinion of the present writer that the clue to this problem might be found in the difference of the behaviors of Si and S as dopants in VPE GaAs, so far as their effects on the creation of "mobility killer" centers are concerned (see Shao Jiuan *et al.*, 1986).

Li Xiqiang *et al.* (1984) of SIM have successfully prepared Si GaAs SOI films of 0.1–0.5 μm thick by sputtering. They have also investigated laser recrystallization of these SOI films by using cw Ar laser beam irradiation. Their preliminary results indicate that the grain size after recrystallization can reach 1 μm, with the Ga/As ratio ~1. Further research is in progress, with a view to increasing the grain size by optimizing the irradiation conditions.

c. *Neutron Transmutation Doping*

The work on neutron transmutation doping of III–V compounds has been meagre. Mo Peigen and Li Shouchun (1984) of SIM and Li Yuexin, Gao Jijin and Li Shiling of the Institute of Atomic Energy have applied this technique to undoped HB GaAs. The relative standard deviation in carrier concentration for a neutron irradiated and properly annealed GaAs sample obtained by the electrochemical C–V method is found to be 4.6%, which is better than the homogeneity determined previously by Mo Peigen and Yang Linbao (1982) of SIM on Si-doped samples. The mobility in the samples annealed at 800°C for a period longer than 2 hours is, however, found to be low, 1820 cm^2/V s for $n = 5.5 \times 10^{17}$ cm^{-3}, in approximate agreement with the results reported by Greene (1979).

17. DIFFUSION AND OXIDATION

a. Diffusion

Yuan Yourong *et al.* (1981) of the Changchun Institute of Physics have studied Zn diffusion in $Ga_{1-x}Al_xAs$ and GaAs. By carrying out diffusion under comparable conditions, these investigators have been able to obtain the difference in diffusion rates for Zn in $Ga_{1-x}Al_xAs$ and GaAs, the ratio being found to be 1:2 in the temperature range 600–700°C for an x value of 0.4. The mechanism for this difference is discussed.

Gong Mengnan and Zhang Fuwen (1981) of the above mentioned Institute have reported on a new diffusion process based on the box method for the diffusion of Zn in $Ga_{1-x}Al_xAs$ and GaAs. This process is similar to the conventional diffusion process used for Si devices. On the basis of the work of previous investigators, Gong Mengnan and Zhang Fuwen have succeeded in obtaining predetermined junction depth, smooth surface and reproducible results. The surface carrier concentration can reach 10^{20} cm^{-3} at a diffusion temperature of 650–750°C. Besides, it is shown that the light-emitting diodes fabricated by using the box diffusion process exhibit reasonably good ohmic-contact behavior, which is tentatively attributed to the incorporation of some gaseous species in the diffusion process. It is the opinion of these investigators that this process can also be applied to GaP and GaAsP LEDs, in addition to those based on GaAs and $Ga_{1-x}Al_xAs$.

The diffusion characteristics for Zn in InP using different Zn sources have been investigated by Zhu Youcai and Gao Dingsan (1982) of JLU. As a new source of Zn, the ternary compound with Zn : In : P = 43.2 : 2.7 : 54.1 has been prepared for the first time. It is shown that the ternary compound is the best so far as the reproducibility of the junction depth and the $X_i - \sqrt{t}$ relationship are concerned.

Pang Yongxiu and Sun Bingyu (1985) of SIM have studied the diffusion of Cd and Zn in InP by the use of eight diffusion sources, including CD, Zn and their compounds. It is found that Cd and, especially, CdP_2 show a relatively slow rate of diffusion, and are, therefore, good sources for diffusion in InP. The experimental results are discussed on the basis of the neutral complex hypothesis suggested by Tien and Miller (1979).

b. Oxidation

The research on oxidation includes process technology, growth characteristics and bulk properties of the oxide layer with interfacial and related properties, discussed in a previous section (13b).

The technology for the growth of the oxide layer on GaAs by anodic oxidation in oxygen plasma has been investigated by Zhang Guansheng *et al.* (1983) of FDU in collaboration with Zhao Guozhen and Niu Chengfa of the

Institute of Materials. These investigators have designed an apparatus based on rf glow discharge, which can be successfully applied to the oxidation of GaAs wafers as large as 40 cm in diameter. The relationship between film thickness and time of oxidation at different temperatures is found to be linear. The films grown are amorphous and uniform. As usual, the in-depth profiles for the As-grown anodic oxide obtained by AES-XPS display three distinct layers with different Ga/As ratios. A relatively homogeneous oxide layer with a thin interface is formed on annealing. The electrical properties of the oxide films are comparable to those reported in recent literature, according to these investigators.

Similar work has been carried out by Chen Zhihao *et al.* (1982) of SIM and Dong Zhiwu (1983) of JLU. The mechanism of oxidation is discussed by Chen Zhihao *et al.* and is shown to be related to mass transport in the oxide by the drift of the mobile ions Ga^{3+} and As^{3+}. Their experimental data are shown to be in agreement with theoretical prediction. In the same paper, Chen Zhihao *et al.* have reported the optical properties of the oxide film obtained from spectroscopic ellipsometry by a method after Aspnes (1977). The composition of the oxide has also been obtained in this way, by adopting the refractive index data on As_2O_3 and Ga_2O_3 reported by Umeno (1979). It is the belief of Chen Zhihao *et al.* (1985) that the usually-observed abnormal C–V characteristics of GaAs MIS structure could originate from deviations from stoichiometry and the presence of elementary As in the insulating film. In order to investigate device applications of the oxide film, Dong Zhiwu of JLU has fabricated *n*-channel, depletion mode MOSFETs by growing a 0.2–0.3 μm thick VPE layer on the SI GaAs substrate and then an oxide layer of 1300 Å on the epilayer. The output characteristics of a typical MOSFET is given, but some abnormal phenomena are found to be present, as expected.

The plasma oxidation of InP is more difficult than that of GaAs, on account of the simultaneous sputtering of the oxide film. Tang Housun *et al.* (1985) of FDU have succeeded in the growth of a satisfactory oxide layer on InP by introducing an intermediate layer of SiO_2 by sputtering, or of Al_2O_3 by evaporation and oxidation. The SiO_2–In_2O_3/P_2O_5 film grown by this technique is found to be dense, stable and quite uniform in thickness over an area of 8 cm^2.

Anodic and thermal oxidation have been studied by Liu Hongdu *et al.* (1982) of BJU. The native oxides grown are evaluated by means of X-ray diffraction and infrared absorption spectra and by measurements on refractive indices, expansion coefficients and Young's moduli. The results indicate that the native oxides grown by these techniques can be applied to planar optical devices and integrated optics, with GaAs as the substrate, while the stress and photo-elastic effect must be taken into consideration in the application of thermal oxide films.

A new anodic oxidation process based on the use of anhydrous tartaric acid–ethylene glycol solution has been developed by Zhou Mian and Wang Weiyuan (1984) of SIM for growing native oxides on GaAs, InP and Al_2O_3 films, with an improvement of the film quality. The use of the non-aqueous solution also helps to stabilize the oxidation process, according to these investigators.

18. Ohmic Contacts

a. GaAs

Wu Dingfen *et al.* of SIM have studied the various aspects of ohmic contact, including composition of the Au/Ge alloy, optimum thickness of the metal layer and effect of different components in the Au–Ge and Au–Ge–Ni systems on *n*-type GaAs. It is found by Wu Dingfen and Chen Fenkou (1979) of SIM that the thickness of the Au–Ge–Ni (eutectic AuGe + 5 w% Ni) layer on *n*-GaAs ($n \sim 10^{17}$ cm^{-3}) should be >1000 Å in order to avoid the rapid rise in specific contact resistance. These investigators (1980) have also found that the optimum Au content of the Au–Ge alloy is around 96 w% Au, while the optimum Ni content is different for different Au/Ge ratios. The use of noneutectic Au/Ge alloy ohmic contact has been investigated by Wu Dingfen *et al.* (1984) for diffused *n*-channel GaAs MESFETs, resulting in a very smooth surface, a specific contact resistance comparable with that reported in the literature and, therefore, also in improved device performances. The composition of the Au/Ge alloy used by these investigators is 150 Å Ge/2000 ~ 2500 Å Au.

Laser alloying of Au–Ge–Ni ohmic contacts on *n*-GaAs has been studied by a number of investigators using both pulse and cw laser beams. Using a modulated Q-ruby pulse laser, Wu Hengxian (1980) of the Shanghai Institute of Optics and Fine Mechanics, in collaboration with Lin Chenglu *et al.* of SIM, has found an optimum laser energy density of 1.5 J cm^{-2} for producing satisfactory ohmic Au–Ge–Ni contacts on Te : *n*-GaAs, with $n = (3 \sim 9) \times 10^{16}$ cm^{-3}. As expected, the contacts have certain advantages over thermally alloyed ones. The pulse laser beam technology has also been used by Yang Qianzhi *et al.* (1981) for producing ohmic contacts on both *n*-GaAs and *p*-GaAlAs or *p*-GaAs of a GaAlAs/GaAs solar cell. An additional advantage of this technology is the possibility of using different laser energy densities for the *n* and *p* sides of the cell so that alloying can take place under optimum conditions for both types of materials. The alloy used for the *p*-type material is Cr–Au, Cr–Au–Zn or Ag–Zn. On the other hand, backside irradiation through the substrate with a cw CO_2 laser has been used by Tsou Shichang *et al.* (1983) of SIM for producing Au–Ge–Ni ohmic contacts on *n*-GaAs with $n = 6 \times 10^{14}$–7×10^{17} cm^{-3}. It is found by these investigators that the laser alloyed contacts show better ohmic behavior and a contact resistance lower than furnace alloyed ones of approximately one order of magnitude for low bulk concentrations ($\sim 5 \times 10^{15}$ cm^{-3}).

b. InP and GaP

Much of the work on ohmic contacts on InP and InGaAsP has been done on *p*-type materials, as the specific resistance of such contacts are much higher than that of the ohmic contacts on *n*-InP. The alloys used for making the contacts consist of Au–Zn, Au + Zn + Au Au–Ge–Ni, Au–Ge–Ni + Au and Ti–Pd–Au. Jing Xinliang *et al.* (1982) of Jiaotong University have used the conventional Au–Zn alloy (10 w% Zn) for producing ohmic contacts on both *p*-InP and *p*-InGaAsP, and found that the contacts on the quaternary alloy have a lower specific contact resistance. Successive layers of Au and Zn to provide a Au + Zn + Au structure have been used by Wang Lijun *et al.* (1982) of JLU to prevent the loss of Zn by evaporation during vacuum deposition. The optimum thickness of the first Au layer is found to be ~300 Å, with the Zn and the top Au layers kept at ~450 Å and ~1200 Å respectively, and with alloying carried out at 420°C for 5 min. Zhang Guicheng and Shui Hailung (1984a) and Zhang Guichang *et al.* (1985a), all of SIM, have, however, chosen to use the Ti–Pd–Au alloy as the ohmic contact material for *p*-InP by sputtering successively 500 Å Ti, 3000 Å Pd and 3000 Å Au layers on In : InP. The use of the Ti and Pd layers is found to have the effect of retarding the mutual diffusion between Au and In at 450°C and to have certain other advantages over the conventional Au–Zn alloy. Therefore, it is the opinion of these investigators that the *p*-InP/TiPdAu electrode should have practical implications in the fabrication of InGaAsP/InP DH lasers.

Zhu Bing *et al.* (1984) of Nanjing University, in collaboration with Pan Maohung of Nanjing No. 19 Factory of Telecommunication Devices and with Mao Paohua and Cheng Yongxi of NSR, have used irradiation with a cw Co_2 laser on both the front and back sides to produce Au-Ge-Ni ohmic contacts on *n*-type InP. It is found that the specific contact resistance is lower in the case of AuGeNi + Au alloys compared with that of AuGeNi alloys. An explanation for this difference is offered on the basis of AES studies.

The formation of ohmic contacts on both *n*- and *p*-type GaP has been investigated by Wang Lijun and Yuan Yuan (1983) of JLU by using Au–Ge–Ni, Ni + Au–Ge–Ni and Au–Sb + Au–Ge–Ni alloys for *n*-GaP and Au–Zn for *p*-GaP. It is found that the use of Ni + Au–Ge–Ni can prevent the balling phenomenon and that the use of Au–Sb(1 w% Sb) + Au–Ge–Ni results in an apppreciable lowering of specific contact resistance.

c. Calculation of Specific Resistance

An ohmic contact model for GaAs and other semiconductors has been devised by Wu Dingfen and Wang Dening (1985) of SIM, using the *n*-type semiconductor as an example. According to this model, the specific resistance ρ_c should be the sum of ρ_{c_1} and ρ_{c_2}, of which the former occurs between the contact metal and the underlying heavily doped semiconductor layer (N_{DC}) after alloying, and the latter is brought about by a barrier arising from the concentration difference between N_{DC} and N_D in the active layer. Either ρ_{c_1} or ρ_{c_2} can be predominant, depending upon the conditions of alloying and other parameters. Under optimum conditions, N_{DC} can be so high that ρ_{c_1} may be neglected, then the usual inverse proportionality between ρ_c and N_D (see, for instance, Wu Dingfen, 1972, and Brasslau, 1981) would be valid if $N_D < N_C$ ($N_C \sim$ effective state density). The experimental results on Au–Ge–Ni ohmic contacts on *n*-GaAs reported in the literature can be explained very well according to these investigators. It is also their opinion that the model can be applied to *p*-type GaAs and InP, and to *p*-type Si as well.

Wu Dingfen and Wang Dening (1986) of SIM have also calculated the specific contact resistance ρ_c for both Au/*n*-InP and Au/*p*-InP, on the basis of carrier transport study of metal-semiconductor systems (see Chang *et al.*, 1971, for Si and GaAs; Lei *et al.*, 1978, for GaP). The experimental data on AuZn/*p*-InP ohmic contacts reported in the literature are interpreted on the basis of theoretical calculations. The problems encountered in the *p*-type ohmic contact experiments of these investigators are also discussed.

19. Etching, Ion Milling and Reactive Ion Etching

a. Etching

The new etching solution of the HCl–H_2O_2–H_2O system suggested by Shaw (1981) has been independently investigated by Wang Limo *et al.* (1982) and Wang Limo (1984) of the Sichuan Solid-State Circuits Research Institute. The residual oxide film left on the GaAs surface etched with this etchant is found to be thin and not contaminated by non-volatile impurities. A ternary diagram of iso-etching-rate contours has been worked out. Besides, etching characteristics and kinetics of the said new etchant have been studied, and an etching rate equation has been derived. The calculated results are found to be in good agreement with experimental data.

A new electrochemical method for selective etching of SI GaAs has been developed by Lu Fengzhen *et al.* (1983) of SIM. This method is based on the use of the citric acid–hydrogen peroxide–water etchant. Their experimental results show that a smooth surface and flat-bottomed holes can be obtained, and the smoothness of the surface is attributed to the effect of a DC current applied in the process.

Jiang Yong and Yan Guisheng (1982) of the Wuhan Research Institute of Posts and Telecommunications have studied the etching characteristics of the etchants of the systems Br–CH_3OH and HCl–H_2O, with a view to meeting the requirements for fine etching needed in the fabrication of GaInAsP/InP DH lasers.

b. Ion Milling and Reactive Ion Etching

Ion milling of InP and GaAs and its application in device fabrication have been studied by Fu Xinding *et al.* (1982) of SIM, with the LK-1 ion-milling machine manufactured by SIM. Their experimental results on GaAs are in agreement with those of Mozzi *et al.* (1979). It is found, however, that the change in color of the InP surface should be avoided by using the proper beam density and ion energy. The use of ion milling for layer stripping of InP is found to be as successful as that of GaAs if such precautions are taken.

Both the Sichuan Institute of Piezoelectric and Acoustooptic Technology and SIM have designed and manufactured reactive ion etching (RIE) machines suitable for application in VLSI technology. The RIE group of the Sichuan Institute, led by Li Qinggui and Li Xingming (1982), has successfully applied this new technique to a number of semiconductor materials, including GaAs and InP. Preliminary attempts have also been made by said group to apply this technique to acoustic surface wave devices and integrated optoelectronic circuits. Yan Jinglong and Zhou Zuyao (1985) of SIM have studied impurity contamination in reactive ion etching of silicon and found that the contamination is very slight and can be easily eliminated.

VIII. Microwave Devices and Integrated Circuits

A review of the work on microwave semiconductor devices in China has been published in Microwave System News (Anonymous, 1981), so only the work carried out since then will be covered here. The preliminary work on HEMTs fabricated from GaAs/GaAlAs superlattice structures and microwave devices based on III–V compounds other than GaAs will not be included in this review, with the exception of InP TEDs.

Recent years have seen wide applications of ion implantation technology in the fabrication of GaAs microwave devices, especially the MESFETs. Devices with gate lengths of 0.3–0.5 μm (Liang Chunguang *et al.*, 1983) and 0.3–0.7 μm (Chen Kejin *et al.*, 1982), which are shorter than the 0.5–0.8 μm cited in the previous paper (Zou Yuanxi, 1981), have been successfully fabricated by the engineers of HSRI and NSR respectively. Besides, computer simulation and analysis have been increasingly used in a number of institutes.

The use of polyimide for surface passivation of GaAs MESFETs and mixer diodes, and as a substitute for SiO_2 in the Ti–Pt–Au/n-GaAs Schottky barrier, has been investigated by Yang Hanpeng *et al.* (1982) and Sun Daoyun *et al.* (1982) of HSRI.

The technology for monolithic GaAs MESFET integrated circuits has been studied by Wang Limo *et al.* (1982) of the Sichuan Solid-State Circuits Research Institute, through the fabrication of buffered FET logic (BFL) integrated circuits.

An isoplanar structure GaAs MESFET using VPE-buffer-SI substrate materials was developed at HSRI in 1976 (Deng Xiancan, 1977). Compared with the conventional mesa-type structure MESFET, the isoplanar structure device is claimed to have certain significant advantages. Moreover, conventional n-type ion-implanted planar GaAs MESFETs and monolithic ICs must be annealed at temperatures higher than 800°C after ion implantation, while the isoplanar structure devices need not. Another advantage of the isoplanar structure is the absence of radiation damage in the active region of the device, as pointed out by Deng Xiancan (1983a).

Han Jihong *et al.* (1984) of NSR have successfully applied a proton-implantation isolation technique to the fabrication of 12 GHz power and low noise GaAs MESFETs, with improved device yield and reproducibility.

At the 1985 National Symposium on Semiconductor Physics, held at Xiamen, Fujian, several papers on the physics and behavior of $Ga_{1-x}Al_xAs$/GaAs heterostructures appeared. One of the papers (Wu Dingfen *et al.*, 1986) reports the cooperative work of SIM and Jiaotong University on the fabrication of both normally-on and normally-off $Ga_{1-x}As$/GaAs HEMTs. The transconductance obtained is much better than that of ion-implanted MESFETs with the same pattern and gate size, and increases by a factor of 1.7 with a decrease in temperature from 300 to 77 K.

20. Microwave Devices

a. MESFETs

The work on MESFETs has been largely concentrated in two institutes: (i) HSRI, and (ii) NSR.

(1) *LN GaAs MESFETs.* Some typical device parameters and performance data for LN GaAs MESFETs are listed in Table XV.

In the field of discrete microwave devices, significant progress has been made in the performance and application of GaAs MESFETs (see Tables XV–XVII). The optimum noise figure for 4 and 12 GHz LN GaAS MESFETs with an associated gain of 10 and 7.5 db has been reduced to 0.6 and 1.4 db

TABLE XV

PERFORMANCE DATA FOR LN GaAs MESFETs

Active layer								
Growth method	Thickness μm	L μm	W μm	f GHz	NF db	Ga db	Year	Author and affiliation
		0.3 ~ 0.5		18 12	4.2 2.6	6.2 10.0	1983	Liang Chunguang *et al.* (HSRI)
VPE				4	1.0	10	1982	NSR[b]
VPE	0.2	0.3 ~ 0.7	280	9.5	1.4	8.1	1982	Chen Kejin *et al.* (NSR)
VPE		0.5		4	0.7	10	1983	HSRI[a]
VPE				18	3.2		1983	HSRI[a]
VPE	0.2	0.5		12	1.4	7.5	1984	Chen Kejin *et al.* (NSR)
MBE	0.22			2	1.5	17	1984	Inst. Phys. and NSR[c]

[a] Private communication from Deng Xiancai of HSRI.
[b] Private communication from Han Jihong of NSR.
[c] Private communication from Zhou Junming of Inst. Phys.

respectively, as reported by Yang Hanpeng *et al.* (1984)[31] of HSRI and Yu Shifa (1984)[32] of NSR. Owing to their high performances, the 4 GHz LN GaAs MESFETs have been widely and satisfactorily applied as high frequency tuners in satellite ground receivers in China (1986, according to a correspondence written by Jin Qiuhua). The tuners operate in the frequency range of 3.7–4.2 GHz and have noise temperatures of 65–85 K. Instead of using VPE materials, the ion implanted MP GaAs MESFETs fabricated by Luo Haiyun *et al.* (1986)[33] of HSRI operate at 15 GHz and have an output power of 140 mW and an associated gain of 4.5 db.

(2) *MP GaAs MESFETs.* Some typical device parameters and performance data for MP GaAs MESFETs are listed in Table XVI.

(3) *Dual gate GaAs MERSFETs.* Some typical device parameters and performance data for dual gate GaAs MESFETs are listed in Table XVII.

[31] *Abstracts, Zhongqing Symp. on Compound Materials, Solid Microwave Devices and Optoelectronic Devices*, p. 118.
[32] *Ibid.*, p. 122.
[33] *Semiconductor Information.* No. 1, p. 1.

TABLE XVI

PERFORMANCE DATA FOR MP GaAs MESFETs

Active layer								
Growth method	Thickness μm	L μm	W μm	f GHz	Power (cw) mW	Power gain db	Year	Author and affiliation
VPE	0.2 ~ 0.5			4	3000	4	1982	Huang Lirong *et al.* (HSRI)
VPE	0.3	1	1200	10	573	4.6	1982	Li Songfa and Lu Haiyun (HSRI)
VPE				4	1500	3.5	1982	NSR[b]
VPE				6	1000	4	1982	NSR[b]
VPE	0.3 ~ 0.5	1	1200	9 11	540 300	3.2 3	1982	Wu Yunhong *et al.* (NSR)
VPE				12	100	4	1982	NSR[b]
VPE			1200	10	710	5	1983	HSRI[a]

[a] Private communication from Deng Xiancai of HSRI.
[b] Private communication from Han Jihong of NSR.

TABLE XVII

PERFORMANCE DATA FOR DUAL GATE GaAs MESFETs

Active layer								
Growth method	Thickness μm	L μm	W μm	f GHz	NF db	Ga db	Year	Author and affiliation
Implanted	0.3	1.5	300	1	0.8	11.5	1981	Wang Weiyuan *et al.* (SIM)
VPE				1 2	0.7 0.8	22 15	1982	Zheng Reiying (NSR)
VPE				1.9	1.5	10	1982	NSR[a]
VPE				4	3.7	16.5	1982	NSR[a]
VPE	0.5	0.4 ~ 0.6		12	3.5	15	1982	Deng Xiancan *et al.* (HSRI)
VPE				12	3.1	14	1983	Deng Xiancan (HSRI)

[a] Private communication from Han Jihong of NSR.

It can easily be seen from Tables XV to XVII that a significant advance has been made in recent years in the performance of GaAs MESFETs in China compared with the data cited previously (Zou Yuanxi, 1981).

It may be added in this connection that an experimental dual gate GaAs MIS SB FET with a heavily doped channel and an anodic oxide film of 70 Å thick, grown by a new technique (see Section 17b), has been fabricated by Zhou Mian and Wang Weiyuan (1984) of SIM. The breakdown voltage of this FET, with gate length 2 μm and width 400 μm, is 4–5 V and the transconductance is 25 mS/mm—both factors being higher than those of the FETs normally made in their laboratory—while the gate-source capacitance is reduced to 2/3 of the normal value.

The simultaneous formation of the heavily doped channel and ohmic contact by a single ion-implantation and the improvement in homogeneity resulting from the use of the heavily doped channel are believed to have practical implications in the fabrication of GaAs ICs.

b. *Other Microwave Devices*

In the following sections, only those microwave devices which show a significant improvement in performance compared with those described in the previous paper (Zou Yuanxi, 1981) will be discussed, along with some new devices.

(1) *GaAs and InP TEDs.* A typical 1.25 cm single modulating GaAs TED fabricated by NSR[34] has the following device performance: modulation band width 800 MHz, working band width 500 MHz when operated at 25 GHz, output power cw 50 mW, pushing figure 2 MHz/V, long term stability, relative frequency variation 5×10^{-5}, power variation 5×10^{-5}.

Gunn diodes fabricated from InP are expected to exhibit better characteristics than GaAs TEDs. Using a multilayer structure grown by the process, Deng Yanmao *et al.* (1982) of NSR have succeeded in fabricating InP TEDs with the device performing as follows: output power 151 mW, converting efficiency 2.44% operated at 50.6 GHz, output power 147 mW, conversion efficiency 2.54% operated at 58.3 GHz.

(2) *GaAs mixer diodes.* The 4–6 mm GaAs SB mixer diodes fabricated by NSR have the following device performance: C_{jo} 0.03 ~ 0.04 pF, R_s 1.7 Ω, conversion loss <5.0 db and NF 6.5 db (including NF_{If} 1.5 db), both at 69 GHz.

[34] Private communication from Han Jihong of NSR.

(3) *GaAs hyperabrupt varactor diodes.* Using a Lo–Hi–Lo multilayer structure grown by the VPE process on Te-doped GaAs substrate, the engineers of NSR[35] have fabricated GaAs hyperabrupt varactor diodes with the device performance as follows: X band, $C_{jo} \sim 1.0$ pF, $V_{BR} > 20$ V, $\gamma_c > 5$–7, $f_{c\text{-}4} > 100$–200 GHz.

(4) *GaAs bipolar microwave transistor.* Su Liman (1983) of the Beijing Electron Tube Factory had reported experimental results on the *npn* GaAs heterojunction microwave bipolar transistor of self-alignment structure. It is claimed that the *npn* GaAs transistor has several advantages compared with Si transistors.

21. Integrated Circuits

The development of GaAs integrated circuits in China is, admittedly, still in its primitive stage; therefore, only a few examples will be given below.

a. Microwave ICs

(1) *Monolithic X band GaAs MESFET oscillators.* Ye Yukang *et al.* (1983) of NSR have successfully fabricated monolithic X band GaAs MESFET oscillators with the performance as follows: output power 30 mW at 10.3 GHz, 40 mW at 8.2 GHz, efficiency 15%. However, the frequency stability is not very satisfactory, according to these investigators.

The materials used for fabricating GaAs digital integrated circuits have been turned to ion-implanted wafers since 1983. An 11-stage GaAs ring oscillator with a delay time of 60–70 ps/gate and frequency dividers with clock frequency of 2.1 GHz has been fabricated by Yong Yaozhong (1985)[36] of HSRI. Si-implanted wafers, after being annealed thermally or rapid thermally, have been used by Lee Binghui and Deng Xiancan (1986)[37] of HSRI. In order to improve the IC performances, Xia Guanqun and Wang Weiyuan (1986, to be presented at 1986 Quanzhou Symp. on Compound Materials, Solid Microwave Devices and Optoelectric Devices) of SIM have successfully used the low-resistance WAg(30%)/W gate in the fabrication of self-aligned GaAs MESFETs.

(2) *Monolithic GaAs amplifiers.* The isoplanar structure developed by Deng Xiancan of HSRI has been adopted for the fabrication of a monolithic, single stage broadband amplifier with 2 db noise figure and 10 db gain in 100 MHz to 1.5 GHz band (Deng Xiancan, 1983a).

[35] *Ibid.*

[36] Abstracts, 1985 Emei Symp., p. 78.

[37] *Semiconductor Information.* No. 3, p. 1.

The GaAs monolithic microwave integrated amplifiers may be taken as an example of the achievements in MMIC. Since 1984, amplifiers with operating frequencies of 10 GHz, band widths of 250 MHz, linear gains over 3 db and output power of 110 mW have been fabricated in a chip 1.84 × 1.32 mm^2. And in 1985, the linear gain of the amplifier operated at 10 GHz and fabricated in a 1.8 × 1.7 mm^2 chip has increased to 5 db in a band width of 800 MHz, according to Tien Erwen (1986)[38].

b. Digital Logic GaAs ICs

A monolithic digital logic GaAs IC with a propagation delay of 60 ps at room temperature has been successfully fabricated by the engineers of HSRI (Deng Xiancan, 1983b), using the isoplanar structure.

IX. Optoelectronic Devices

Besides the three Symposia mentioned in Part I, two Symposia on Optical Communications were held in Wuhan in 1979 and 1983. Some of the papers presented at the Second Symposium on Optical Fibre Communications are also included in this review, along with the results on related topics reported in recent periodicals.

22. Laser Diodes

a. GaAlAs/GaAs DH and Related Lasers

(1) *Attempts to improve the performance and/or yield of the lasers.* Although the laboratory life of GaAlAs/GaAs DH lasers has exceeded 10^4 hours (The DH Laser Group of IS, 1981; Shan Zhenguo of the Shanghai Institute of Optics and Fine Mechanics, 1981), it is the opinion of Shan Zhenguo (1983) that much work remains to be done in order to raise the yield, control the mode and improve the performance of the lasers. For this purpose, the infrared transmission technique has been used by Shan Zhenguo to observe directly the luminous position in the DH lasers, so that a reliable basis for improving the technology used in the fabrication of the lasers can be obtained.

A method for measuring the relaxation oscillation effect in proton-bombarded stripe single-mode, as well as in multimode GaAlAs/GaAs DH lasers, has been suggested by Zhao Liqing *et al.* (1983) of IS. The photon lifetime τ_p and the absorption loss coefficient in the lasing region are calculated and discussed. Moreover, the influence of such parameters as

[38] *SSERP.* **6**, p. 88.

spontaneous radiation, superluminescence and direct forward bias on the relaxation oscillation in multimode lasers, and their depression capability have been investigated.

Zhang Jingming and Zheng Baozhen (1982) of IS have investigated the near field spectra of GaAlAs/GaAs DH lasers and concluded that the kink behavior of the L–I curve and multiple groups of the longitudinal mode can be attributed to nonuniform distribution of Al in the active layer. A similar conclusion regarding the effect of the nonuniform distribution of Al on the phenomenon of self-sustained pulsation of GaAlAs/GaAs DH lasers has been obtained by Zhao Liqing and Wang Qiming (1982) of IS, in collaboration with Zhang Cunshan and Wu Zhenqiu of Hebei University, by measuring the time resolution spectra and the transient near field distribution.

Wang Shouwu *et al.* (1982) of IS have derived an expression for the dependence of the electrooptical delay time on the injection pulse current and the front edge rise time in DH lasers. The measured results of these investigators on a typical stripe-type DH laser are found to be in agreement with the calculated values. The practical implication of this work lies in the possibility of increasing the pulse modulation rate, according to these investigators.

A generalized formula has been deduced by Feng Zhechuan and Liv Hongdu (1983) of BJU for the radius of curvature and layer stresses originated from thermal strain in semiconductor multilayer structures. Various approximate expressions, including the useful forms for the thick substrate, have been derived and used by these investigators to calculate the active stress and its dependence on layer thickness and Al fraction for conventional GaAlAs/GaAs DH lasers, as well as for visible buffered GaAlAs DH lasers. An analytical expression for the vanishing of the active layer stress has also been given.

The phenomenon of facet erosion has been investigated by Zhuang Wanru and Yang Peisheng (1983) of IS. Their results indicate that facet oxidation is caused by a photochemical reaction in air, O_2 or wet N_2, while a dry nitrogen atmosphere has practically no influence on the facet parameters.

Liao Xianbing (1984) of the Yongchuan Institute of Photoelectric Technology has investigated the effects of sputtered Al_2O_3 and Si_3N_4 coatings on the lifetime of GaAlAs/GaAs DH lasers. It is found that Si_3N_4 is more easily grown and provides an effective facet coating for the DH lasers.

(2) *Visible-infrared GaAlAs lasers.* Infrared GaAlAs lasers with lasing wavelength ~8800 Å have been successfully fabricated by Zheng Guangfu *et al.* (1983) of the Yongchuan Institute of Photoelectric Technology. The room temperature cw life is of the order of 10^4 hours, and the cw operating ambient temperature is as high as 150°C.

Some preliminary work on Visible GaAlAs lasers has been done by several investigators. Both Zheng Guangfu *et al.* (1983) and Deng Ximin *et al.* (1982, 1983) of JLU have measured the effect of Al content in the melt on the lasing wavelength under their respective experimental conditions. Visible GaAlAs lasers with cw operation at room temperature and lasing wavelength of ~7600 Å have been obtained by Zheng Guangfu *et al.*, although much remains to be done on improving the lifetime of the experimental lasers.

b. InGaAsP/InP DH Lasers

A series of efforts have been made by the scientists of IS on the fabrication of 1.3 μm InGaAsP/InP DH lasers. Preliminary success in the fabrication of proton-defined stripe geometry, 1.3 μm InGaAsP/InP DH lasers has been achieved by Zhu Longde *et al.* (1981), resulting in two diodes capable of room temperature cw operation. By adopting a series of measures for improving the design of the device structure and epitaxial growth of heterojunctions, as well as for optimizing the doping concentrations, proton bombardment conditions, etc., Zhu Longde *et al.* (1982) have succeeded in fabricating proton-defined stripe geometry, 1.3 μm InGaAsP/InP DH lasers with a room temperature cw operating life of over 8000 hours, good reproducibility and satisfactory laser characteristics. In the meanwhile, Zhu Longde *et al.* (1984) have investigated the absorption characteristics and PL properties of InP and InGaAsP after proton bombardment under different conditions, and the subsequent annealing treatment. It is found by these investigators that lasers having low threshold current density, high quantum efficiency, higher kink point in the output current characteristics and capable of being operated in the nearly single longitudinal mode can be attained under optimum bombardment conditions. Additionally, Duan Shukun *et al.* (1983) have fabricated InGaAsP/InP DH lasers with chemically etched-mirrors. It is found that the average threshold current density of one etched-mirror laser is similar to that of the standard cleaved-mirror devices fabricated from the same wafer, and that the etched-mirror lasers have a higher threshold current density.

Proton-defined stripe geometry, 1.5 μm InGaAsP/InP DH lasers have been fabricated by both the Laser Group of the Third Department of HSRI (1982) and Chen Jun and Sun Kechang (1984) of the Yongchuan Institute of Photoelectric Technology. Both of these research groups adopt the method of growing an additional anti-backmelting layer as a means to prevent the melting of the active layer. The latter investigators have been able to fabricate the said DH lasers with a room temperature cw operating life of over 3000 hours.

Certain techniques in the fabrication and application of InGaAsP/InP DH lasers, including the use of low temperature-deposited SiO_2 film as mask for deep Zn diffusion, fine etching for InP and InGaAsP materials, and coupling between the InGaAsP/InP DH laser and the single mode optical fibre, have been studied by scientists of the Wuhan Research Institute of Posts and Telecommunications (Yang Guisheng *et al.*, 1982); Jiang Yong and Yang Guisheng, 1982; Ran Chongzhu and Wang Yuzhang, 1983).

Miao Zhongli *et al.* (1982) of JLU have investigated the filamentary light-emitting behavior of electrode-defined stripe geometry InGaAsP/InP DH lasers, with a view to attaining stable transverse mode operating diodes. Another work of JLU is that of Wang Xianren *et al.* (1985), who have succeeded in measuring the crystalline perfectness of the active layer of an InGaAsP/InP DH laser by means of an ordinary X-ray diffraction, instead of the double-crystal diffraction unit.

The effect of certain structural parameters on the threshold current density for planar stripe geometry InGaAsP/InP DH lasers has been studied by Meng Shile *et al.* (1983) of Jiaotong University.

Since 1983, considerable research and development on the 1.3 μm InGaAsP/InP index-guided, single-mode laser diodes used for long wavelength fiber optical communication have been carried out in China. Several structures of low threshold InGaAsP/InP lasers have been successfully fabricated on a laboratory scale. Zhao Songshaw *et al.* (1986)[39] of the Wuhan Research Institute of Posts and Telecommunications have reported that 1.3 μm InGaAsP/InP double-channel planar buried heterostructure (DC-PBH) lasers have been fabricated with super-cooling growth techniques by a two-step growth process. The minimum threshold current of the lasers at room temperature is 15 mA and a typical value is 20 mA. The highest cw operation temperature is 80°C, and the output power is 2 mW. The stable single longitudinal mode is obtained at four times the threshold current. Low threshold current, high output power, fundamental mode 1.3 μm InGaAsP/InP BH lasers have been fabricated by Wang Wei *et al.* (1986)[40] of IS. The cw threshold current is as low as 10 mA at room temperature. The linear output power is as high as 20 mW. The highest operating temperature is 100°C. The stable fundamental transverse mode operation can be obtained at a 2.5-fold DC threshold current. Moreover, the Yongchuan Institute of Photoelectric Technology has recently declared that it can provide GJ322 type 1.3 μm InGaAsP/InP DC-PBH laser modules as a commercial product.[41] These modules have been used in experimental fiber optical digital transmission.

[39] *CJS.* 7, p. 324.
[40] *CJS.* 7, p. 337.
[41] Private communication from Pan Huizhen of SIM.

c. *GaAlAs/GaAs DH Lasers with Special Functions and Integrated Optical Circuits*

Being a three terminal device, a GaAlAs/GaAs DH laser with two sections in a common cavity has some special functions, including longitudinal mode selectivity, optical bistability and extrashort pulse emitting. Wang Qiming *et al.* (1985) and Huang Xi *et al.* (1983) of IS have measured the stable-state characteristics of such DH lasers. The threshold current density, operating characteristics and the upward shift of the "kinks" are discussed.

Research work on integrated optics has been carried out almost exclusively at SIM. A monolithic integrated device consisting of a twin-mesa GaAlAs/GaAs laser, a waveguide and an optoelectronic detector or amplifier has been fabricated by Pan Huizhen *et al.* (1981a). Xiao Zhongyao *et al.* (1982, 1983) have reported on a twin-mesa GaAlAs/GaAs laser integrated with a passive waveguide, capable of oscillating in single-longitudinal and fundamental-transverse modes, due to the action of optical injection locking of the frequency. This device has, however, a high threshold current density. To overcome this drawback, Xiao Zhongyao *et al.* (1984) have devised an integrated twin-planar-stripe GaAlAs/GaAs laser coupled with a passive waveguide. The variation of mode behavior with driving current and ambient temperature and the optical injection lock-in action of the two active cavities are examined and discussed. An integrated laser consisting of two active cavities coupled with a convex passive waveguide buried in the substrate has been fabricated by Pan Huizhen *et al.* (1983a). In order to lower the threshold current and develop single-mode fibre optical communication, Pan Huizhen *et al.* (1983b) have also designed and fabricated an integrated channel-substrate planar-stripe GaAs/GaAlAs laser coupled with a two-dimensional waveguide. The longitudinal mode selectivity of this device is discussed, along with the relation between the active layer thickness on the one hand, and the mode power transfer efficiency and threshold current density on the other. Computer simulation has been used in the design of this device by Wang Dening and Pan Huizhen (1984). A buried crescent structure has been adopted on the basis of the result of this simulation, leading to a significant lowering of the threshold current (Pan Huizhen *et al.*, 1985).

23. Light Emitting Diodes

a. *GaAlAs/GaAs and InGaAsP/InP DH LEDs*

Scientists of SIM have reported on two types of GaAlAs/GaAs DH LEDs for use in optical fibre communication. Pan Huizhen *et al.* (1981b) have successfully fabricated small area high radiance surface-emitting LEDs of the Burrus type having radiation power greater than 100 W/sr. cm^2, an output power of the tail fibre (inner diameter 60 μm, N.A. 0.17) of 200 μW and an extrapolated life of 10^5 hr. These performance parameters are, according to these investigators, comparable with those of similar devices fabricated by a number of companies in both the U.S.A. and Japan. Moreover, these diodes have already, as pointed out by these investigators, come into use in the 1.8 km, 8.448 Mb/s, PCM-120 route experimental optic-fibre telephone communication system erected in Shanghai. As another type of LED useful in optical fibre communication, high speed and high radiance edge-emitting LEDs have been fabricated by Pang Yongxiu *et al.* (1982, 1983). The devices have an optical waveguide structure and a moderate doping level. The performance parameters are: frequency response >60 MHz, structure radiance 1017 W/sr. cm^2, output power of the tail fibre (core diameter 60 μm, N.A. 0.17) 118 μW and coupling efficiency 11%. Through optimization of the stripe geometry, Pang Yongxiu *et al.* (1984) have been able to obtain coupled power of >100 μW (as high as 300 μW) with the standard graded index fibre (N.A. 0.2, core diameter 50 μm).

Similar work has been done by engineers of the Yongchuan Institute of Photoelectric Technology (Zhan Suzhen, 1984; Yang Delin, 1986). In order to increase the optical output power and modulation frequency, Yang Delin has fabricated edge-emitting GaAlAs/GaAs LEDs coupled to optical fibres with output monitoring. A Si–PIN detector is used for this purpose.

The relative merits of the Burrus surface-emitting, restricted edge-emitting and the *p–n* junction, isolated stripe geometry edge-emitting GaAlAs/LEDs are discussed by Pang Yongxiu *et al.* (1983b) of SIM, on the basis of their experimental results.

The corresponding InGaAsP/InP DH LEDs have also been fabricated by scientists of SIM (Shui Hailong *et al.*, 1982a,b). For the Burrus type surface-emitting DH LEDs, the device output has attained >1 mW at 100 mA driving current, with the highest value being ~2 mW. The output power of the tail fibre (core diameter 60 μm, N.A. 0.23) is over 50 μW. The characteristics of the LEDs are described and discussed from the viewpoint of output saturation and degradation. Edge-emitting InGaAsP/InP DH LEDs have been fabricated, and the effects of certain material parameters on the device characteristics investigated by Zhang Guicheng *et al.* (1985b). It is found that an active layer thickness of ~0.2 μm and the location of the *p–n* junction in the active layer have some important advantages. Both Xu Shaohua *et al.* (1983) of SIM and Zhang Wansheng *et al.* (1983) of HSRI have fabricated InGaAsP/InP DH LEDs with a monolithically integrated spherical lens shaped up by etching for improving the coupling efficiency into the optical fibre. It is found by Xu Shaohua *et al.* that a coupling efficiency of 6% can be obtained, and, accordingly, the output power can be increased by 30%.

b. *Orange $GaAs_{0.35}P_{0.65}$: N, Yellow $GaAs_{0.15}P_{0.85}$: N LEDs and Green GaP LEDs*

LEDs based on the ternary alloy GaAsP have been fabricated and studied by Fang Zhilie of FDU and his collaborators (Fang Zhilie *et al.*, 1982; Fang Zhilie and Bai Xiaodong, 1983). The epitaxial materials are prepared by the As–PCl_3–H_2–Ga process, with the introduction of NH_3 as the source of N doping. Yellow and orange LEDs using the epitaxial materials $GaAs_{0.15}P_{0.85}$: N/GaP and $GaAs_{0.35}P_{0.65}$: N/GaP are fabricated, and their optoelectrical characteristics measured. The peak wavelengths are determined as 5900 and 6300 Å respectively. It is found that the luminescence efficiency drops considerably for the yellow LEDs as the carrier concentration exceeds 3×10^{17} cm^{-3}, and the optimum carrier concentration for the orange LEDs is slightly less than 1×10^{17} cm^{-3}.

Ding Zuchang *et al.* (1982, 1983) of Zhejiang University have successfully fabricated GaP : N green LEDs with a good reproducibility. The following results on device performance have been obtained: peak of radiative spectrum ~5600 Å with red peak practically absent, total photon flux >10 mlm with a maximum of 20 mlm at a current density of 12 A cm^{-2}.

High radiance planar LEDs with a light-emitting area more than 5 times that of ordinary LEDs have been designed and fabricated by Fan Zhilie and Xu Jianzhong (1984) of FDU. It is claimed by these investigators that these devices have already found some important practical applications.

24. Degradation of DH Lasers and Related LEDs

a. *A Model for Slow Degradation of GaAlAs/GaAs DH Lasers and Related LEDs*

The mechanism for slow degradation of GaAlAs/GaAs DH lasers has yet to be elucidated, although it is generally believed to be related to an increase in nonradiative recombination centers in the active region during their long period operation (Miller *et al.*, 1977b). The nature of the recombination centers involved, therefore, merits our serious attention.

Hämmerling and Huber (1977) have studied the degradation of $GaAs_{0.6}P_{0.4}$: Se LEDs by plotting the figure of merit B/J (B—brightness, J—current density) versus As_2 pressure during Zn diffusion, as is shown in Fig. 17. In addition, Zou Yuanxi (1980) has plotted the ratio (Δ light output power)/(initial light output power) versus carrier concentration n on the basis of the heat treatment data of Hämmerling and Huber, as is shown in Fig. 18. It can be seen from Figs. 17 and 18 that linear relationships exist in both cases, with the slopes equal to $-1/2$ and $1/2$ respectively. Since $[V_{Ga}]$ is proportional to $(P_{As_2})^{1/2}$, it follows from Figs. 17 and 18 that the

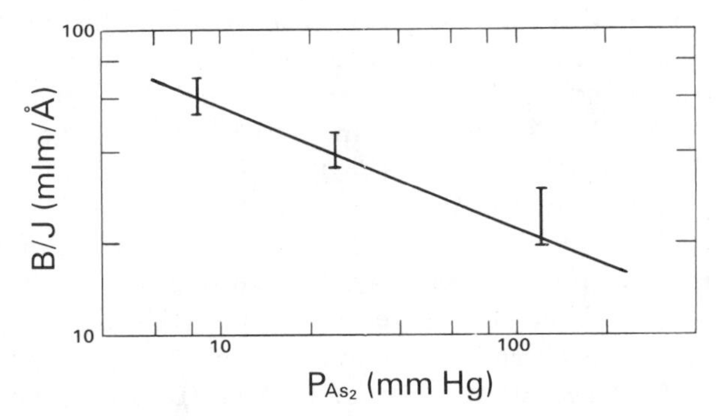

FIG. 17. Effect of arsenic pressure during Zn diffusion on figure of merit of $GaAs_{0.6}P_{0.4}$: Se LEDs. [From Hämmerling and Huber (1977), reproduced with permission from *Journal of Electronics Materials*, Vol. **6**, No. 6, 1977, a publication of The Metallurgical Society of AIME, Warrendale, PA, U.S.A.]

concentration of the recombination centers involved should be proportional to $[V_{Ga}]$ and $n^{1/2}$. It follows immediately that the recombination centers responsible for the degradation of $GaAs_{0.6}P_{0.4}$: Se LEDs could be the A hole trap observed by Lang and Logan (1975) in LPE GaAs, in accordance with the A-B-C model discussed in Section 15a, if the latter can be applied to the ternary compound as well. The A trap is chosen in preference to the B in view of the influence of the former on MCDL in *n*-GaAs, as mentioned in Section 11.

As pointed out by a number of investigators (Dixon and Hartman, 1977; Hartman and Dixon, 1976; Petroff *et al.*, 1976), there exist some similarities between the degradation of GaAlAs/GaAs DH lasers and certain related LEDs. Therefore, Zou Yuanxi (1980) has suggested that the nonradiative recombination centers involved in the degradation of GaAlAs/GaAs DH lasers could also be the A trap mentioned above. It is gratifying to note

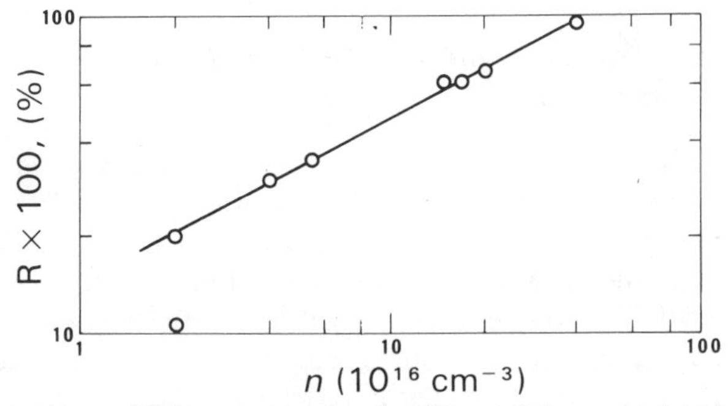

FIG. 18. Dependence of light output attenuation (R = △ light output/initial light output) for $GaAs_{0.6}P_{0.4}$: Se LEDs based on data of Hämmerling and Huber (1977).

in this connection that Lopez *et al.* (1977) have found, by DLTS measurements, a rapid increase in hole traps during the degradation of $GaAs_{0.6}P_{0.4}$ LEDs, and that Kondo *et al.* (1982) have reported similar findings in GaAlAs LEDs, with $p \sim 2 \times 10^{16}\ cm^{-3}$ in the active layer during long period operation. The probable sources of $(V_{Ga})_2$ for the formation of A and B traps have been discussed by Zou Yuanxi (1980).

According to this model for slow degradation, a decrease in the concentration of the A trap in the active layer should lengthen the lifetime of the GaAlAs/GaAs DH lasers. It is interesting to note in this connection that such a decrease has actually been observed by Zhou Jicheng *et al.* (1985) in a recent study of the long period operation of GaAs LEDs with a homogeneous junction.

Based on experimental data on the temperature dependence of the degradation coefficient for their GaAlAs/GaAs LEDs of the Burrus type, Pang Yongxiu *et al.* (1983b) of SIM have been able to derive an extra-polated room temperature half-life time of 1×10^6 h and an apparent activation energy of 0.6 eV. As the slow degradation of GaAlAs/GaAs LEDs has been observed to take place within the GaAlAs layer (Yamakaoshi, 1977), this value for the activation energy is believed to be valid also for edge-emitting GaAlAs/GaAs LEDs by Pang Yongxiu *et al.*

According to Gao Jilin *et al.* (1985) of IS, the deep level LE2 observed by them in the n-$Ga_{0.7}Al_{0.3}As$ (Sn) layer of a four-layer structure, located at 0.37 eV below the conduction band, should be responsible for slow degradation of GaAlAs/GaAs DH LDs, during which a significant increase in LE2 is observed. Moreover, these investigators have been able to show definitely that this level is related to oxygen. It is our suspicion, therefore, that this level could also be related to V_{III} formed by the reaction between Al and oxygen. The diffusion of this level into the active layer will eventually result in an increase of the A trap, which might be the level directly responsible for slow degradition of DH laser diodes.

b. Degradation Studies on InGaAsP/InP DH LEDs, LDs and Other LEDs

Zhang Guicheng and Shui Hailong (1984b) of SIM have studied the room temperature (15–30°C) degradation characteristics of InGaAsP/InP DH lasers in air, including $(P/P_o) - t$ relationship, EL pattern of the light-emitting region and spectrum and I–V characteristics. The stability of the spectrum and I–V characteristics is ascribed to the absence of new dark structures during long period operation. The degradation mechanism is discussed on the basis of the change in τ_{nr} by using the extended defect model (Horikoshi *et al.*, 1979).

Lu Bingmei *et al.* (1982) of HSRI have studied degradation of InGaAsP/InP DH lasers at both room and elevated temperatures. The room temperature operating life has approached 10^4 hours, while that extrapolated to room temperature has exceeded 10^8 hours. It is the opinion of these investigators that slow degradation is caused by the creation of nonradiative recombination centers through the recombination-enhanced rediffusion of zinc, so as to make the *p–n* junction move away from the active region.

Zhou Bizhong *et al.* (1985) of Xiamen University have studied the degradation of green GaP : N LEDs after electron irradiation at 1 MeV. It is found that the green light intensity decreases by about two orders of magnitude as a result of electron irradiation. This deleterious effect is attributed to the creation of a number of deep levels, to which the defects V_{Ga}, V_P, P_{Ga}, Ga_p and their complexes are tentatively assigned.

25. Solar Cells and Optoelectronic Detectors

a. GaAlAs/GaAs Solar Cells

The research effort in the manufacture of high efficiency GaAlAs/GaAs solar cells has diminished considerably in favor of the development of amorphous silicon solar cells. Although research work is still being carried on at SIM, it has been limited to the conventional LPE GaAlAs/GaAs solar cells so far.

The AM1 efficiency of GaAlAs/GaAs solar cells having an area of 0.2–1 cm^2 fabricated by Guan Limin *et al.* (1986a) of SIM, using the LPE material provided by Yang Qianzhi *et al.* (1985), has attained >20% with the use of double-layer anti-reflection coating. This efficiency represents a significant increase over the 16–18% reported in the previous paper (Zou Yuanxi, 1981). As pointed out by these investigators, this improvement in AM1 efficiency has been obtained by adopting a series of measures, including adoption of a specially designed graphite boat, improvement in the background purity of the LPE layer, careful control of layer thickness and doping and investigation of MCDL in both types of GaAs material as a function of carrier concentration. The dark I–V characteristics of the cells measured by Guan Limin *et al.* (1986b) of SIM indicate that further improvement in AM1 efficiency can be expected by minimizing recombination via deep levels in the epitaxial material.

b. $In_{0.53}Ga_{0.47}As$ Long Wavelength Detectors and Other Optoelectronic Devices

Recent years have seen only a limited amount of published work in China on $In_{0.53}Ga_{0.,47}As$ long wavelength detectors. Briefly, both LPE and VPE layers have been grown, but the carrier concentration is not sufficiently low (see Section 4b). Pan Yungang (1983) of the Yongchuan Institute of Photoelectric Technology and Wang Shutang *et al.* (1984) of IS have reported on the growth of material by VPE and LPE respectively, as well as on fabrication of the detector. In general, the performance of the detectors has yet to be improved, presumably through further purification of the epitaxial material. At present, the dark current is smaller for the devices using the LPE material.

Li Shihai *et al.* (1985) of NSR have met with preliminary success in the fabrication of $In_{0.53}Ga_{0.47}As/InP$ photodiodes. Some typical device parameters are: I_D 28 nA (−1.5 V), C_{jo} 7 pF, η47% ($\lambda = 1.3\ \mu$m).

Guo Kangjin *et al.* (1982) of SIM have fabricated Pt SB avalanche photodetectors based on GaAs, with a view to meeting the requirements for monolithic integrated optic circuits. It is found that the mesa structure is the best of the four different structures studied. The multiplication factor of the mesa-type detector can be higher than 1000, while the dark current density is as low as 1.13×10^{-4} A/cm^{-2} at 0.9 V_B. The response speed is <1 ns.

Theoretical and experimental investigations of planar GaAs Gunn photodetectors have been reported by Wei Cejun of IS and Klein and Beneking of Aachen University (1982).

A back illuminated heterostructure $In_{0.53}Ga_{0.47}As/InP$ PIN photodiode has been worked out by Wang Shutang *et al.* (1986, to be published) of IS. The coincidence between the distribution of the generated rate of photocarriers and the electric field in the depletion region has been found to lead to much better device performances.

X. Concluding Remarks

Recent work on III–V compounds in China has been reviewed. Compared with the paper presented at the Eighth International Symposium on GaAs and Related Compounds, held in Vienna in 1980 (Zou Yuanxi, 1981), this review reflects a big stride forward on practically all the frontiers of research in the field of III–V compounds. However, it can easily be seen that there are still new technologies, as well as sophisticated devices, on which the work done has been of a preliminary nature only. It will suffice to cite here a few examples, such as MBE and MOVPE on the materials side, and HEMT and GaAs IC from the device point of view.

It is well known that China has plenty of gallium resources, as well as adequate production facilities. The same situation happens to exist in the case of indium too. Therefore, China is in an advantageous position to develop her III–V compound industry in the forseeable future. The establishment of such an industry depends, however, upon certain other conditions, including intensive research in various phases discussed in this review. Needless to say, the realization of the goal of erecting such an industry will also result in shortening the distance between the silicon and III–V compound industries.

While it is the responsibility of Chinese scientists and engineers working in this field to make their contributions to the establishment of such an industry, international cooperation and intercourse are expected to play an important role in the advancement of our knowledge in this broad area. If the work done in China should prove useful to scientists and engineers of other countries working in the field of III–V compounds, the purpose of the present review would be accomplished.

Acknowledgments

The writer wishes to thank all the related authors for their consent to including their work in this review and Professors Wang Houji, Wang Weiyuan, Peng Ruiwu, Zhou Jicheng, Pan Huizhen and Shui Hailong and Drs. Li Aizhen, Mo Peigen, Zhou Binglin, Wang Rongjuan, Zhu Jinfeng and Wu Zhuoqing for their valuable suggestions. The assistance of Ms. Hanna Kiang, Li Beilan and Shi Zhenjuan; Messrs, Yu Haisheng, Zhang Yeping, Wu Jisen, Wu Ju and Liang Bingwen and many other colleagues at the Shanghai Institute of Metallurgy in preparing this manuscript is also greatly appreciated.

References

For the sake of brevity, the following abbreviations will be used:

(1) 1977 Liuzhou Symp.—The 1977 National Symposium on Gallium Arsenide and Related Compounds, Liuzhou.

(2) 1981 Shanghai Symp.—The 1983 National Symposium on Gallium Arsenide and Related Compounds, Shanghai.

(3) 1982 Kunming Symp.—The 1982 National Symposium on Semiconducting Compound Materials, Solid Microwave Devices and Optoelectronic Devices, Kunming.

(4) 1982 Changzhou Symp.—The 1982 National Symposium on Electron, Ion and Photon Beams and Micro-Fabrication Technology, Changzhou.

(5) 1983 Nanjing Symp.—The 1983 National Symposium on Semiconductor Physics, Nanjing.

(6) 1983 Wuhan Symp.—The 1983 National Symposium on Optical Fibre Communications, Wuhan.

(7) AES—Acta Electronica Sinica.

(8) APS—Acta Physica Sinica.

(9) CJS—Chinese Journal of Semiconductors.

(10) JAS—Journal of Applied Sciences (in Chinese).

(11) LDD—Luminescence and Display Devices (in Chinese).

(12) SSERP—Solid State Electronics Research and Progress (in Chinese).

(13) STC—Science and Technology Communications (in Chinese).

(14) XYJS—Xi You Jin Shu (Rare Metals).

Anonymous (1981). *Microwave System News*. **11**, 90.

Aspnes, D. E., Schwartz, B., Studna, A. A., Derick, L., and Kosi, L. A. (1977). *J. Appl. Phys.* **48**, 3510.

Balagurov, L. A., Gimel'farb, F. A., Karataev, V. V., Nemtsova, G. A., and Omel'yamovskii, E. M. (1976). *Kristallografiya* **21**, 1242.

Bao Qingcheng, Wang Qiming, Peng Huaide, Zhu Longde and Gao Jilin (1982). *In* "1982 Kunming Symp.," p. 56.

Barnes, C. E., and Samara, G. A. (1983). *Appl. Phys. Lett.* **43,** 677.

Bonner, W. A. (1980). *Mater. Res. Bull.* **15,** 63.

Braslau, N. (1981). *J. Vac. Sci. Technol.* **19,** 803.

Cadoret, R., Hollan, L., Loyau, J. B., Oberlin, M., and Oberlin, A. (1975). *J. Cryst. Growth* **29,** 187.

Chai Xiyun, Liu Xunlang, Jiao Jinghua, Zhao Jianqun, Cao Huimei, Sun Wenrong and Xie Shizhong (1983). *In* "Abstracts, 1983 Shanghai Symp.," p. 5.

Chandra, A., and Eastman, L. F. (1979). *Solid State Electron.* **22,** 645.

Chang, C. Y., Fang, Y. K., and Sze, S. M. (1971). *Solid State Electron.* **14,** 541.

Chen Chao and Shen Qihua (1983). *J. Xiamen Univ. (Natur. Sci. Ed.)* **22,** 20.

Chen Chao and Zhang Qi (1983). *In* "1983 Shangai Symp.," p. 368.

Chen Gaoting, Fang Zujie and Qiu Rongshang (1981). *In* "1981 Shanghai Symp.," p. 244.

Chen Jun and Sun Kechang (1984). *DIANZI KEXUE (Electron. Sci).* No. 1, 9.

Chen Kejin, Chen Shiyang, Yu Shifa and Liu Qiming (1982). *In* "1982 Kunming Symp.," p. 121.

Chen Kejin, Chen Shiyang and Yu Shifa (1984). *SSERP* **4** (2), 1.

Chen Keming, Qiu Lanhua, Cui Yude, Chen Weide and Jiang Pingqing (1982). *CJS* **3,** 233.

Chen Keming, Qiu Lanhua and Chen Weide (1985). To be published in *CJS.*

Chen Tingjie, Wu Lingxi, Xu Shouding, Meng Qinghui, Yun Kun and Li Yongkang (1982). *CJS* **3,** 169.

Chen Zhihao, Yu Kaisheng and Wang Weiyuan (1982). *KEXUE TONGBAO (Science)* **28,** 1180.

Chen Zhihao, Zhu Wenyu, Zhu Weiwen and Wang Weiyuan (1985). *In* "Abstracts, 1985 Emei Symp. on GaAs and Related Compounds," p. 60.

Chen Ziyao, Shao Yongfu, Zhu Fuying and Peng Ruiwu (1985). *CJS* **6,** 181.

Cheng Qixiang (1981). *In* "1981 Shanghai Symp.," p. 146.

Cheng Qixiang (1982). *In* "Semi-Insulating III–V Materials" (2nd Int. Conf., Shiva, England), p. 70.

Chu Yiming, He Hongjia, Cao Funian, Bai Yuke, Fei Xueying and Wang Fenglian (1981). *CJS* **2,** 86.

Cockayne, B., Macewan, W. R., and Brown, G. T. (1980). *J. Mater. Sci.* **15,** 2785.

Conwell, E. M., and Vessell, M. O. (1968). *Phys. Rev.* **166,** 797.

Cooperative Group for MBE System (1981). *CJS* **2,** 167.

Dai Daoxuan and Tang Houshun (1985). *Vac. Sci. Technol.* (in Chinese) **5** (3), 1.

Dai Daoxuan and Zou Huiliang (1985). Unpublished.

Dai Daoxuan, Tang Houshun and Jiang Guobao (1985). *CJS* **6,** 212.

Deng Xiancan (1977). *In* "1977 Shanghai Symp.," p. 102.

Deng Xiancan (1983a). *Nucl. Instrum. Methods* **209/210,** 657.

Deng Xiancan (1983b). *In* "1983 Shanghai Symp.," p. 336.

Deng Xiancan, Ma Zhenchang, Yang Yaozhong, Yang Guoan and Zhang Mian (1982). *In* "1982 Kunming Symp.," p. 114.

Deng Ximin, Jin Baofu, Du Guotong, Yang Jian, Ren Linfu, Yang Wenyan and Cao Dingsan (1982). *In* "1982 Kunming Symp.," p. 271.

Deng Ximin, Du Guotong, Yang Jian, Ren Linfu, Yang Wenyan and Gao Dingsan (1983). *In* "Abstracts, 1983 Wuhan Symp.," p. 60.

Deng Yanmao, Zhang Hongzhi, Sheng Yongxi and Fang Jingzhi (1982). *In* "1982 IEEE MTT-S Int. Microwave Symp. Digest," p. 516.

DH Laser Group of IS (1981). *Laser J.* (in Chinese) **8** (10), 16.

DiLorenzo, J. V. (1972). *J. Cryst. Growth* **17,** 189.

Ding Moyuan, Shi Yihe, Fu Tao, Wang Yan and Li Shuangxi (1982). *CJS* **3,** 162.

Ding Moyuan, Shi Yihe, Wang Yan, Li Shuangxi and Fu Tao (1984). *AES* **4,** 108.
Ding Xunmin, Yang Shu, Dong Guosheng and Wang Xun (1984). *CJS* **5,** 162.
Ding Zuchang (1982). *J. Zhejiang Univ.* No. 2, 62.
Ding Zuchang, Hua Weimin, Cao Guangsheng and Zhu Guoying (1983). *In* "1983 Shaghai Symp.," p. 111.
Dixon, R. W., and Hartman, R. L. (1977). *J. Appl. Phys.* **48,** 3225.
Dong Zhiwu (1983). *Acta Scientiarum Naturalium Univ. Jilinensis.* No. 1, 76.
Duan Shukun, Shi Zhongcheng, Li Jingran, Lu Hui, Wang Liming and Sun Furong (1983). *CJS* **4,** 291.
Eisen, F. H., Figgins, J. A., Immorlica, A. A., Kuyas, R. L., Ludington, B. W., Welch, B. M., Wen, C. P., and Zucca, R. (1975). *AD* A018090.
Fairman, R. D., Chen, R. T., Oliver, J. R., and Chen, D. R. (1981). *IEEE Trans.* **ED-28,** 135.
Fan Guanghan, Hu Guang, Fu Shuqing, Xie Jiangfeng and Liu Hanzhi (1981). *In* "1981 Shanghai Symp.," p. 183.
Fan Maolan (1983). *In* "1983 Shanghai Symp.," p. 170.
Fang Dungfu, Wang Xiangxi, Xu Yongquan, Miao Hanying and Mou Panjian (1983). *JAS* **1,** 235.
Fang Dungfu, Wang Xiangxi, Xu Yongquan and Tan Litong (1984). *J. Cryst. Growth* **66,** 327.
Fang Zhilie and Bai Xiaodong (1983). *Shanghai Semicond.* No. 4, 30.
Fang Zhilie and Li Chungji (1983). *In* "1983 Shanghai Symp.," p. 108.
Fang Zhilie and Xu Jianzhong (1984). *Electron. Sci. Technol.* (in Chinese). No. 2, 4.
Fang Zhilie, Wang Mansheng and Ni Linfu (1982). *LDD.* No. 1, 40.
Favenec, P. N. (1976). *J. Appl. Phys.* **47,** 2532.
Feng Yougang, Sun Henghui, Fang Zhilie and Sheng Chi (1983). *J. Fudan Univ. (Natur. Sci. Ed.)* **22,** 367.
Feng Zhechuan and Liu Hongdu (1983). *CJS* **4,** 171.
Fu Shuqing, Xie Jiangfeng and Liu Hanzhi (1981). *In* "1981 Shanghai Symp.," p. 182.
Fu Xinding, Chen Guomin, Ren Congxin, Fang Hongli, Yang Jie, Shao Yongfu and Wang Guangyu (1982). *STC.* No. 3, 20.
Gao Jilin, Wu Ronghan, Li Zhaoyin and Gao Shufen (1985). *CJS* **6,** 245.
Gong Mengnan and Zhang Fuwen (1981). *In* "1983 Shanghai Symp.," p. 333.
Gong Xiuying and Zhao Jianshe (1981). *XYJS.* No. 5, 40.
Gong Jishu, Xu Zhongying, Xu Junying, Li Yuzhang, Chen Lianghui, Shi Zhongcheng and Du Baoxun (1982). *CJS* **3,** 474.
Greene, P. D. (1973). *J. Phys.* **D6,** 1550.
Greene, P. D. (1979). *Solid State Commun.* **32,** 325.
Guan Limin, Sui Zhaowen and Xu Hong (1986b). *J. Solar Energy* (in Chinese) **7,** 86.
Guan Limin, Sui Zhaowen, Chen Fengkou and Yu Wenjie (1986a). Presented at 2nd Int. Photovoltaic Sci. and Engrg. Conf., Beijing, Aug. 19–21, 1986.
Guo Kangjin, Hu Weiyang, Yao Wenlan, Chen Liangyang and Shui Hailong (1982). *In* "1982 Kunming Symp.," p. 302.
Hämmerling, N., and Huber, D. (1977). *J. Electron. Mater.* **6,** 581.
Han Jihong, Wu Luxun, Shao Zhenya and Xiao Dejian (1982). *SSERP* **2** (4), 65.
Han Jihong, Wu Luxun and Shao Zhenya (1984). *AES* **12** (6), 97.
Hartman, R. L., and Dixon, R. W. (1976). *Appl. Phys. Lett.* **26,** 233.
Haegawa, F., and Majerfeld, A. (1976). *Electron Lett.* **12,** 52.
He Jianhua, Hong Liansheng, Wei Xiaoyi and Pan Chengheng (1982). *In* "1982 Kunming Symp.," p. 73.
Hilsum, C., and Rose-Innes, A. C. (1961). *Semiconducting III–V Compounds*, Pergamon Press, New York.

Hobgood, H. M., Eldridge, G. W., Barrett, D. L., and Thomas, R. N. (1981). *IEEE Trans.* **ED-28,** 140.

Hollis, J. E. L., Choo, S. C., and Heasell, E. L. (1967). *J. Appl. Phys.* **38,** 1626.

Horikoshi, Y., Kobayashi, T., and Furukawa, Y. (1979). *Jpn. J. Appl. Phys.* **18,** 2237.

Hou Xiaoyuan, Yu Mingren and Wang Xun (1984), *CJS* **5,** 171.

Huang Lirong, Huang Songzhang and Guan Pikai (1982). *In* "1982 Kunming Symp.," p. 145.

Huang Qisheng, Grimmeiss, H. G., and Samuelson, L. (1985). *J. Appl. Phys.* **58,** 3068.

Huang Shangxiang (1981). *In* "1981 Shanghai Symp.," p. 172.

Huang Shangxiang (1983). *CJS* **4,** 197.

Huang Xi, Wang Qiming and Du Baoxun (1983). *CJS* **4,** 449.

Hulme, K. F., and Mullin, J. B. (1962). *Solid State Electron.* **5,** 211.

Ikoma, T., Takikawa, M., and Taniguchi, M. (1982). *In* "1982 Symp. on Gallium Arsenide and Related Compounds 1981 (Inst. Phys. Ser. No. 63)," p. 191.

Jiang Xuezhao (1985). *Rare Metals* (in Chinese) **4** (3), 35.

Jiang Yong and Yang Guisheng (1982). *In* "1982 Kunming Symp.," p. 233.

Jiao Pengfei, Chen Weixi, Liu Hongxun and Yu Lisheng (1983). *In* "1983 Shanghai Symp.," p. 252.

Jin Gang, Ma Kejun, Yu Zhengzhong and Xu Ping (1980). *Infrared Phys. Technol.* (in Chinese) **9** (5), 17.

Jin Gang, Yu Zhengzhong, Ma Kejun, Cao Juying and Xu Ping (1981). *Infrared Phys. Technol.* (in Chinese) **10** (4), 12.

Jing Xinliang, Xu Xinhui, Zhu Qimu, Gong Xiaocheng and Chen Yixin (1982). *Shanghai Metals (Nonferros Ed.)* **3** (3), 37.

Kaminska, M., Lagowski, J., Parsey, J., and Gatos, H. C. (1981). Presented at 3rd Lund Conf. on Deep-Level Impurities in Semiconductors, U.S.A.

Kaminska, M., Lagowski, J., Parsey, J., and Gatos, H. C. (1982). *In* "1982 Gallium Arsenide and Related Compounds 1981 (Inst. Phys. Ser. No. 63)," p. 197.

Kondo, K., Isozumi, S., Yamakoshi, S., and Kotani, T. (1982). *In* "1982 Gallium Arsenide and Related Compounds 1981 (Inst. Phys. Ser. No. 63)," p. 227.

Koshiga, F., and Sugano, T. (1979). *Thin Solid Films* **56,** 39.

Kong Meiyin, Sun Dianzhao, Huang Yunheng, Liang Jiben, Chen Zhonggui and Li Qiwang (1984). *CJS* **5,** 226.

Lagowski, J., Gatos, H. C., Parsey, J. M., Wada, K., Kaminska, M., and Walukiewicz, W. (1982). *Appl. Phys. Lett.* **40,** 342.

Lagowski, J., Lin, D. G., Aoyama, T., and Gatos, H. C. (1984). *Appl. Phys. Lett.* **44,** 336.

Laithwaite, K., Newman, R. C., Angress, J. F., and Gledhill, G. A. (1977). *In* "1977 Gallium Arsenide and Related Compounds 1976 (Inst. Phys. Ser. No. 33a)," p. 133.

Lang, D. V., and Logan, R. A. (1975). *J. Electron. Mater.* **4,** 1053.

Lang, D. V., and Logan, R. A. (1979). *Phys. Rev.* **B19,** 1015.

Lang, D. V., Hartman, R. L., and Shumaker, N. E. (1976). *J. Appl. Phys.* **47,** 4986.

Laporet, J. L., Cadoret, M., and Cadoret, R. (1980). *J. Cryst. Growth* **50,** 663.

Laser Group of 3rd Dept. of HSRI (1982). *In* "1982 Kunming Symp.," p. 217.

Lavinson, M., and Temkin, H. (1983). *Appl. Phys. Lett.* **42,** 604.

Lei, T. F., Lee, C. L., and Chang, C. Y. (1978). *Solid State Electron.* **21,** 385.

Li Aizhen and Milnes, A. G. (1983). *J. Cryst. Growth* **62,** 95.

Li Aizhen, Guo Kangjin, Feng Jiangsheng, Shao Qirong and Zou Yuanxi (1965). Unpublished work, SIM.

Li Aizhen, Xin Shangheng and Milnes, A. G. (1983a). *J. Electron. Mater.* **12,** 71.

Li Aizhen, Cheng, H., and Milnes, A. G. (1983b). *J. Electrochem. Soc.* **130,** 2072.

Li Aizhen, Shen Dexin and Qiu Jianhua (1984). Presented at 3rd National Symp. on Surface Phys., Xian, China.
Li Aizhen, Shen Dexin, Qiu Jianhua and Yang Yuefei (1985a). *J. Vac. Sci. Technol.* **B3**, 746.
Li Aizhen, Milnes, A. G., Chen Zhiyao, Shao Yongfu and Wang Shoubo (1985b). *J. Vac. Sci. Technol.* **B3**, 629.
Li Changshi, Du Yinlan, Xie Zhicheng and Qu Yunxia (1982). *Anal. Chem.* (in Chinese). No. 9, 541.
Li Mingdi, Gong Xiaocheng, Zhu Rengmu and Chen Yixin (1983). *JAS* **1**, 327.
Li Qinggui and Li Xingming (1982). *In* "Abstracts, 1982 Changzhou Symp.," p. 148.
Li Shihai, Zhang Shangqiong, Chen Peiqiu, Zhang Yanbin, Zhang Chuongrin and Chen Xiaojian (1985). *SSERP* **5**, 189.
Li Songfa and Lu Haiyun (1982). *In* "1982 Kunming Symp.," p. 124.
Li Xiqiang, Chen Zhiyao, Lin Chenglu and Wang Weiyuan (1984). *In Energy Beam-Solid Interactions and Transient Thermal Processing* (John C. C. Fan and W. M. Johnson, eds.), p. 621. North-Holland, New York.
Liang Chunguang, Chen Xiaoze, Yang Hanpeng, Wang Shujun and Cao Yulu (1983). *DIANZI KEXUE (Electron. Sci.).* No. 3, 1.
Liang Jingguo, Feng Sunqi, Pan Guiming and Chen Weixi (1981). *CJS* **2**, 273.
Liao Xianbing (1984). *Semicond. Photoelectron.* (in Chinese) **5** (1), 20.
Lin Lanyin, Fang Zhaoqiang, Zhou Bojun, Zhu Suzhen, Xiang Xianbi and Wu Rongyan (1982a). *J. Cryst. Growth* **56**, 533.
Lin Lanying, Lin Yaowang, Zhong Xingru, Zhong Yanyun and Li Xiulan (1982b). *J. Cryst. Growth* **56**, 344.
Lin Shiming, Wang Qiming, Du Baoxun, Shi Zhongcheng and Gao Jilin (1982). *CJS* **3**, 175.
Lin Yaowang, Zhang Yanyun, Li Xiulan and Lin Lanying (1984). *CJS* **5**, 547.
Lin Zhaohui and Zhang Lizhu (1983). *In* "Abstracts, 1983 Shanghai Symp.," p. 42.
Lin Zhaohui, Zhang Lizhu, Ren Hong, Xu Huiying, Zhang Berui, Qin Guogang and Cui Yucheng (1984). *CJS* **5**, 497.
Lin Zhaohui, Qin Guogang, Zhang Lizhu, Chen Kun, Xu Huiying, Zhang Borui and Yin Qingmin (1985). *LDD* **6** (1), 7.
Lin Zhongpeng (1977). *Anal. Chem.* (in Chinese). No. 4, 268.
Lin Zhongpeng (1981). *Anal. Chem.* (in Chinese). No. 3, 253.
Lin Hongdu, Chen Weixi and Feng Zhechuan (1982). *CJS* **3**, 359.
Liu Hongxun, Chen Weixi and Yu Lisheng (1983). *CJS* **4**, 93.
Liu Qiongen, Wu Zhuoqing and Zhou Meixhen (1980). *STC* (Anal. Chem. Special Issue) **44**.
Liu Xitian, Deng Zhijie and Yu Bincai (1983). *In* "1983 Shanghai Symp.," p. 67.
Logan, R. M., and Hurle, D. T. J. (1971). *J. Phys. Chem. Solids* **32**, 1739.
Lopez, C. Garcia, A., and Munoz, E. (1977). *Electron. Lett.* **13**, 460.
Lu Bingmei, Liu Hanxun and Chen Jiwen (1982). *In* "1982 Kunming Symp.," p. 217.
Lu Dacheng (1982a). *CJS* **3**, 329.
Lu Dacheng (1982b). *In* "1982 Kunming Symp.," p. 12.
Lu Fengzhen, Wang Jiakuan, Xu Chenmei and Ding Yongqing (1981). *In* "1981 Shanghai Symp.," p. 134.
Lu Fengzhen, Ding Yongqing and Peng Ruiwu (1983). *JAS* **1**, 243.
Lu Jiangcai and Chen Jun (1983). *In* "1983 Shanghai Symp.," p. 280.
Ma Jinglin, Li Guchui, Yan Fengzheng, Liu Yili, Zhou Ruiying, Li Zhizhong and Din Xinhua (1984). *CJS* **5**, 676.
Ma Shimang and Tan Jingzi (1984). *SSERP* **4** (1), 33.
Mao Yuguo, Tang Bingrong, Dai Li, Tao Xinqing and Kung Yuzhen (1984). *SSERP* **4** (2), 41.

Martin, G. M., Jacob, G., Poiblaud, G., Goltzene, A., and Schwab, C. (1980). *In* "Proc. 11th Int. Conf. on Defects and Radiation Effects in Semiconductors (Inst. Phys. Ser. No. 59)," p. 281.
Meng Shile, Zhu Qimu, Gong Xiaocheng and Chen Yixin (1983). *SSERP* **3** (4), 26.
Miao Zhongli, Wang Jing and Gao Dingsan (1982). *In* "1982 Kunming Symp." (R. A. Huggins, ed.), p. 221.
Miller, G. L., Lang, D. V., and Kimmerling, L. C. (1977b). *Ann. Rev. of Mater. Sci.*, Vol. **7**, p. 377. Ann. Rev. Inc., Palo Alto, California.
Miller, M. D., Oelsen, G. H., and Ettenberg, M. (1977a). *Appl. Phys. Lett.* **31**, 358.
Milnes, A. G. (1983). *In* "Advances in Electron Physics" (P. W. Hankes, ed.), Vol. **61**, p. 63. Academic Press, New York, New York.
Mitonneau, A., Mircea, A., Martin, G. M., and Pons, D. (1979). *Rev. Phys. Appl.* **14**, 853.
Miyazawa, S., Mizutani, T., and Yamazaki, H. (1982). *Jpn. J. Appl. Phys. Lett.* **21**, L542.
Mo Jinji, Zheng Yanlan and Jiang Wenda (1984). *XYJS* **8** (2), 8.
Mo Peigen (1985). *JAS* **3**, 38.
Mo Peigen and Yang Linbao (1982). *XYJS* **6** (2), 34.
Mo Peigen, Yang Jinhua, Li Shouchun, Jiang Dawei, Zho Huifang and Zhang Guoming (1985a). *JAS* **3**, 355.
Mo Peigen, Li Shouchun, Li Yuexin, Gao Jinling and Li Shiling (1984). *CJS* **5**, 562.
Mo Peigen, Wu Ju, Zou Yuanxi, Yang Jinhua and Li Shouchun (1985b). *In* "Semi-Insulating III–V Materials, Kah-nee-ta (Shiva)," p. 134.
Morizane, K., and Mori, Y. (1978). *J. Cryst. Growth* **45**, 164.
Mozzi, R. L., Fabian, W., and Piekarski, F. J. (1979). *Appl. Phys. Lett.* **35**, 337.
Nanishi, Y., Yamazaki, H., Mizutani, T., and Miyazawa, S. (1982). *In* "1982 Gallium Arsenide and Related Compounds 1981 (Inst. Phys. Ser. No. 63)," p. 7.
Nishizawa, J., Shih Yijing, Suto, K., and Koike, M. (1982). *J. Appl. Phys.* **53**, 3878.
Oliver, J. R., Fairman, R. D., Chen, R. T., and Yu, P. W. (1981). *Electron. Lett.* **17**, 839.
Ozeki, M., Kommeno, J., Shibatomi, A., and Okawa, S. (1979). *J. Appl. Phys.* **50**, 4808.
Pan Huizhen, Xiao Zongyao, Shen Pengnian, Chen Zongquan, Fu Xiaomei, Zhu liming, Wu Guanqun and Shui Hailong (1981a). *KEXUE TONGBAO (Science)* **26**, 1141.
Pan Huizhen, Zhang Guicheng, Xu Shaohua, Pang Yongxiu, Cheng Zongquan, Fu Xiaomei, Zhu Liming and Hu Daosan (1981b). *J. Electron* (in Chinese) **3** (1), 22.
Pan Huizhen, Xu Guohua, Xiao Zongyao, Cheng Zongquan, Shen Pengnian, Fu Xiaomei and Wang Huiming (1983a). *Acta Optica Sinica* **3**, 785.
Pan Huizhen, Xu Guohua, Xiao Zongyao and Wang Dening (1983b). Presented at 4th Int. Conf. on Integrated Optics and Opt. Commun., Tokyo.
Pan Huizhen, Yang Yi, Xiao Zhongyao and Wang Dening (1985). *In* "1985 Symp. on Gallium Arsenide and Related Compounds (Inst. Phys. Ser. No. 74)," p. 485.
Pan Yungang (1983). *In* "1983 Wuhan Symp.," p. 70.
Pang Yongxiu and Sun Bingyu (1985). To be published in *DIANZI KEXUE (Electron. Sci)* **7**, 297.
Pang Yongxiu, Cheng Zongquan, Fu Xiaomei, Zhu Liming, Wang Huimin and Pan Huizhen (1982). *J. Ch. Inst. Commun.* **3** (2), 86.
Pang Yongxiu, Cheng Zongquan, Fu Xiaomei, Zhu Liming, Wang Huimin and Pan Huizhen (1983a). *IEEE Trans.* **ED-30**, 348.
Pang Yongxiu, Pang Huizhen, Gong Liangen, Wu Guanqun, Wang Dening and Wu Zhen (1984). Accepted for presentation at 9th IEEE Int. Semicond. Laser Conf., Rio de Janeiro, Brazil.
Pang Yongxiu, Zhang Guicheng, Chen Qiyu, Xiao Zongyao and Pan Huizhen (1983b). *Optical Fibre Commun.* (in Chinese). No. 1, 57.
Partin, D. L., Chen, J. W., Milnes, A. G., and Vassamillet, L. F. (1979). *J. Appl. Phys.* **50**, 6845.

Peng Ruiwu (1980). *Acta Metallurgica Sinica* **16,** 308.
Peng Ruiwu (1982). *J. Cryst. Growth* **56,** 351.
Peng Ruiwu (1984). Presented at 6th Amer. Conf. on Cryst. Growth and 6th Int. Conf. on Vapor Growth and Epitaxy, Atlantic City, New Jersey, U.S.A.
Peng Ruiwu, Jiang Wenda, Sun Shangzhu, Xu Chenmei, Sheng Songhua and Cai Yang (1983). Presented at 7th Int. Conf. on Cryst. Growth, Stuttgart, F.G.R.
Peng Zhenfu, Xu Jiachi, Wu Shangzhen and Wang Shanxiao (1982), *SSERP* **2** (2) 54.
Petroff, P. M., Lorimar, O. G., and Ralston, J. M. (1976). *J. Appl. Phys.* **47,** 583.
Qian Jiayu and Lu Jinyuan (1984). *Rare Metals* (in Chinese) **3,** 106.
Qiao Yong, Lu Jianguo, Lu Chaowei, Shao Yongfu and Wang Weiyuan (1983). *CJS* **4,** 561.
Qiu Sicho (1984). *J. Central China Inst.* **12,** 73.
Ran Chongzhu and Wang Yuzhang (1983). *In* "Abstracts, 1983 Wuhan Symp.," p. 79.
Sadana, D. K., Washburn, J., and Zee, T. (1982). *In* "1982 Symp. on Gallium Arsenide and Related Compounds 1981 (Inst. Phys. Ser. No 63)," p. 359.
Schwartz, G. P. (1980). *J. Electrochem. Soc.* **127,** 2269.
Shan Zhenguo (1981). *Laser J.* (in Chinese) **8** (6), 23.
Shan Zhengguo (1983). *CJS* **4,** 265.
Shao Jiuan, Zou Yuanxi and Peng Ruiwu (1986). *JAS* **4,** 16.
Shao Yongfu, Wang Guangyu, Chen Zhiyao, Liang Qi, Li Cuiyun, Zhu Fuying, Tan Wenling, Fu Xinding and Peng Ruiwu (1981). *In* "1981 Shanghai Symp.," p. 239.
Shao Yongfu, Chen Zhiyao and Peng Ruiwu (1982). *CJS* **3,** 215.
Shaw, D. W. (1981). *J. Electrochem. Soc.* **128,** 874.
Shen Chi and Zhan Qianbao (1985). *JAS* **3,** 52.
Shen Houyuan, Liang Junwu and Chu Yiming (1984). *CJS* **5,** 233.
Shen Linkang (1975). *Microelectronics* (in Chinese). No. 1, 1.
Shen Linkang and Jiang Jinyi (1982). *Microelectronics* (in Chinese). No. 1, 15.
Shen Linkang, Jiang Jinyi, Zhang Jinxin and Xiao Qixin (1982). *XYJS* **6** (6), 1.
Shen Linkang, Jiang Jinyi, Xiao Qixin and Zhang Jinxin (1983). *In* "Abstracts 1983 Shanghai Symp.," p. 28.
Sheng Songhua, Sun Shagzhu, Su Jiannong and Peng Ruiwu (1983). *In* "1983 Shanghai Symp.," p. 121.
Shi Huiying, Yu Haisheng, Ren Yaocheng, Zou Yuanxi, Jiang Lingdi, Sun Qiuxia and Hu Jian (1983). *Rare Metals* (in Chinese) **2** (1), 39.
Shi Huiying, Yu Haisheng, Zhou Binglin, Ren Yaocheng, Zou Yuanxi, Chen Zhenxiu and Chiang Yulang (1984). *Mater. Lett.* **2,** 313.
Shi Huiying, Yu Haisheng, Ren Yaocheng, Zou Yuanxi, Chiang Yulan, Sun Qiuxia and Hu Jian (1986). To be published.
Shi Yijing (1982). *In* "1982 Kunming Symp.," p. 283.
Shi Zhiwen, Ma Guorong, Yang Perisheng, Zuang Wanru, Hu Tiandou, Li Jingran and Lu Hui (1982). *In* "1982 Kunming Symp.," p. 104.
Shi Zhiyi and Zhou Qiumin (1984). *Rare Metals* (in Chinese) **3** (1), 98.
Shui Hailong, Zhang Guicheng, Wu Xiangsheng, Chen Qiyu, Xu Shaohua, Yang Yi, Chen Ruizhang and Hu Daosan (1982a). *LDD* **3** (3), 1.
Shui Hailong, Zhang Guicheng, Wu Xiangsheng, Chen Qiyu, Xu Shaohua, Yang Yi, Chen Ruizhang and Hu Daosan (1982b). *J. Electron.* (in Chinese) **4,** 286.
Stringfellow, G. B. (1970). *J. Electrochem. Soc.* **117,** 1301.
Stringfellow, G. B. (1980). *Appl. Phys. Lett.* **36,** 1.
Stupp, E. H., and Milch, A. (1977). *J. Appl. Phys.* **48,** 282.
Su Liman (1983). *In* "1983 Symp. on Gallium Arsenide and Related Compounds 1982 (Inst. Phys. Ser. No. 65)," p. 423.

Sun Daoyun, Tong Zhengsheng, Li Daping and Liu Weixin (1982). *In* "1982 Kunming Symp.," p. 203.
Sun Dianzhao, Chen Zhonggui, Liang Jiben, Huang Yunheng and Kong Meiying (1984). *In* "Abstracts, 3rd Int. MBE Conf., San Francisco, California," p. 87.
Sun Guiru (1983). *In* "1983 Shaghai Symp.," p. 236.
Sun Guiru and Liu Ansheng (1983). *In* "1983 Shanghai Symp.," p. 231.
Sun Guiru and Tan Lifang (1983). *XYJS* **7** (4), 28.
Sun Huiling, Wang Peida, Li Xiuqiong, Tang Gefei, Xu Jiadong and He Caixia (1983). *Micro-Fabrication Technology* (in Chinese). No. 2, 33.
Sun Qing, Zou Yuanxi and Shi Huiying (1983). *Mater. Lett.* **2,** 79.
Sun Qing, Zou Yuanxi, Shi Huiying and Wu Wenhai (1984). *JAS* **2,** 195.
Sun Shangzhu, Sheng Songhua and Peng Ruiwu (1981). *In* "1981 Shanghai Symp.," p. 139.
Sun Tongnian, Liu Silin and Gao Shuzeng (1982). *In* "Semi-Insulating III–V Materials" (2nd Int. Conf.), p. 61. Shiva Publishing Limited, England.
Tan Lifang, Liao Liying, Tan Huizu, Lu Qidong and Zou Yuanxi (1982). *In* "Semi-Insulating III–V Materials" (2nd Int. Conf.), p. 248. Shiva Publishing Limited, England.
Tan Litong, Zhang Minquan, Hu Yusheng and Fang Dunfu (1982). *XYJS* **6** (6), 1.
Tang Houshun, Yu Xitong, Mi Wei, Xiong Pei and Cheng Changan (1984). *Semicond. J.* (in Chinese) **9** (1), 49.
Tang Houshun, Yu Xitong, Li Chuan, Zhang Guansheng, Gao Huimin and Chen Guanmeng (1985). *J. Fudan Univ. (Natur. Sci. Ed.)* **24,** 101.
Taniguchi, M., and Ikoma, T. (1983a). *In* "1983 Symp. on Gallium Arsenide and Related Compounds 1982 (Inst. Phys. Ser. No. 65)," p. 65.
Taniguchi, M., and Ikoma, T. (1983b). *J. Appl. Phys.* **54,** 6448.
Thompson, F., and Newman, R. C. (1972). *J. Phys.* **C5,** 1999.
Tien, P. K., and Miller, B. I. (1979). *Appl. Phys. Lett.* **34,** 701.
Tsou Shichang, Wang Weiyuan, Lin Chenglu and Xia Guanqun (1983). *AES* **11** (1), 104.
Tu Xiangzheng, Ge Yuru, Wang Weiming and He Denglung (1982). *APS* **31,** 78.
Umeno, M., and Ameniya, Y. (1979). *Surface Sci.* **86,** 314.
Van Vechten, J. A. (1975). *J. Electrochem. Soc.* **122,** 423.
Wada, O., Majerfeld, A., and Choudhury, A. N. M. M. (1980). *J. Appl. Phys.* **51,** 423.
Walukiewicz, W., Lagowski, J., and Gatos, H. C. (1981). *J. Appl. Phys.* **52,** 5853.
Wan Qun, Li Yuzhen and Sun Guiru (1984). *Rare Metals* (in Chinese) **3** (2), 13.
Wang Caihao, Zhou Xiuda, Guo Yanming, Bian Jinhui, Shi Juyuan, Wu Wenhai, Wang Rongjuan, Zou Yuanxi and Xiong Qinghua (1982). *J. Shanghai Univ. Sci. Technol.* No. 1, 1.
Wang Dening, Cheng Zhaonian and Wang Weiyuan (1980). *APS* **29,** 1452.
Wang Dening and Pan Huizhen (1984). *Acta Optica Sinica* **4,** 126.
Wang Dingguo (1977). *In* "1977 Liuzhou Symp.," p. 452.
Wang Dingguo and Wang Quansheng (1984). *XYJS* **8** (3), 39.
Wang Dingguo, Wang Quansheng, Xiong Yinyu, Liu Yongfang,, Zhao Xuyuan, Sun Yanhua and Zhou Minfu (1981). in "1981 Shanghai Symp.," p. 86.
Wang Feng and Zhang Ruihua (1981). *In* "1981 Shanghai Symp.," p. 166.
Wang Feng, Zhong Ruihua, Gao Lixin and Li Aiqin (1983). *In* "1983 Shanghai Symp.," p. 168.
Wang Fuxian and Zuo Qipei (1983). *In* "1983 Shanghai Symp.," p. 275.
Wang Guangyu and Ding Yongqing (1985). *Rare Metals* (in Chinese) **4,** 11.
Wang Guangyu and Peng Ruiwu (1985). Unpublished.
Wang Guanyu, Zou Yuanxi, Peng Ruiwu and Shao Yongfu (1982). *Rare Metals* (in Chinese) **1** (1), 4.
Wang Houchi (1982). *Fresenius Z. Anal. Chem.* **313,** 385.

Wang Lijun and Yuan Yuan (1983). *In* "1983 Shanghai Symp.," p. 327.
Wang Lijun, Gao Dingsan and Miao Zongli (1982). *In* "1982 Kunming Symp.," p. 252.
Wang Limo (1984). *CJS* **5**, 56.
Wang Limo, Luo Haoping, Li Qizhong, Yang Fengchen, Jian Fong, Qu Zhiren, Chen Lingyun and Ma Hongfang (1982). *CJS* **3**, 222.
Wang Qiming, Du Baoxun, Huang Xi and Zhao Liqing (1985). *J. Ch. Inst. Commun.* **6** (1), 65.
Wang Rongjuan and Xiong Qinghua (1980). *STC* (Anal. Chem. Special Issue) **40.**
Wang Shaobo, Wu Ruidi and Xue Zhongfa (1981). *In* "1981 Shanghai Symp.," p. 216.
Wang Shaobo, Wu Ruidi and Xue Zhongfa (1984). *JAS* **2**, 267.
Wang Shouwu, Wang Zhongming and Xu Jizong (1981). *CJS* **2**, 189.
Wang Shouwu, Zhao Liqing, Zhang Cunshan, Deng Shenggui (1982). *CJS* **3**, 113.
Wang Shutang, Pan Rongjun and Zeng Jing (1984). *CJS* **5**, 214.
Wang Tiancheng (1981). *In* "1981 Shanghai Symp.," p. 155.
Wang Tiancheng, Fan Wenjiang and Ma Guichen (1983). *In* "1983 Shanghai Symp.," p. 185.
Wang Weiyuan, Xu Jingyang, Ni Qimin, Tan Ruhuan, Liu Yueqin and Qiu Yueying (1979). *APS* **28**, 684.
Wang Weiyuan, Qiao Yong, Lin Chenglu, Luo Chaowei and Zhou Yongquan (1982). *APS* **31**, 71.
Wang Weiyuan, Wang Dening, Qiao Yong, Cheng Zhaonian, Xia Guanqun and Xu Jingyang (1983). *In* "1983 Shanghai Symp.," p. 291.
Wang Xianren, Yan Weiping, Lan Tian and Wang Shaoping (1985). *AES* **13** (6), 31.
Wang Yiping, Guo Haizhou, Mao Yuguo, Lin Shuzhi and Tao Xinqin (1983). *In* "1983 Shanghai Symp.," p. 100.
Wang Yunsheng and Zhuang Enyou (1981). *In* "1981 Shanghai Symp.," p. 123.
Wang Zhangguo, Ledebo, L. Å., and Grimmeiss, H. G. (1984). *J. Phys.* **C17**, 259.
Wang Zhongchang, Pan Yanwan, Zhang Guifang, Yang Zhengquan, Qi Xinguo, Luan Hongfa, Liu Zhenqi and Gen Yanchang (1983). *In* "1983 Shanghai Symp.," p. 85.
Wei Cejun, Klein, J. H., and Beneking, H. (1982). *IEEE Trans.* **ED-29**, 1442.
Weimann, G. (1979). *Thin Solid Films* **56**, 173.
Weiner, M. E. (1972). *J. Electrochem Soc.* **119**, 496.
Weiner, M. E., and Jordon, S. E. (1972). *J. Appl. Phys.* **43**, 1767.
Weisberg, L. R., and Blanc, J. (1960). *In* "Proc. Int. Conf. on Semicond., Prague," p. 940.
Williams, E. W., Elder, W., Astles, M. G., Webb, M., Mullin, J. E., Staugan, B., and Tufton, P. J. (1973). *J. Electrochem. Soc.* **120**, 1741.
Wu Dingfen (1972). *In* "1972 Shanghai Symp.," p. 275.
Wu Dingfen and Chen Fenkou (1979). *STC.* No. 2, 27.
Wu Dingfen and Chen Fenkou (1980). *CJS* **1**, 100.
Wu Dingfen and Wang Dening (1985). *APS* **34**, 331.
Wu Dingfen and Wang Dening (1986). To be published in *JAS*.
Wu Dingfen, Daembkes, H., and Heime, K. (1984). *CJS* **5**, 99.
Wu Fumin, Dong Ruixin and Zhang Xueshu (1982). *In* "1982 Kunming Symp.," p. 70.
Wu Hengxian, Lin Chenglu, Wu Dingfen, Liu Xianghuai, Chen Fenkuo and Wu Huijuan (1980). *Nature J.* (in Chinese) **3** (6), 474.
Wu Jinying, Mou Cuiping, Qu Yunxia and Feng Zhenzao (1986). *Anal. Chem.* (in Chinese) **14**, 304.
Wu Jisen, Guan Anmin, Sun Yilin and Zou Yuanxi (1984b). *Mater. Lett.* **2**, 504.
Wu Jisen, Zou Yuanxi and Mo Peigen (1984a). *JAS* **2**, 1.
Wu Lingxi, Liu Xunlang, Xie Shizhong, Meng Qinhui and Li Yongkang (1984). *CJS* **2**, 132.
Wu Saijuan, Chou Lanhua, Lu Dacheng, Yu Ping and Wang Ruilin (1982). *CJS* **3**, 155.
Wu Xiangsheng, Yang Yi and Li Yunping (1982a). *LDD.* No. 2, 71.

Wu Xiangsheng, Yang Yi, Li Yunping, Tang Qiangmei and Shui Hailong (1982b). *CJS* **3**, 162.
Wu Xiangsheng, Li Yunping and Yang Yi (1983). *LDD.* No. 1, 1.
Wu Yunhong, Gong Bangrui and Lu Zhenzhong (1982). *In* "1982 Kunming Symp.," p. 126.
Wu Zhuoqing and Zhou Meizhen (1980). *STC* (Anal. Chem. Special Issue) **64**.
Wu Zhuoqing, Zhou Meizhen and Huang Huiming (1982). *In* "1982 National Symp. on Atomic Absorption Spectroscopy," Paper No. 029, p. 1.
Xiang Xianbi, Yang Xiquan, Fang Zhaoqiang, Zhu Suzhen, Wu Rangyuan and Lin Lanying (1985). *CJS* **6**, 386.
Xiao Zongyao, Shen Pengnian and Pan Huizhen (1982). *In* "Technical Digest, 6th Topical Meeting on Integrated and Guided Wave Optics," Pacific Grove, California, U.S.A., p. Th66-1
Xiao Zongyao, Shen Pengnian and Pan Huizhen (1983). *Acta Optica Sinica* **3**, 107.
Xiao Zongyao, Pan Huizhen and Chen Lianyong (1984). *Acta Optica Sinica* **4**, 499.
Xie Baozhen, Liu Shijie, Zhang Jingping, Yin Shiduan, Gu Quan and Dai Aiping (1984). *CJS* **5**, 81.
Xie Boxing, Lin Youshen, Duan Weili, Yan Guang and Yin Wen (1983). *In* "Abstracts, 1983 Nanjing Symp.," p. 76.
Xin Shangheng and Eastman, L. F. (1983). *In* "1983 Shanghai Symp.," p. 152.
Xin Shangheng, Schaff, W. J., Wood, C. E. C., and Eastman, L. F. (1982). *Appl. Phys. Lett.* **41**, 742.
Xin Shangheng, Wood, C. E. C., De Simone, D., Palmateer, S., and Eastman, L. F. (1982). *Electron. Lett.* **18**, 3.
Xiong Qinghua and Wang Rongjuan (1983). *Anal. Chem.* (in Chinese) **12**, 377.
Xu Chenmei, Li Cuiyun, Wang Bohong and Peng Ruiwu (1985). *J. Electronics* (in Chinese) **7**, 369.
Xu Guochua, Pan Huizhen, Wang Huimin and Fu Xiaomei (1982). *Shanghai Metals (Nonferrous Ed.)* **3** (4), 26.
Xu Hongda and Lin Lanying (1966). *KEXUE TONGBAO (Science)* **17**, 115.
Xu Junying, Chen Lianghui, Gong Jishu, Xu Zhongying, Zhuang Weihua, Li Yuzhang, Xu Jingzong and Wu Lingxi (1984). *J. Luminescence* **31 & 32**, 454.
Xu Shaohua, Yao Wenlan, Dong Rongkang and Wu Xiangsheng (1983). *In* "1983 Shanghai Symp.," p. 348.
Xu Shouding (1983). *In* "1983 Shanghai Symp.," p. 248.
Xu Yongquan, Wang Xiangxi, Mou Panjang and Fang Dunfu (1984). *Rare Metals* (in Chinese) **3**, 29.
Yamakaoshi, S. (1977). *Appl. Phys. Lett.* **31**, 627.
Yan Jinlong and Zhou Zuyao (1985). To be published in *Vac. Sci. and Technol.* (in Chinese).
Yan Shulan and Zhou Shuxing (1982). *In* "Abstracts, 1982 Changzhou Symp.," p. 98.
Yang Dejia (1981). *In* "1981 Shanghai Symp.," p. 211.
Yang Delin (1984). *Semicond. Photoelectron* (in Chinese) **5** (1), 12.
Yang Guisheng, Pan Zhonghao, Xiao Min, Jiang Yong, Chen Minggan, Zheng Yunsheng and Fu Yisheng (1982). *In* "1982 Kunming Symp.," p. 217.
Yang Hanpeng, Liu Zongmu, Liu Zonglin, Qin Ganming (1982). *In* "1982 Kunming Symp.," p. 199.
Yang Hengqing, Jiang Guoqing, Chen Yujin and Bao Zongming (1984). *CJS* **5**, 48.
Yang Linbao, Gong Liangen and Jiang Lindi (1984). *LDD* **5** (1), 59.
Yang Qianzhi, Wu Dingfen and Guan Limin (1981). *J. Solar Energy* (in Chinese) **2**, 125.
Yang Qianzhi, Min Huifang, Miao Zhongxi, Xu Yucang, Wang Zhenying and Wang Xingda (1985). Preliminary Program, 18th IEEE Photovoltaic Specialists Conf., Las Vegas, Navada, Oct. 21–25, 1985, p. 92.

Yang Yi, Wu Xiangsheng, Zhu Runshen, Wu Ruhuan and Li Yunpin (1983). *STC*. No. 3, 29.
Yang Yinhui, Liu Chenglin and Xu Juxiang (1981). *In* "1981 Shanghai Symp.," p. 160.
Yang Yinhui, Liu Chenglin, Xu Juxiang and Li Baoliang (1985). *XYJS* **9** (1), 11.
Yao Ganzhao, Huang Chaohong, Wang Hegui, Feng Jinheng and Zhen Xianfu (1983). *In* "Abstracts, 1983 Shanghai Symp.," p. 92.
Ye Yukang, Wang Fuchen and Lou Jienion (1983). *AES* **11,** 18.
Yin Qingmin, Jiang Yongfang and Li Yulu (1981). *In* "1981 Shanghai Symp.," p. 20.
Yokoyama, N., Mimura, T., Odani, K., and Fukuta, M. (1978). *Appl. Phys. Lett.* **32,** 58.
Yu, P. W., and Walters, D. C. (1982). *Appl. Phys. Lett.* **41,** 863.
Yu Guangao, Wang Cuilian and Huang Enfu (1983). *SSERP* **3** (1), 69.
Yu Jinzhong, Ju Jingli and Chen Qiong (1982). *In* "1982 Kunming Symp.," p. 92.
Yu Jinzhong, Iwai, S., Aoyagi, Y., and Toyoda, K. (1985). *CJS* **6,** 123.
Yu Kun, Meng Qinghui, Li Yongkang, Wu Lingxi, Chen Tingjie and Xu Shouding (1981). *CJS* **2,** 243.
Yu Lisheng and Wang Gunda (1982). *CJS* **3,** 465.
Yu Lisheng, Chen Weixi, Jiao Pengfei and Liu Xongxun (1983). *In* "1983 Shanghai Symp.," p. 290.
Yu Lisheng, Liu Hongxun, Chen Weixi, He Hui, Sun Yi and Wu Lei (1984). *CJS* **5,** 339.
Yu Mingren, Yang Guang and Wang Xun (1983). *APS* **32,** 799.
Yu Mingren, Wang Hongchuan, Fang Jilie, Hou Xiaoyuan and Wang Xun (1984). *APS* **33,** 1713.
Yu Suping and Huang Qisheng (1983). *J. Xiamen Univ. (Natur. Sci. Ed.)* **22,** 10.
Yu Zhenzhong, Jin Gang, Chen Xinjiang and Ma Kejun (1980a). *APS* **29,** 11.
Yu Zhenzhong, Jin Gang, Chen Xinjiang and Ma Kejun (1980b). *APS* **29,** 19.
Yu Zhenzhong, Ma Kejun and Jin Gang (1983). *CJS* **4,** 407.
Yuan Renkuan and Xu Junmin (1984). *CJS* **5,** 661.
Yuan Yourong, Meng Shuhua and Li Yuehua (1981). *In* "1981 Shanghai Symp.," p. 280.
Zeisse, C. R. (1977). *J. Vac. Sci. Technol.* **14,** 957.
Zeng Xianfu and Chung, D. D. L. (1982). *J. Vac. Sci. Technol.* **21,** 611.
Zhan Suzhen (1984). *Ch. J. Lasers* **11,** 283.
Zhang Fujia and Sun Da (1985). To be published in *J. Lanzhou University.*
Zhang Guansheng, Tang Houshun, Huang Dusen, Su Jiuling and Yu Xitong (1981a). *In* "1981 Shanghai Symp.," p. 259.
Zhang Guansheng, Huang Dusen, Tang Houshun and Yu Xitong (1981b). *In* "1981 Shanghai Symp.," p. 272.
Zhang Guansheng, Tang Houshun, Huang Dusen, Yu Xitong, Zhao Guozhen and Niu Chengfa (1983). *CJS* **4,** 181.
Zhang Guicheng and Shui Hailong (1984a). *DIANZI KEXUE (Electron. Sci.)* **6,** 172.
Zhang Guicheng and Shui Hailong (1984b). *LDD* **5** (2), 65.
Zhang Guicheng, Cheng Zongquan and Yu Zhizhong (1985a). *Vac. Sci. and Technol.* (in Chinese) **5** (6), 33.
Zhang Guicheng, Shao Yongfu and Gong Liangen (1985b). *LDD* **5** (4), 1.
Zhang Jingming and Zheng Baozhen (1982). *CJS* **3,** 1.
Zhang Lizhu, lin Zhaohui, Zhou Hao, Xu Huiying, Zhang Borui and Yin Qingmin (1983). *In* "1983 Shanghai Symp.," p. 250.
Zhang Minquan, Zou Yuanxi and Fang Dunfu (1983). Unpublished work, SIM.
Zhang Minquan, Zou Yuanxi and Fang Dunfu (1985). *Acta Metallurgica Sinica* **21,** A81.
Zhang Taishou (1979). Unpublished work, GINM.
Zhang Wangsheng, Li Shengrui, Ma Chunyuan, Liu Hanxun and Yao Xiaoe (1983). *In* "Abstracts, 1983 Wuhan Symp.," p. 59.
Zhang Wenjun, Ai Guangqin and Song Shuqin (1981). *In* "1981 Shanghai Symp.," p. 147.

Zhang Xueshu, Wu Fumin, Zhao Weiping and Yu Lixia (1983). *In* "Abstracts, 1983 Shanghai Symp.," p. 18.
Zhang Yunqin and Tang Daiwei (1985). To be published in *J. Instrum. Mater.* (in Chinese).
Zhang Yunsan and Gao Jilin (1983). *In* "Abstracts, 1983 Nanjing Symp.," p. 53.
Zhao Liqing, Wang Qiming, Zhang Cunshan and Wu Zhenqiu (1982). *CJS* **3**, 312.
Zhao Liqing, Zhang Cunshan, Tang Zheng and Wang Qiming (1983). *CJS* **4**, 239.
Zheng Guangfu, Liao Xianbing, Zheng Xianming and Hu Enzhi (1983). *CJS* **4**, 200.
Zhou Binglin and Chen Zhenxiu (1985). *CJS* **6**, 100.
Zhou Binglin, Ploog, K., Gmelin, E., Zheng, X. Q., and Shultz, M. (1982). *Appl. Phys.* **A28**, 223.
Zhou Bizhong (1984). *J. Xiamen Univ. (Natur. Sci. Ed.)* **23**, 173.
Zhou Bizhong, Tsai Yaoqiang and Xu Yu (1985). *J. Xiaman Univ. (Natur. Sci. Ed.)* **24**, 330.
Zhou Jicheng, Lu Yian, Guan Limin and Chen Huanji (1985). *In* "1985 Emei Symp. on GaAs and Related Compounds," p. 54.
Zhou Junming, Bao Changlin, Huang Yi, Yang Zhongxing and Lin Zhangda (1983). Presented at 1983 Shanghai Symp. in English.
Zhou Mian and Wang Weiyuan (1984). *CJS* **5**, 577.
Zhou Min (1982). *In* "1982 Kunming Symp.," p. 95.
Zhu Bing, Bao Ximao, Li Hesheng, Pan Hongmao, Mao Baohua and Sheng Yongxi (1984). *CJS* **5**, 554.
Zhu Longde, Zhang Shenglian, Wang Xiaojie, Wang Li, Gao Shufen and Device Technol. Group (1981). *CJS* **2**, 212.
Zhu Longde, Wang Xiaojie, Wang Li, Hu Xiongwei, Gao Shufen, Sun Furong, Zhang Fengjun, Wang Liming, Zhang Shenglian, Li Jingran and Lu Hui (1982). *In* "1982 Kunming Symp.," p. 213.
Zhu Longde, Wang Li and Hu Xiongwei (1984). *CJS* **5**, 66.
Zhu Youcai and Gao Dingsan (1982). *In* "1982 Kunming Symp.," p. 249.
Zhuang Wanru and Yang Peisheng (1983). *In* "Abstracts, 1983 Wuhan Symp.," p. 70.
Zhuang Weihua, Chen Peili, Zheng Baozheng, Xu Junying, Li Yuzhang and Xu Jizong (1984a). Presented at 10th European Conf. on Opt. Commun., Stuttgart, F.R.G.
Zhuang Weihua, Chen Peili, Zheng Baozheng, Xu Junying, Li Yuzhang and Xu Peili (1985). *IEEE Quantum Electronics* **QE21**, 712.
Zhuang Weihua, Zheng Baozhen, Xu Junying, Li Yuzhang, Xu Jizong and Chen Peili (1984b). *CJS* **5**, 538.
Zou Yuanxi (1972). *In* "1972 Shanghai Symp.," p. 22.
Zou Yuanxi (1973). Presented at National Conf. on Infrared Mater. and Technol., Shanghai, China.
Zou Yuanxi (1974). Presented at National Conf. on Solid State Microwave Technol., Shi Jia Zhuang, Hebei, China.
Zou Yuanxi as the group leader (1976). *Acta Metallurgica Sinica* **12**, 154.
Zou Yuanxi (1977). *XYJS*. No. 2, 5.
Zou Yuanxi (1980). *AES*. No. 1, 60.
Zou Yuanxi (1981). *In* "1981 Gallium Arsenide and Related Compounds 1980 (Inst. Phys. Ser. No. 56)," p. 213.
Zou Yuanxi (1982). *In* "1982 Gallium Arsenide and Related Compounds 1981 (Inst. Phys. Ser. No. 63)," p. 185.
Zou Yuanxi (1983a). *In* "1983 Shanghai Symp.," p. 189.
Zou Yuanxi (1983b). *JAS* **1**, 107.
Zou Yuanxi and Lu Naikun (1977). *In* "1977 Liuzhou Symp.," p. 173.

Zou Yuanxi, Zhou Jicheng, Mo Peigen, Lu Fengzhen, Li Liansheng, Shao Jiuan and Huang Lei (1983). *In* "1983 Gallium Arsenide and Related Compounds 1982 (Inst. Phys. Ser. No. 65)," p. 49.

Zou Yuanxi, Zhou Jicheng, Lu Yian, Wang Guangyu, Hu Binghua, Lu Bingfang, Li Cuncai, Li Liansheng, Shao Jiuan and Sheng Chi (1984). *In* "Proc. 13th Int. Conf. on Defects in Semiconductors, Coronado, USA," p. 1021.

Zou Yuanxi, Mo Peigen, Min Huifang, Yang Qingzhi, Zhou Jicheng and Zhang (Guicheng). *In* "Materials Research Society Symposia Proc., San Francisco," **48**, 385.

CHAPTER 2

InAs-Alloyed GaAs Substrates for Direct Ion Implantation

Harvey Winston, Andrew T. Hunter, Hiroshi Kimura and Robert E. Lee

HUGHES RESEARCH LABORATORIES
MALIBU, CALIFORNIA

I. Introduction

Semi-insulating GaAs substrates for direct ion implantation fabrication of field effect transistors and integrated circuits have become increasingly available in the last few years. The generally preferred process for preparing semi-insulating GaAs is liquid encapsulated Czochralski (LEC) growth without the intentional addition of impurities, and many commercial suppliers are offering substrate wafers made from LEC material. While these LEC substrates are in general use, they suffer from one significant limitation: the FETs, which are fabricated on a wafer by the formation of Schottky barrier gates over the active *n*-type channels formed by ion implantation doping with donor impurities, do not have identical electrical behavior. In terms of FET threshold voltage, standard deviations of the order of 100 mV are observed for FETs on a single wafer, even when great care has been taken to provide for uniform implantation fluence and unchanging device geometry over the wafer. Even with this degree of nonuniformity, useful circuits can be fabricated, but lower standard deviations of threshold voltage (and better uniformity of other device characteristics as well) would improve the yields of simple GaAs ICs and make more complex ICs feasible. In this

ISBN 0-12-752126-7

chapter, we will discuss an approach to better uniformity based on eliminating dislocations from GaAs substrate crystals by replacing a small fraction of the Ga atoms with the isoelectronic substituent In.

In the direct ion implantation process for preparing active channels, ions of a selected donor species, such as Si or Se, impinge on a semi-insulating GaAs substrate. The orientation of the substrate with respect to the beam is chosen so as to minimize deep penetration by channeling; the energy of the ions is selected to yield the desired profile in depth; and the fluence is adjusted to produce the desired threshold voltage. Immediately after implantation, the channel region contains a great deal of lattice damage caused by interaction with the beam of energetic ions as they lose their kinetic energy and come to rest. This damage is annealed out in a post-implant thermal treatment, during which the GaAs surface must be protected against decomposition by a capping layer or by a suitable ambient gas. Ideally, after anneal, the implanted donor density in the channel region would be independent of the lateral coordinates and depend only on the depth coordinate, thus ensuring uniform threshold voltages for FETs fabricated on the wafer. Threshold voltage is here taken to be the gate voltage at which the conduction electrons in the channel are fully depleted and the source-drain impedance rises abruptly. The voltage needed to deplete the channel depends on the net donor density and its variation with depth; in the simplified case of a step-function profile with a uniform density N extending to a depth L, the solution of Poisson's equation leads to the well-known equation for threshold voltage

$$V_T = V_B - \frac{eNL^2}{2\varepsilon},$$

where e is the electronic charge, ε is the substrate dielectric constant, and V_B is the barrier height of the Schottky barrier formed by the gate metal on the semiconductor surface—about 0.75 V for GaAs. Large values of N and L lead to negative values of V_T; this case corresponds to depletion-mode (or normally-on) FETs, which require negative gate voltage to cut off the channel conduction. Enhancement-mode (or normally-off) FETs are those with such low values of N and L that V_T is positive. For positive voltages larger than V_T, the applied voltage overcomes the effect of the built-in voltage of the gate, and the channel begins to conduct. Since N is the net donor density, V_T depends not only on the implanted ion density, but also on the activation efficiency—that is, the fraction of implanted ions that becomes active as donors less any fraction activating as acceptors. Variations of activation efficiency across a wafer can thus cause variations in threshold voltage. In fact, threshold voltage is not constant across a wafer, and this lack of uniformity is a severe limitation on the complexity of circuits that can be fabricated with reasonable yields.

As circuit complexity increases, progressively tighter tolerances are placed on the individual elements making up integrated circuits, such as the inverter in digital logic or the memory cell in random access memories. The inverters or memory cells must all have closely similar switching threshold voltages for the elements to interact with each other and produce correct functioning of the circuit. In MIMICs (microwave and millimeter-wave integrated circuits), the basic element is an analog gain or control unit that must have closely controlled properties in relation to other basic elements, so that the desired gain, pass band, power, noise level, or impedance match can be achieved. Thus, both digital and analog GaAs integrated circuits require a high degree of element matching to achieve circuit functionality. The number of fully functional circuits per wafer is a major factor in manufacturing yield; it is adversely affected by random defects, such as discontinuous metal interconnects, metal shorts between adjacent structures, or particulates on the wafer surface (which can be reduced by careful attention to device fabrication processing) and by variations in individual FET parameters. The FET threshold voltage, for example, can be influenced by systematic variations across the wafer in the angle of incidence of the implanted ions (Kasahara *et al.*, 1985b) or by variations in gate length caused by nonuniform photolithography (Lee, 1984). These variations can also be reduced by care in the device processing. However, there remain variations originating in the substrate material itself; these can be eliminated only during crystal growth and pre-implant processing, and these provide the motivation for efforts to improve substrate materials.

The degree of parameter control required by various circuits depends on the circuit type, the complexity level, and the speed requirements, as well as individual details of the circuit and logic designs. It is impossible to state definite uniformity requirements, except in relation to specific circuit designs, and it is an over-simplification to emphasize the uniformity of a single parameter, such as threshold voltage. However, general rule-of-thumb guidelines, as shown in Table I, provide a useful starting point. Microwave circuits have the widest latitude in threshold voltage because gain can be

TABLE I

SUBSTRATE UNIFORMITY REQUIREMENTS

Circuit	Threshold voltage standard deviation	Comments
MIMIC circuit	50–250 mV	Multi-stage amplifier
Logic	25–40 mV	16 × 16 multiplier
Memory	15–35 mV	16 K SRAM
A/D converter	5–10 mV	8-bit high speed

achieved with FETs having a wide range of threshold voltage. However, microwave circuits are subject to other uniformity requirements related to the *s*-parameters of the individual FETs at their operating points. Logic and memory have more stringent threshold uniformity requirements, depending on circuit complexity and logic family. Finally, analog-to-digital circuits require the highest degree of threshold voltage matching in order to attain high resolution of the analog signal. It is important to emphasize that the threshold voltage tolerance depends significantly on the circuit design and logic family employed, and that some designs have wider uniformity latitudes than others. As an illustration of this, a study on a 16-Kbit SRAM circuit employing a total of 10^5 enhancement/depletion mode FET devices showed full operation only when FET threshold scatter was less than 15 mV (Hirayama *et al.*, 1986). A similar study for the same circuit based on a complementary J-FET memory cell (Zuleeg) projected full operation with a 35 mV threshold variation.

Some of the substrate-related nonuniformity observed in practice is the result of subsurface sawing, polishing damage, or contamination during handling, but when these potential sources of nonuniformity are reduced or eliminated, there remains an often unacceptable degree of device nonuniformity, the cause of which is to be sought ultimately in the substrate wafer itself. Many investigations have shown variations across unimplanted substrate wafers of their resistivity, mobility, defect and impurity concentrations, dislocation density, and luminescence behavior, and there have been many attempts to establish correlations between the distributions of substrate properties and the distributions of carrier concentration (or of threshold voltages and saturation current of FETs) in the active layers fabricated on the substrates by direct implantation. Nanishi *et al.* (1982) demonstrated a correlation between the distribution of dislocations in substrates and the distributions of threshold voltage and saturation current of FETs fabricated on them. In a later study, Nanishi *et al.* (1983) presented striking evidence of a micro-scale correlation between lineage-like dislocations and the saturation current of nearby FETs. The importance of dislocation-free wafers in silicon IC technology, no doubt, contributed as well to an increasing demand by groups developing GaAs integrated circuits for substrates with lower dislocation densities, which were expected to lead to more uniform FET characteristics across a wafer.

One useful technique for reducing or eliminating dislocations in GaAs while retaining the advantages of LEC growth is the substitution of In for part of the Ga. The resulting material is formally an alloy of InAs and GaAs, $In_xGa_{1-x}As$, with x usually smaller than 0.01. (Alloys with x close to 0.5, lattice-matched to InP and used in opto-electronic applications, are prepared by different methods and are not treated here.) It seems best to designate these dilute alloys as InAs-alloyed GaAs in order to emphasize that the ratio of In + Ga to As is essentially unity. However, most papers in the field refer to In-doped or In-alloyed GaAs. We prefer to avoid calling the material In-doped GaAs, because the In, as an isoelectronic substituent on the Ga site, has no electrical doping effect. In this chapter on the InAs-alloyed GaAs, we will discuss its growth, metallurgical and crystallographic behavior, electrical and optical properties, and effectiveness as a substrate for achieving uniform device properties.

II. Crystal Growth

The crystal growth of InAs-alloyed GaAs by the LEC method has been guided by the theoretical explanations provided by Jordan *et al.* (1980, 1985) and by Russian workers (Avdonin *et al.*, 1972; Mil'vidskii and Bochkarev, 1978; Mil'vidsky *et al.*, 1981) for the observed distribution of dislocations in LEC GaAs. During the growth process, as the crystal is pulled from the melt and emerges through the boric oxide encapsulant layer into the ambient atmosphere of the growth furnace, it experiences temperature distributions and associated thermoelastic stresses in excess of the critical resolved shear stress (CRSS) for dislocation formation. In many cases, the radial distribution and axial symmetry of dislocations observed in LEC GaAs have been satisfactorily explained by assuming that the dislocation density depends on the amount by which the local thermoelastic stress exceeds the CRSS. The stresses on narrow crystals are not very large, in accordance with the observation that dislocation-free GaAs crystals about a cm in diameter have been prepared. However, it appeared impossible to grow LEC crystals with a low dislocation density and useful diameters (at least two inches) for substrate purposes, because of the thermoelastic stresses generated in the wider crystals by the thermal gradients in the usual LEC growth furnace designs. However, these theoretical considerations provided a basis for understanding why horizontal Bridgman crystals, which are not exposed to high thermal gradients, generally have lower dislocation densities than LEC crystals. These considerations also suggested that reducing the thermal gradients in LEC furnaces would at least reduce the dislocation density, even if the requirements of melt stability and diameter control precluded a thermal gradient low enough to eliminate dislocations entirely. In a verification of the effectiveness of reduced thermal gradients in LEC growth, Matsumoto *et al.* (1984) described modifications to a standard LEC furnace that reduced the gradients to 30–60°C cm^{-1} from the original 100–150°C cm^{-1}; the dislocation density in 3 in wafers grown in the modified furnace was 1–4 $\times 10^4$ cm^{-2}, in contrast to values several times as high for wafers from crystals grown in the unmodified gradient.

Since the reduction of thermal gradients proved inadequate to reduce the thermoelastic stress below the CRSS and eliminate dislocations, several workers proposed increasing the CRSS by adding impurities. Mil'vidskii and Bochkarev (1978) summarized previous Russian work on the reduction of dislocation density in LEC semiconductor crystals by the addition of dopants, and Mil'vidsky, Osvensky, and Shifrin (1981) included, in a general discussion of doping effects on dislocations, a plot specifically showing the reduction of the dislocation density in LEC crystals of GaAs containing various concentrations of In. They suggested a mechanism for the increase in CRSS with impurity concentration based on the blocking of dislocation movement by the formation of Cottrell impurity atmospheres around the dislocations. According to this model, In should be an effective impurity for increasing CRSS, and Mil'vidsky *et al.* indicate that, at 10^{19} In cm^{-3}, a 25 mm diameter LEC crystal has a dislocation density of 100 cm^{-2}, in contrast to 10^4 cm^{-2} for pure material.

The Russian workers noted that In, as an isoelectronic impurity, should not act as a donor or acceptor, and they reported Czochralski-grown In-containing GaAs that was semi-insulating (Solov'eva *et al.*, 1981) and material that was both semi-insulating and of reduced dislocation density (Solov'eva *et al.*, 1981). A French group at LEP (Jacob, 1982; Jacob *et al.*, 1983) was the first to report the growth of an InAs-alloyed GaAs up to 35 mm in diameter that was semi-insulating as well as largely dislocation-free, and had electrical properties essentially identical to those of unalloyed crystals, In the 1982 paper, Jacob also reported the growth of semi-insulating and largely dislocation-free InAs-alloyed GaAs by the very low thermal gradient liquid-encapsulated Kyropoulos (LEK) process. Jacob's work clearly showed the feasibility of obtaining fairly large, low-dislocation semi-insulating wafers of alloyed material suitable for device fabrication.

Several subsequent papers (M. Duseaux and S. Martin, 1984; Hobgood *et al.*, 1984; Barrett *et al.*, 1984; Kimura *et al.*, 1984; Tada *et al.*, 1984; Foulkes *et al.*, 1984; Elliott *et al.*, 1984) gave further accounts of the growth of low-dislocation, InAs-alloyed semi-insulating GaAs suitable for device substrates, and of the influence of growth conditions on the crystal properties. We conclude from these studies that, with certain significant exceptions, the usual methods (Kirkpatrick *et al.*, 1984; Thomas *et al.*, 1984; Stolte, 1984) for growing GaAs crystals for device substrates apply for alloyed material. The exceptions arise from the change in the concentration of In in the melt as it is rejected from the growing crystal (the distribution coefficient of In is smaller than unity).

Both high-pressure and low-pressure LEC growers may be used. In the low-pressure case, the starting material is pre-synthesized GaAs, and the In is supplied as InAs. With high-pressure growth furnaces, the starting charge may be pre-synthesized or it may consist of elemental Ga, In and As; the crucible material is usually pyrolytic boron nitride. The electrical behavior of the alloy crystal will depend, as usual, on maintaining a slight stoichiometric excess of As over the Group III elements in the melt—in this case, the total of Ga plus In. If the initial melt is As-deficient, at some stage of the growth the melt will become depleted of As to the extent that the rest of the crystal will be *p*-type, with its electrical behavior dominated by residual acceptor impurities and native acceptor defects inherent to growth from an As-poor melt. With As-rich melts, the crystal is semi-insulating, with its electrical behavior controlled by the native defect EL2. The principal acceptor impurity is C, and its concentration can be controlled by the selection of the moisture content of the boric oxide encapsulant (Hunter *et al.*, 1984).

As the crystal grows, the melt becomes richer in In, as shown in Fig. 1. Analysis of this plot leads to a value of the distribution coefficient k of 0.12. The increasing concentration of In in the melt will eventually create the conditions for constitutional supercooling and cellular growth. To defer the onset of constitutional supercooling as long as possible, the growth rate must be kept low, but low growth rates mean very long growth times—up to several

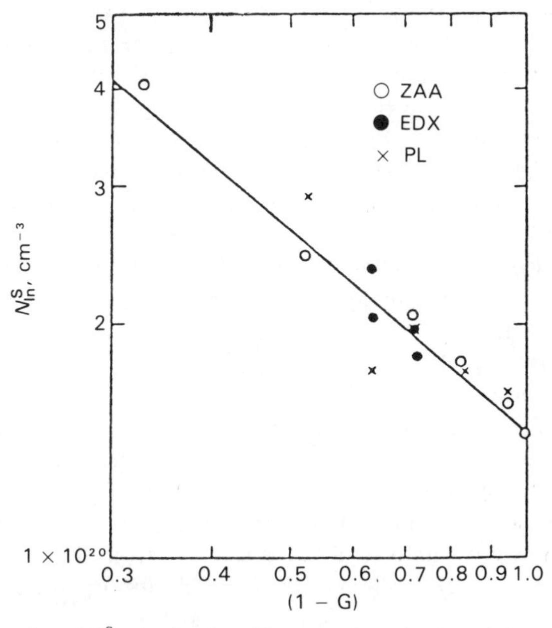

FIG. 1. Log–log plot of $N_{\mathrm{In}}^{\mathrm{S}}$, the density of In atoms in an In-alloyed GaAs crystal, versus (1 − G), the fraction of the original melt still in the liquid phase. The indium density was measured by Zeeman atomic absorption (ZAA) analysis, energy-dispersive X-ray (EDX) analysis, and photoluminescence (PL) determination of bandgap shift.

days—for the long crystals that are needed to supply large numbers of wafers. In turn, the long growth times may allow the excessive escape of As from the melt, and interfere with the proper stoichiometry for semi-insulating behavior. One practical expedient is to program the growth rate to be rapid near the seed end of the crystal and to slow down at the stage of growth for which the melt concentration of In is at the constitutional supercooling limit for the original rapid rate (McGuigan, 1986). Growth rates used have ranged from about 1 cm/hr, appropriate for pure GaAs, to 4 mm/hr.

Of course, the onset of constitutional supercooling can also be deferred to a later stage of crystal growth by using a smaller initial concentration of In. The In concentration necessary to reduce the dislocation density will be smaller when the thermal gradients in the growth furnace and the associated thermoelastic stresses are reduced, although lower thermal gradients also increase the susceptibility to constitutional supercooling. A comparison of different experiments confirms that essentially dislocation-free material occurs at lower In concentrations for growers with lower thermal gradients. The trade-offs between growth rate, In concentration, and thermal gradients are significant issues in growing dislocation-free alloy material. An important economic consideration is the high cost of In relative to the other starting materials; because of the low value of k, the melt concentration must be much higher than the desired initial crystal concentration of In, and the cost of the In approaches that of the Ga. Furthermore, some part of the crystal will be unusable, because the In concentration has become too high and cellular growth or second-phase precipitation has occurred. Any large-scale production of alloy crystals will have to include recovery of the rejected In.

As we will see in the next section, there is often a region of high dislocation density around the periphery of otherwise low dislocation alloy crystals, probably resulting from the loss of As as the crystal emerges from the boric oxide encapsulant. Also, many LEC growth furnaces support quite large temperature fluctuations in the melt close to the growth interface, resulting from melt convection and leading to striations. In crystals containing In, the striations are particularly prominent in X-ray topographs, and the associated fluctuations of the In concentration represent an undesirable nonuniformity. To solve these problems, and to allow for a particularly low value of thermal gradient in a high-pressure LEC furnace, Kohda *et al.* (1985) have described a fully encapsulated Czochralski process with applied vertical magnetic field (VM-FEC). The growing crystal never emerges from the thick boric oxide layer, so that it is never exposed to the ambient atmosphere of the furnace, and the thick encapsulant also impedes the loss of As from the melt. The thick encapsulant layer, in conjunction with two separately controllable heater elements, ensures a low thermal gradient, and the magnetic field stabilizes convection in the melt so as to eliminate large temperature excursions near the interface. In addition, the rotation rates for the crystal and crucible were chosen to eliminate rotational striations related to radially asymmetric melt temperature distributions and the steady component of melt convection that is not eliminated by the magnetic field (Osaka and Hoshikawa, 1984). The resulting crystals are completely dislocation-free and striation-free, and contain 10^{20} In cm^{-3} in the tail end, corresponding to crystallization of half the melt. The corresponding value of x is 0.004. In this experiment, there were no threading dislocations at the center of the crystal, a fact which the authors attribute to the use of a dislocation-free seed crystal and optimized conditions for seeding and necking. Presumably, the dislocation-free seed was itself cut from an InAs-alloyed GaAs ingot rather than from an unalloyed ingot, so that its lattice constant would be a closer match to the alloy material deposited from the melt. Most of the crystals grown by other groups contained dislocations at the crystal center, even when attempts were made to eliminate them by necking in the seed.

In alloying is beneficial in reducing dislocation density in horizontal Bridgman growth as well, as shown by Inoue *et al.* (1986). Consistent with the low thermal gradients in Bridgman growth, In concentrations in the crystal of $1.5–4 \times 10^{19}\,cm^{-3}$ resulted in dislocation densities below $1000\,cm^{-2}$, as compared to unalloyed material, which, in the same equipment, had dislocation densities from 3000 to $6000\,cm^{-2}$.

III. Crystallography and Structure

Interpretations of the behavior of InAs-alloyed GaAs assume that the In atoms occupy Ga lattice sites. Observations of Rutherford backscattering (RBS) and channeling by Kuriyama *et al.* (1986) show that the minimum backscattering yields of 1.5 MeV He ions along the [100] axis are consistent with this assumption. The increase in backscattering yields caused by the presence of 10^{20} In cm^{-3} corresponds to an interstitial fraction of 0.016. Since the In–As bond is 6% longer than the Ga–As bond, and the bond lengths remain essentially unchanged in InAs–GaAs alloys (Mikkelsen, 1983), the four As nearest neighbors of each In atom will be displaced from their lattice positions and contribute to the interstitial fraction. The total number of displaced nearest-neighbor As atoms is, thus, $4 \times 10^{20}\,cm^{-3}$, or a fraction of about 0.01; since the displacements must extend somewhat beyond the nearest neighbor shell, the observed value of 0.016 is not unexpected. Other RBS studies show that over 94% of the In is in substitutional sites on the Ga sublattice in GaAs containing 1 atom % In (McGuigan, 1966). Further qualitative evidence of the strain caused by the lattice misfit of In in a similar alloy crystal appears in nuclear magnetic resonance experiments that show that half of the As sites are strained. A final piece of evidence for the lattice location of In is provided by the examination of the influence of In on the quasi-forbidden X-ray reflections of GaAs (Fujimoto *et al.*, 1986). Because of the small difference in structure factor between Ga and As, certain formally allowed X-ray reflections in GaAs have very low intensity. Calculations show that if In is on the Ga sublattice, increasing its average scattering power toward that of the As sublattice, the intensities of the quasi-forbidden 200 and 600 reflections decrease still further. The reported observations are in accord with the predictions. (Incorporation of Al, which occupies Ga sites and lowers the average scattering power of the Ga sublattice, results in an increase of intensity in these reflections, also in full accord with calculations.)

The residual grown-in stresses associated with these strains in alloy material manifest themselves as brittleness and fragility of the crystals, Cracking during the preparation of wafers from ingots is an unfortunately frequent occurrence. However, McGuigan *et al.* (1986) point out that an anneal at 950°C for 18 hours relieves the stress; after anneal, the wafer yields from ingots of InAs-alloyed and unalloyed GaAs are identical. The same workers report that the alloy is, in some sense, harder than unalloyed GaAs, in that more than twice as much time is needed for polishing. After polishing, an alloyed wafer shows only about 1/15 as much subsurface damage as unalloyed material, according to a photon backscattering technique, approaching high-quality silicon wafers in this respect.

X-ray topographs of almost dislocation-free LEC alloy crystals containing about 10^{20} In cm^{-3} (Kitano *et al.*, 1985; Ishikawa *et al.*, 1986) reveal striations related to local fluctuations in In content, estimated at 15% of its average value, as well as a region highly distorted, probably as the result of In segregation. The observed X-ray contrast is caused by variations in lattice constant, with contrast from lattice orientation absent except in slipped regions near the wafer periphery. Markedly different conclusions are drawn in the same papers about an unalloyed LEC crystal with a dislocation density reported as 4000 cm^{-3}. In this case, the X-ray contrast is due entirely to variations in lattice orientation over a range of 320 arc seconds. For the alloyed sample the range of lattice orientation is only 5 arc seconds. Figure 2 is an X-ray topograph of an alloy wafer from our laboratory (Kimura, 1984a). The appearance of contrast across the whole wafer indicates the small variation in orientation; dislocations are evident in slippage at the periphery and in a narrow central core, but are not readily apparent over most of the wafer area. The irregular striation patterns in this crystal, grown in a high-pressure LEC furnace with a high temperature gradient, suggests a bumpy growth interface, rather than one clearly convex or concave.

The strain in as-grown alloy crystals is a consequence of the success of alloying in reducing dislocation content. Stresses that in unalloyed crystals

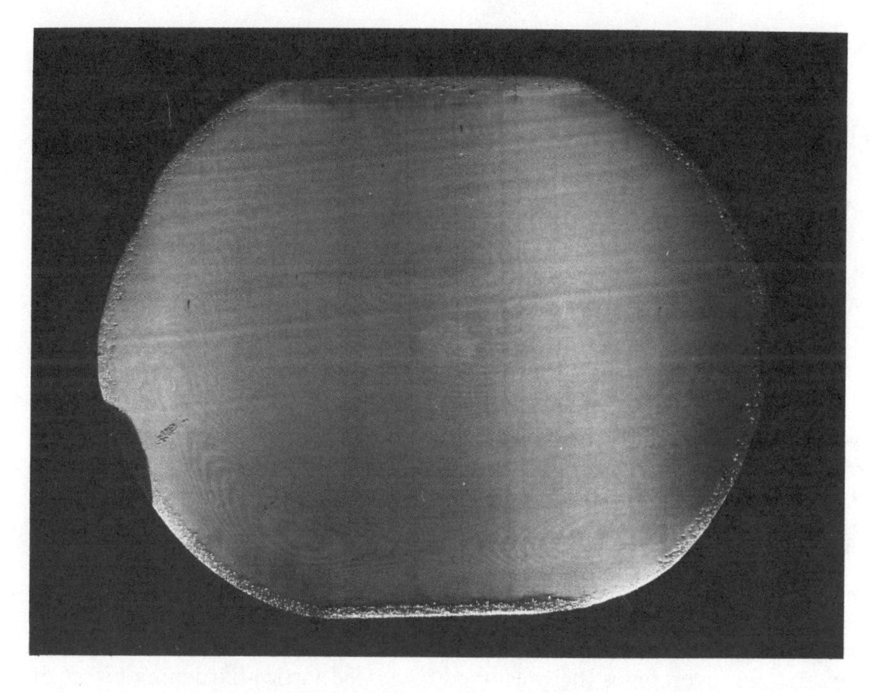

FIG 2. X-ray topograph of a (100) wafer of In-alloyed GaAs.

are relieved by dislocations are frozen into the alloy crystals. More direct evidence of the dislocation reduction comes from X-ray topography and, most definitely, from the counting of KOH etch pits. Figure 3 shows the appearance of etched (001) surfaces of wafers cut normal to the [001] growth axis from different parts of an alloy ingot with increasing In content, while Fig. 4 shows the etch pit densities (EPD) at different positions along a [110] diameter (Kimura, 1984a). Near the seed end, the In concentration is presumably not high enough to prevent the formation of some dislocations, which appear in a rosette pattern aligned along ⟨100⟩ directions, but there are still sizable dislocation-free areas. The middle wafer, with more In, is almost dislocation-free; however, the region at the original periphery was removed in grinding the ingot to a 2-inch diameter, and there were probably dislocations in this region, just as in the case of the seed-end wafer. These peripheral dislocations may result from high thermal stresses near the outer diameter of the crystal, in accord with the thermoelastic theory, or they may be generated as As loss from the growing crystal emerging from the boric oxide encapsulant leaves Ga droplets that can act as dislocation sources. The wafer richest in In shows arrays of dislocations aligned along ⟨110⟩ directions and has less dislocation-free area than the middle wafer. This may

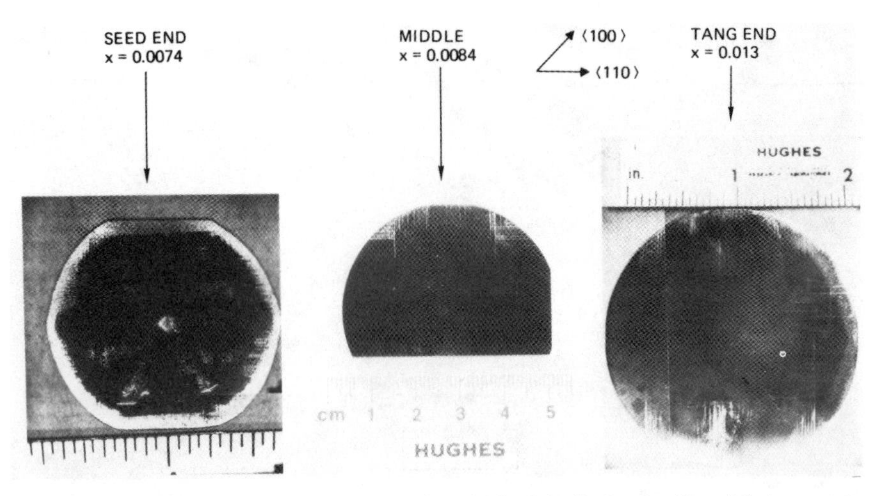

FIG. 3. KOH-etched (100) slices from the seed, middle and tail of an In-alloyed GaAs crystal. The composition of each slice is given by the value of x in the formula $Ga_{1-x}In_xAs$.

indicate a change in the thermal gradients as the crystal grows longer, or perhaps a change in growth conditions, such as the near-onset of cellular growth, that overcomes the presumed increased lattice-hardening effect of a higher In concentration. In any event, these results are in agreement with many other studies showing that $In_xGa_{1-x}As$, with x near 0.01 or even lower, exhibits remarkably reduced dislocation densities. Also, the network arrangement of dislocations arrayed into cell walls surrounding dislocation-free regions a tenth or so of a millimeter across, characteristic of much unalloyed LEC GaAs (Ishii *et al.*, 1984), is absent.

All the wafers in Fig. 3 have dislocations at their center, threading along the axis. These may be caused, or perhaps just confined to the center, by the central maximum in thermoelastic stress predicted theoretically (Jordan, 1984). On the other hand, the central dislocations are sometimes not observed at all (Kohda, 1985). In our own laboratory (Kimura, 1986) the formation of a long thin neck on the seed has prevented the appearance of the central dislocations even in a high gradient furnace. Jacob (1982) has suggested that a suitably misoriented seed might diminish the generation of central dislocations. Scott *et al.* (1985), applying X-ray topography to alloy crystals grown in a low-gradient environment with 2×10^{19} In cm^{-3}, found that the central dislocations along the growth axis are not simply extensions of those in the seed. Dislocations in the neck that are clearly connected to ones in the seed have Burgers vectors along [110] and [111] types of directions, while the straight dislocations passing on into the crystal have Burgers vectors of the type [010]. There are also helicoidal dislocations that,

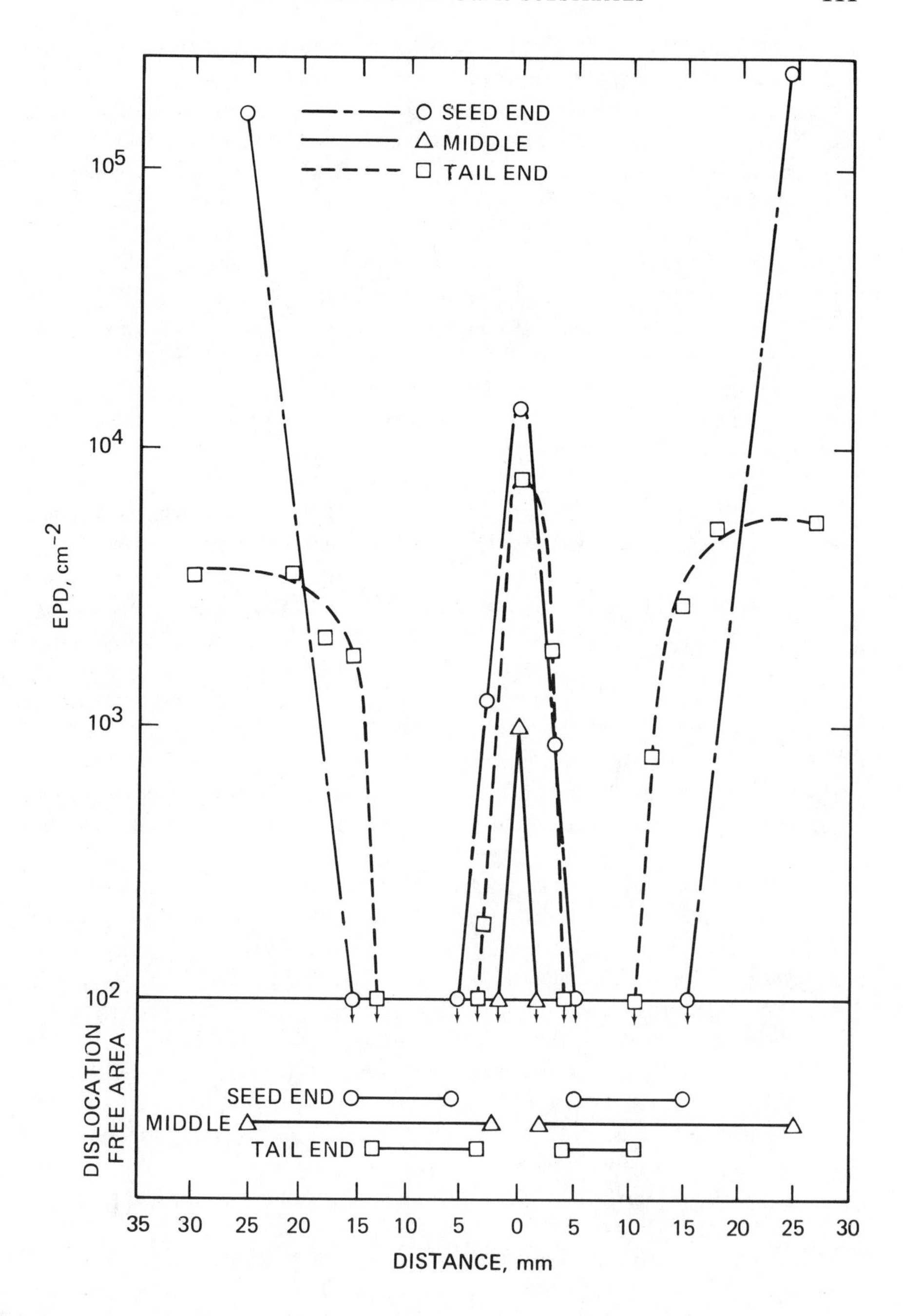

FIG. 4. Plots of etch pit density along a [110] diameter for the three wafers shown in Fig. 3.

presumably as the result of heavy decoration, never go out of contrast, and thus cannot be assigned Burgers vectors. Both straight and helicoidal dislocations propagate normal to the growth striations that are indicated by a strong periodic contrast and that are related to the segregation of In. For In concentrations less than $2 \times 10^{19}\,\mathrm{cm}^{-3}$, the crystals have dislocation densities of about $3000\,\mathrm{cm}^{-2}$, without the central straight and helicoidal dislocations. The authors contend that the central dislocations cannot be explained by thermoelastic stress arguments alone, and emphasize the importance of controlling stoichiometry and native defect concentrations in any attempt to eliminate the central dislocations. Similar detailed X-ray topographic observations of dislocations in alloyed crystals are reported by Pichaud *et al.* (1985) and Burle-Durbec *et al.* (1986). Their work shows that most of the dislocations remaining in alloy crystals have Burgers vectors of the type $\langle 110 \rangle$; it should be noted that the thermoelastic theories assume that slip occurs in the $\langle 110 \rangle \{111\}$ glide systems. However, the topographs of alloyed crystals reveal dislocations with other Burgers vectors as well. There are growth striations with a very regular $20\,\mu\mathrm{m}$ spacing, apparently not related to dislocations, and Pichaud *et al.* invoke a longer-period In content variation to explain some of the details of the distribution of dislocations along the length of the crystal. Of course, the practical interest in the dislocations remaining in InAs-alloyed GaAs is in finding methods to eliminate them, so that the entire wafer area can be used for device fabrication.

The addition of In to the melt has, indeed, led to greatly reduced dislocation density, and, in a general way, the predictions of the thermoelastic theory seem to have been confirmed. For example, in low-thermal-gradient growth, lower concentrations of In are needed to reduce dislocation density than in high-gradient situations. However, actual measurements of CRSS on low-dislocation InAs-alloyed GaAs fail to reveal significant increases from alloying. Hobgood *et al.* (1986) found no discernible increase in CRSS between unalloyed and InAs-alloyed GaAs grown by LEC methods from pyrolytic boron nitride crucibles up to 600°C, although the CRSS values for LEC material are modestly higher than those for non-LEC crystals. These observations are particularly surprising because (1) application of an improved thermoelastic stress model requires a 28-fold increase in CRSS for the alloy (with $8 \times 10^{19}\,\mathrm{In\,cm}^{-3}$) to account for the observed reduction in dislocations (McGuigan, 1986), and (2) a solution-hardening model based on the resistance to dislocation motion offered by the $InAs_4$ units (21% greater in volume than $GaAs_4$ units because of the longer bonds) predicts appreciable strenthening (Ehrenreich and Hirth, 1985). Matsui and Yokoyama (1986) found, however, that the hindrance to dislocation motion by the presence of In atoms predicted from the Ehrenreich and Hirth theory does not occur. Measurements by Tabache *et al.* (1986) on alloyed and unalloyed GaAs (presumably grown by low-gradient LEC methods from quartz crucibles, in contrast to Hobgood's material) do show a doubling of the CRSS due to the presence of 2.9×10^{19} In cm^{-3}. This increase is nowhere near enough to explain the dislocation reductions that are observed. The problem is not new, however, since Duseaux and Jacob (1982) found a similar discrepancy on analyzing the thermal stresses experienced by 15 mm diameter unalloyed GaAs crystals they grew by LEC. The maximum stresses were from 30 to 200 times as high as experimentally available values of CRSS; nevertheless, the crystals were dislocation-free. As Duseaux and Jacob point out, experimental values of CRSS in many materials are from one to four orders of magnitude smaller than the theroretical values based on the shear modulus, and the discrepancy is not well understood. There is clearly a need for better understanding of the effects in both unalloyed and alloyed GaAs.

Yonenaga *et al.* (1986) have reported a significant difference in dislocation properties between alloyed and unalloyed GaAs. At 450°C, dislocations of the α type (the atoms in the dislocation core with dangling bonds are Ga or In) in GaAs containing 2×10^{20} In cm^{-3} are immovable at stresses below 10 MPa; β dislocations in the alloy and both α and β dislocations in unalloyed GaAs (boat-grown material reported as containing 10^{13} Si cm^{-3}) move under any stress down to 1 MPa. Yonenaga *et al.* proposed that the pinning of the α dislocations in the alloy impedes dislocation formation by preventing dislocation multiplication by the Frank–Read mechanism, which requires the mobility of all the types of dislocations involved in forming a loop, including the α type. However, Matsui and Yokoyama (1986) reported a smaller difference (about 50%, instead of a factor of ten) in the threshold stresses to start α and β dislocations moving at 300°C; though they found differences in threshold stresses and dislocation velocities between the α and β types of dislocations, the results were the same whether they examined InAs-alloyed or unalloyed LEC samples.

It is well known that unalloyed LEC GaAs crystals contain inclusions of various sizes. Bol'sheva *et al.* (1982) suggested that these can correspond to As rejected from the GaAs phase as the crystal cools. Surfaces cleaved from LEC GaAs (Gant *et al.*, 1983) show 50 μm diameter zones of As related to inclusions in the bulk revealed by cleavage. Earlier, As precipitates of 1000 Å dimensions attached to dislocations had been observed (Cullis *et al.*, 1980). Barrett *et al.* (1984) reported precipitates near the edges of wafers of InAs-alloyed GaAs grown from As-rich melts, revealed by scanning transmission electron microscopy. They presented a figure of one such precipitate, about 1000 Å in diameter, apparently associated with a dislocation line, and shown by energy dispersive X-ray analysis to be As-rich. However, they also reported observing As-rich precipitates in dislocation-free regions of alloy substrates. In their crystals, grown from 1 atom % In melts, they found no evidence in extensive transmission electron microscopy studies for In precipitates. Ogawa (1986) has shown, by infrared light scattering tomography and absorption imaging, that scattering aggregates that may be As are strung along dislocations lines in InAs-alloyed GaAs, but he did not report any other scatterers not connected with dislocations. Cornier *et al.* (1985) found, by TEM analysis, particles about 1000 Å in size in unalloyed LEC material, mostly associated with dislocations. They interpreted these particles as polycrystalline insular GaAs grains. However, they found no such particles in InAs-alloyed GaAs. Earlier, Duseaux *et al.* (1983) found no TEM evidence for microdefects in bulk alloy material. The absence of microdefects is, of course, difficult to prove, but it appears that the only ones observed in the alloy have been associated with dislocations, and there is no evidence for them in dislocation-free regions. We may speculate that in GaAs locally supersaturated with As after cooling from the growth temperature, dislocations serve as nuclei for the formation of As aggregates. When no dislocation is nearby, as in the alloy material, the excess As remains dispersed in solid solution, perhaps forming clusters of a few atoms that are too small to be detected directly. These considerations may be important in connection with the appearance of As-rich regions in the vicinity of dislocations, to be discussed in the next section.

There are reports of other kinds of compositional or crystallographic nonuniformities in alloy material that may be significant in understanding electrical nonuniformity. In-rich inclusions of alloy single crystals with the same crystallographic orientation as the bulk are found in heavily alloyed GaAs (4–5% In) grown by the LEC method (Ono *et al.*, 1986). Even though such In-rich alloys are not likely to be used as substrates, the mechanism of formation of the In-rich inclusions, thought to be constitutional fluctuations in the super-cooled melt, may also be effective at lower In concentrations. With respect to another kind of nonuniformity, careful etching studies by Lessoff (1985) show that the InAs-alloyed GaAs (as well as unalloyed and non-LEC material) shows raised structures on [100] surfaces, ranging in size from 2 to 70 μm, after exposure to a NaOH–KOH etchant. The orientation of the usual dislocation-related etch pits on these raised structures differs by 90° from that in the main part of the surface, suggesting that the raised structures are anti-phase domains.

According to Ozawa *et al.* (1986), InAs-alloyed GaAs, etched in a eutectic hydroxide etch (Lessoff and Gorman, 1984) that melts at a lower temperature than the usual KOH etch, show a fine background of rectangular etch pits that are not associated with dislocations. In contrast, unalloyed, almost dislocation-free GaAs, grown in an ultra-low thermal gradient of 20°C cm^{-1} with a controlled As atmosphere to prevent As loss from the crystal, does not show such a background etch pit pattern. Ozawa *et al.* suggests that the fine background pattern indicates strain or a microscopic nonuniformity of stoichiometry. This observation may be an indication that the InAs-alloyed GaAs, though it has few dislocations, may contain nonuniformities of other kinds.

IV. Electrical and Optical Behavior

The electrical and optical properties of InAs-alloyed GaAs are only slightly different from those of unalloyed GaAs, except insofar as these properties are influenced by the presence of dislocations. In a sense, the properties of dislocation-free alloy material are simpler than those of normal dislocated LEC GaAs.

The value of x in $In_xGa_{1-x}As$ is large enough so that various direct analytical methods can be used to evaluate it. However, there is a small, but easily measurable, shift in the bandgap that shows up as a shift in the energy of the donor-bound-exciton or acceptor-bound-exciton lines in photoluminescence. The relation between x and the line shift at 6 K was determined in our laboratory (Kimura *et al.*, 1984b) as

$$\Delta E\,(\text{meV}) = 1.45 \times 10^3 x.$$

Kitahara *et al.* (1985) determined the coefficient to be 1.505. They emphasized the importance of properly identifying the exciton lines in the alloy material, and showed how this can be done by determining the slope of the intensity of the photoluminescence plotted against the excitation intensity. The measurement is best made at a very low temperature to ensure the sharpness of the exciton lines; the precision of the determination of x is limited by the photoluminescence linewidth, and is estimated to be 0.0005. These authors also identified the photoluminescence lines corresponding to band-to-C-acceptor and donor-to-C-acceptor transitions in the alloy. There were no other acceptor-related transitions, showing that the dominant shallow-acceptor impurity in the alloy sample studied was C, as is very frequently the case in unalloyed material.

Room temperature measurements in our laboratory (Kimura *et al.*, 1984b) of Hall effect and resistivity on seed and tail slices of an alloy crystal that remained semi-insulating throughout its lengths are shown in Table II. The values of resistivity, carrier concentration, and mobility given in the table are averages of four samples for the seed slice and seven samples for the tail slice. For the seed slice, all four samples had mobilities between 6220 and 6340 $cm^2/V\,s$ and resistivities above 10^8 ohm-cm; the seven tail samples ranged in mobility from 6250 to 7140 $cm^2/V\,s$ and were all above 7×10^7 ohm-cm in resistivity. The mobilities and resistivities of all these samples were within the range observed for unalloyed GaAs, although the seed-end mobilities in unalloyed material were often somewhat lower.

TABLE II

ELECTRICAL PROPERTIES OF $In_xGa_{1-x}As$

x	Resistivity (ohm-cm)	Carrier density (cm^{-3})	Mobility (cm^2/V s)
0.0063	2.11_8	5.05_6	6264
0.013	9.45_7	1.01_7	6652

Temperature-dependent Hall measurements on samples from the same semi-insulating ingot can be interpreted as showing a slightly lower activation energy than 0.773 eV, the energy measured in unalloyed GaAs (Kimura *et al.*, 1984b). The shift in activation energy is about 75% of the shift in the bandgap energy for the corresponding samples, as measured by photoluminescence line shift. This result suggests that the dominant deep donor level in the alloyed samples is essentially the same as the EL2 level in unalloyed material, but with its separation from the conduction band edge slightly decreased. Studies of the main electron trap (subsequently identified as EL2) in vapor phase epitaxial films of $In_xGa_{1-x}As$ by Mircea *et al.* (1977) suggest that for small values of x the decrease in activation energy is half that of the bandgap. Measurements of optical absorption at a wavelength of 1.0 μm provided a measure of the EL2 concentrations in the same ingot, 1.6 to 1.9×10^{16} cm^{-3} near the seed and 1.1×10^{16} near the tail. These values are in the normal range for unalloyed semi-insulating GaAs. The melt from which the ingot was grown had an initial mole fraction of As of 0.51, and it might have been expected that the As fraction of the melt would become even higher toward the tail, which usually corresponds to a higher value of EL2 concentration. However, the low growth rate used (4 mm/h) to avoid constitutional supercooling effects caused the growth period to be extended longer than usual, which may have led to the loss of some As from the melt during the growth. The characteristic local vibrational mode absorption of C is found at almost the same wavelength in the alloy material as in unalloyed GaAs, and its intensity corresponds to a concentration of C at the seed of 4×10^{15} cm^{-3} and no more than 1×10^{15} cm^{-3} at the tail. Carbon appears to be the predominant shallow acceptor in this ingot. Thus, the semi-insulating behavior of the alloy crystal is governed, just as in the unalloyed material, by the compensation of carbon or perhaps other shallow acceptors (less any shallow donors that may be present) by the deep donor EL2, a native defect present in GaAs grown from As-rich melts. There is general agreement that the physical basis of the EL2 level is an As antisite, As_{Ga}, but it is not yet resolved whether other native defects are also involved.

Other ingots evaluated in the report by Kimura *et al.* were grown from melts that were As-deficient at the outset and that, therefore, became even more so as essentially stoichiometric $In_xGa_{1-x}As$ crystallized and (possibly) As evaporated from the melt during the long growth runs. These ingots were semi-insulating at the seed end, but at some stage of their growth the melt became sufficiently As-poor that the remainder of the ingots were *p*-type. Temperature-dependent Hall measurements on the *p*-type samples revealed behavior again very similar to that of unalloyed GaAs grown from As-deficient melts. In addition to Hall slopes showing the presence of C acceptors, there is evidence for levels 0.07 and 0.2 eV from the valence band, just as in unalloyed GaAs. Photoluminescence measurements in Fig. 5 (Hunter, unpublished) show the band-to-C and donor-to-C transitions, as well as the transition to the 0.07 eV level, all duly shifted by an amount corresponding to the contraction of the bandgap caused by alloying.

The electrical resistivity is determined by the mobility and the carrier concentration, which itself is determined by the EL2 and shallow acceptor and donor concentrations. A knowledge of the local values of these concentrations and of the mobility would permit a calculation of the resistivity

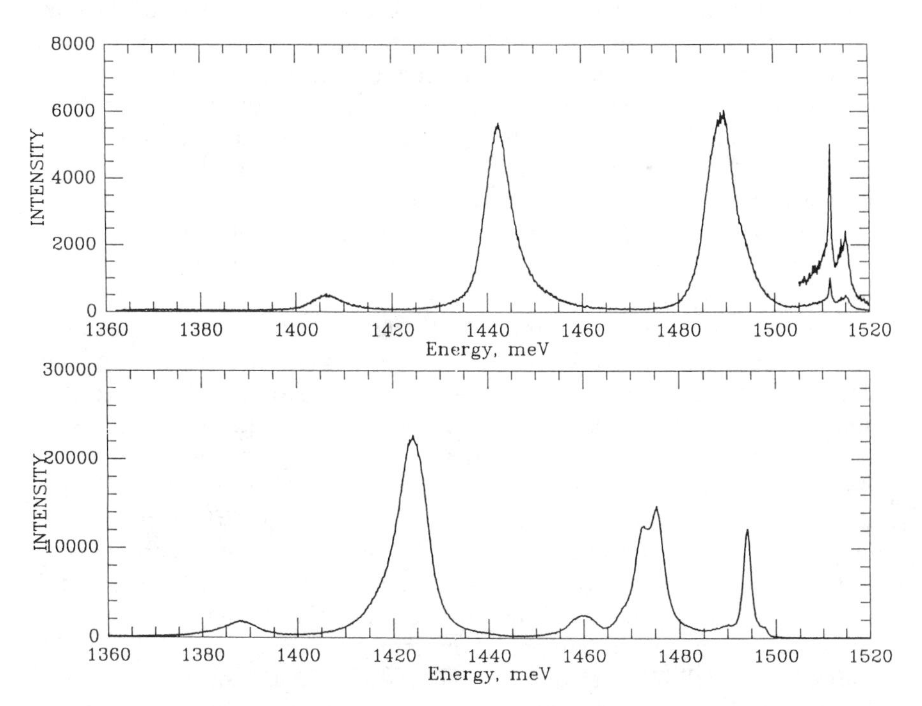

FIG. 5. Photoluminescence spectra of *p*-type GaAs (upper trace) and *p*-type $Ga_{0.988}In_{0.012}$ (lower trace), showing 17 meV shift of major luminescence bands in the In-alloyed material.

variation throughout a crystal. A direct measurement of local resistivity, however, provides a good assessment of bulk material uniformity. The usual methods of resistivity measurement do not have high enough resolution to track the short-range spatial variations that occur in GaAs. Using the three-electrode guard method, Matsumura *et al.* (1985) showed, for undoped and unalloyed semi-insulating LEC GaAs, that, at the cell walls of dislocation networks and at slip lines, the resistivity is more than an order of magnitude smaller than in nearby undislocated areas surrounded by the dislocation cell walls. The 70-μm resolution of the method is necessary to distinguish the cell walls from the dislocation-free regions. Matsumura *et al.* also observed smaller changes in resistivity associated with growth striations (with a 150–200 μm period produced by growth in an applied magnetic field to suppress irregular growth striations). In similar experiments (Miyairi *et al.*, 1985) with 280-μm resolution on InAs-alloyed GaAs grown from a stoichiometric melt and completely free of dislocations and slip lines, the resistivity across the wafers varies only slightly and smoothly, except at the edges. However, crystals grown from either As-rich or As-deficient melts showed almost as much resistivity variation as the unalloyed crystal, in spite of being relatively free of dislocations. It appears that dislocations are not the only causes of resistivity nonuniformity; composition fluctuations related to growth from non-stoichiometric melts can also degrade uniformity.

A very slight decrease in the peak wave number of the carbon local vibrational mode absorption, amounting to 1 cm^{-1} for an In concentration of 2.7×10^{20} cm^{-3} (at 77 K), was reported by Kitagawara *et al.* (1986). The peak also broadens very remarkably. It is known that the location of the C atom on the As site in GaAs leads to a splitting of the local vibrational mode absorption (detectable at liquid He temperature), caused by the random occurrence of the natural isotopes Ga^{69} and Ga^{71} in the random distribution of the naturally occurring isotopes Ga^{69} and Ga^{71} on the nearest neighbor Ga sites (Theis *et al.*, 1982). The presence of In allows for other possible configurations of nearest neighbors around the C atom, leading to further splittings and a down-shift in frequency because of the higher mass of the In atom. This effect may account, at least in part, for the observed broadening and frequency shift in the unresolved absorption at 77 K, though it may also be necessary to take into account the detailed differences in bond lengths and force constants around the C atom when In atoms are present.

Takebe *et al.* (1986) have conducted careful determinations of EL2 and C concentrations in unalloyed and InAs-alloyed GaAs and compared the results with values of the compensation ratio (EL2 concentration divided by the excess of shallow acceptor over shallow donor concentrations) evaluated from temperature-dependent Hall measurements. Measurements on different samples from a given wafer show that the compensation ratio is independent of dislocation density below 1000 cm^{-2}, and increases at higher dislocation densities. The authors report that the observed values of compensation ratio cannot be explained by variations in EL2 and C alone, and that variations in shallow donor densities must be invoked as well.

Observations of a correlation between the variation of dislocation density and the variation of EL2 concentration, measured by near infrared absorption across LEC wafers, have suggested a close relationship between dislocations and EL2. Just such a relationship has been confirmed by several workers (Foulkes *et al.*, 1984; Stirland *et al.*, 1985; Hunter *et al.*, 1986) in samples of InAs-alloyed GaAs in which isolated dislocations could be observed. Dislocations, identified by etching, are surrounded by rod-like absorbing regions with characteristic EL2 absorption. The rods are about 100 μm in diameter and exhibit an EL2 concentration of about 10^{17} cm^{-3}. A further characteristic of EL2 exhibited by these rods is that the absorption can be bleached at cryogenic temperatures by strong illumination. Each dislocation appears to be associated with surrounding EL2; however, there is a background of EL2 not associated with dislocations. The background concentrations vary, but are often as high as those in unalloyed semi-insulating GaAs. Other workers have made detailed maps of the distribution of EL2 in both unalloyed and InAs-alloyed LEC wafers, with the intention of relating the properties of ion-implanted layers formed subsequently on the same wafers to the EL2 distributions (Blakemore and Dobrilla, 1985; Dobrilla *et al.*, 1985; Dobrilla and Blakemore, 1986). The EL2 distributions sometimes, but not always, correlated with dislocation density distribution; when there was a difference between EL2 and dislocation distributions, the properties of implanted layers correlated better with EL2 than with dislocations. Modifications of the EL2 mapping measurement allowed a mapping of stress in the wafer as well. The overall EL2 distribution does not appear to be a direct consequence either of stress or of dislocations.

Photoluminescence and cathodoluminescence intensities are known to be correlated with dislocations in unalloyed GaAs, and the uniformity of luminescence has been suggested as a test for the uniformity of substrates for ion implantation. InAs-alloyed material, with its isolated dislocations, allows a detailed study of luminescence as affected by dislocations. Hunter (1985) has shown that there is a very large contrast of the near-band-edge band-to-acceptor photoluminescence intensity near isolated dislocations in semi-insulating LEC InAs-alloyed GaAs. Bright rings, about 150 μm in radius, surround 100 μm dark spots centered on dislocations. Far from dislocations, the luminescence intensity is several orders of magnitude lower than in the rings around dislocations. The spectra from bright and dark regions do not indicate a major difference in In content. The magnitude of the luminescence contrast suggests that it is the lifetime of free carriers, rather than the concentration of acceptors involved in the luminescence, that directly determines the contrast. Similar cathodoluminescence rings around isolated dislocations in InAs-alloyed GaAs were reported by Winston *et al.* (1985). Although these studies show that luminescence is much brighter in the rings around dislocations than in nearby dislocation-free regions, other work (Noto *et al.*, 1986) on similar InAs-alloyed GaAs suggests that the near-bandgap luminescence is brighter in the dislocation-free regions than in the dislocated areas. The explanation for the discrepancy lies in the spot size of the exciting laser beam. Hunter, using a 20 μm spot, could bring out the luminescence contrast on a micro-scale around dislocations and detect the very bright luminescence in the rings. Noto *et al.*, working with a 2 mm spot, recorded the luminescence intensities averaged over a relatively large area; in their measurements, the very bright rings and the dark areas near dislocations averaged out to an intensity lower than the more uniform intermediate intensity in the dislocation-free regions.

Mapping of EL2 absorption and near-bandgap luminescence intensity across wafers and along ingots of As-grown material reveals, as we have seen, a considerable degree of non-uniformity. As useful as the observed spatial variations in these properties may be as a guide to understanding the effects of growth and post-growth processes on the constitution of the crystal, wafers used as ion implantation substrates always undergo a post-implant anneal (for activating the implanted ions and removing the bombardment damage) that may override the variations observed in virgin material. It is well known that anneals of various kinds improve the electrical properties of unimplanted semi-insulating GaAs (Rumsby *et al.*, 1983, 1984), the uniformity of the 1 μm absorption (Holmes *et al.*, 1984), and the uniformity of the near-bandgap luminescence (Miyazawa *et al.*, 1984; Yokogawa *et al.*, 1984). The InAs-alloyed GaAs also exhibits changes in these properties after annealing, as we shall describe in the next paragraphs.

Martin *et al.* (1985) annealed 4 mm thick slices of almost dislocation-free alloy material for 60 hours at 900°C and found that the EL2 concentration across the slice had become fully uniform. These authors reported that their annealing conditions produced uniform EL2 concentrations of the same value, 1.2×10^{16} cm^{-3}, not only for InAs-alloyed dislocation-free material, but also for normal dislocated LEC GaAs and Bridgman-grown GaAs. They suggested that the common final concentration after annealing corresponded to a thermodynamic equilibrium at the anneal temperature.

Near-bandgap luminescence intensities also become much more uniform in nearly dislocation-free alloy material after annealing. Kimura *et al.* (1986) annealed samples several cm on a side, 3 mm thick, coated with 1000 Å of SiO_2, under conditions similar to typical post-implant anneals, such as 20 min at 830°C. A very pronounced region of increased and uniform luminescence (revealed in both photoluminescence and cathodoluminescence experiments) moved in about 250 μm from the polished and SiO_2-capped surfaces in 20 min, and about 1 mm from a sawed edge of the sample. In a 20 min anneal, the nonradiating areas near dislocations remain unchanged, but the originally nonradiating background far from dislocations increases greatly in intensity. The changes in luminescence are consistent with the diffusion of As vacancies from the surfaces, as suggested by Chin *et al.* (1984) for unalloyed GaAs. Presumably, the role of the As vacancies is to interact with nonradiative recombination centers that lower the carrier lifetime and, thus, reduce luminescence efficiency. Near the surface, As vacancies would interact with the As_{Ga} antisite defects related to the EL2 deep level, to produce an apparent out-diffusion to a depth of perhaps 10 μm, as reported by Makram-Ebeid *et al.* (1982); however, at several hundred μm distance from the surface, it must be another defect that is removed by As vacancy interaction to increase the carrier lifetime. Noto *et al.* (1986) reported improved uniformity of photoluminescence intensity across InAs-alloyed GaAs wafers after annealing. The absence of major changes in EL2 and C concentrations led them also to conclude that the changes in photoluminescence intensity are caused by a decrease in nonradiative recombination centers as a result of the anneal. Additionally, they found that the seed-to-tail hundred-fold drop in luminescence intensity found in as-grown crystals was eliminated by the annealing of either separate wafers or of a whole block of material.

The variation of photoluminescence intensity in the As-grown crystal reflects a difference in thermal history; as the crystal is withdrawn from the melt and passes through the liquid encapsulant and into the atmosphere of the growth furnace, it undergoes an anneal that becomes progressively shorter for the parts of the crystal nearer the tail. The fact that both wafer and block annealing lead to similar increases in luminescence intensity at the tail end to make the axial luminescence profile uniform suggested to Noto *et al.* that contamination introduced during the anneal is not a dominant effect. However, Kimura *et al.* (1986) did find possible evidence for the presence of Cu in the regions of increased luminescence intensity in their experiments, in the form of a 1.34 eV luminescence peak that has been ascribed to Cu, possibly in association with an As vacancy.

The increased uniformity of EL2 concentration and near-bandgap luminescence intensity after anneal in In-alloyed GaAs is quite similar to what is observed in unalloyed GaAs. The in-diffusion of As vacancies from the surface to eliminate thus far unidentified recombination centers appears to account for the annealing effects on photoluminescence. The improved uniformity of EL2 after annealing is probably the result of the presence in both dislocated and dislocation-free material of local reservoirs of As.

We have noted that dislocations serve as nuclei for identifiable As precipitates, while in dislocation-free material, excess As is most likely distributed throughout the bulk in clusters too small to be postively identified as a second phase. As is surely present in as-grown material at room temperature in excess of the thermodynamic solubility of As in solid GaAs, since the existence region of GaAs is wider near the melting point than at lower temperatures, and growth from an As-rich solution will lead to a solid phase with a composition at the As-rich edge of the existence region. At lower temperatures, the existence region of the solid phase will be narrower, so that during the cool-down from the growth temperature, the one-phase solid solution of As in GaAs will become metastable with respect to the equilibrium solid solution and a second phase of As. Its structure at room temperature will depend on the relative rates of nucleation of the As phase and of the diffusion in the crystal of the native defects embodying the As excess. Ikoma and Mochizuki (1985) have proposed that there may be several kinds of EL2, corresponding to different degrees of aggregation of As in solid solution in GaAs. The simplest entity suggested for EL2, the As antisite As_{Ga}, is already a cluster of five As atoms in adjacent substitutional sites; additional interstitial or substitutional As would produce larger clusters that could correspond to the microdefects in InAs-alloyed GaAs mentioned earlier. Still larger clusters then would be observable as a second phase. It is notable that a nearly dislocation-free GaAs grown from a melt kept very close to stoichiometric proportions (Ozawa *et al.*, 1986), so that the growing crystal does not contain a large excess of either component, does not show the fine background etch pattern supposed to reveal microdefects. In any event, for both dislocated and dislocation-free material, an anneal allows redistribution of any excess As, either from actual second-phase precipitates along dislocations or from "colloidal" clusters throughout the bulk, into a more uniform dispersion. If the surface condition of the piece of crystal undergoing the anneal allows the generation of As vacancies, these may also diffuse in to react with excess As and help to achieve an equilibrium distribution of As in the crystal, appropriate to the annealing temperature. Such an equilibrium distribution would imply a uniform distribution of As antisites and, hence, of EL2.

We conclude this section by mentioning experiments on deep levels in InAs-alloyed GaAs. Mitchel and Yu (1985) found a peak at 0.63 eV in the 4.2 K luminescence spectrum similar to, but weaker than, luminescence in GaAs, as well as a 0.77 eV activation energy of the carrier concentration measured by Hall effect. Kitagawara *et al.* (1986) reported deep level transient spectroscopy (DLTS) results on InAs-alloyed GaAs prepared with *n*-type doping to allow the DLTS measurements. They found five levels, four of them corresponding closely in energy and cross section to the well-known traps EL2, EL3, EL5 and EL6 in GaAs, and the fifth exhibiting an energy of 0.26 eV and a cross section of 6.4×10^{-14} cm^2. All these levels, except EL2, are reduced drastically in concentration by an anneal in flowing nitrogen at 850°C for 5 hr, using face-to-face proximity capping by GaAs wafers. The authors noted that under the same annealing conditions, the concentration of the level EL5 in unalloyed GaAs did not decrease.

V. InAs-Alloyed GaAs as a Substrate for Ion Implantation

The underlying reason for the development of InAs-alloyed GaAs was to provide substrates for ion implantation free of dislocations and their effects, whatever they might be, on the activation of implanted dopants. While dislocation-free GaAs had been available for many years, its size was limited, and there seems to be no published record of any systematic studies of ion implantation or device fabrication on dislocation-free material. The alloy material, with large dislocation-free areas, is now obtainable in the same physical dimensions and format as ordinary dislocated LEC material, with wafer diameters of two or three inches, suitable for the fabrication of integrated circuits. We have seen in the preceding sections that the properties of low dislocation density InAs-alloyed GaAs do indeed differ from those of ordinary dislocated material, with the range of influence of isolated dislocations extending for many μm. We have also noted that the properties of the alloyed material can be changed by annealing treatments, often in ways that override the effects of dislocations. Since ion implantation doping always includes a post-implant anneal to heal radiation damage and activate the donor or acceptor function of the implanted dopant, it is reasonable to ask whether the properties of as-grown, pre-annealed substrate wafers in the vicinity of dislocations are relevant to the properties of the active layer produced by implantation and annealing. At the very least, it is necessary to take account of the influence of the thermal history of a substrate sample in assessing the effects of dislocations on dopant activation.

The original experiments that showed a correlation between distributions of dislocations and of FET properties (Nanishi *et al.*, 1982) and specific micro-scale effects of dislocations on nearby FETs (Nanishi *et al.*, 1983) prompted a great deal of further work aimed at understanding the mechanisms and reducing or eliminating the dislocation-related variations in FET properties. The approach in most of these investigations was to implant Si ions into the semi-insulating GaAs substrate material under study, and then conduct a post-implant anneal with an encapsulant or atmosphere intended to prevent decomposition of the GaAs surface. The implantations were conducted in equipment that had been designed to produce uniform ion fluences across the wafer, so that non-uniformities in the distribution of the net donor activity of the implanted Si dopant observed in the active layers were to be ascribed to non-uniformities of activation related to non-uniformities in the substrate itself. The spatial resolution with which the donor concentration was measured needed to be no better than the resolution of the techniques by which substrate properties were observed. Thus, Hall effect measurements of carrier concentration (equivalent to net donor concentration) on Hall samples several millimeters square, cut from the implanted and annealed wafer, were adequate for correlations with dislocation densities revealed by measurement every few millimeters across the wafer of the etch pit densities produced by exposure to molten KOH. The etch pit densities could be measured on the Hall samples themselves or on adjacent wafers from the same ingot. Using these methods, correlations between implanted donor concentration and etch pit density were observed by Bonnet *et al.* (1982) and Honda *et al.* (1983). This was, of course, not surprising, since the FET threshold voltage, known to be a measure of active layer net donor concentration, had already been shown to be correlated with etch pit density. Honda *et al.* noted that more uniform etch pit distributions corresponded to more uniform carrier concentration distributions, and commented that substrates with uniform dislocation density distributions would be highly desirable for the manufacture of GaAs ICs by direct implantation.

The spatial resolution of the Hall effect methods, however, was inadequate to observe the influence of individual dislocations on nearby FETs. Measurements of the spreading resistance of the active layer would have provided a resolution of the order of the diameter of the probe, a few μm, but, unfortunately, barrier effects dominate the impedance of a metal probe on an n-type GaAs surface and prevent the use of this method. The key to higher resolution was the method, introduced by Nanishi *et al.* (1983), of forming an array of closely-spaced FETs (in their experiments, the FETs had $1\,\mu m \times 5\,\mu m$ gates and were laid out on a $200\,\mu m$ square grid). They had observed a significantly larger saturation current for a line of such FETs that happened to be within a few tens of μm of a line of etch pits. Using such an array of closely spaced FETs, Miyazawa *et al.* (1983) were able to determine the distance of each gate from the nearest etch pit and, thus, associate the previously measured FET threshold voltage with the distance to the nearest dislocation. The etch pit density was such that there were tens or hundreds of pits within the $200\,\mu m$ square area around each FET, with the gate-to-nearest-pit spacings, measured in an optical microscope, ranging from 0 to about $100\,\mu m$. They interpreted their results, shown in Fig. 6, as revealing a proximity effect that caused a greater donor activation and correspondingly more negative threshold voltage in the vicinity of a dislocation. FETs more than 20–30 μm from a dislocation exhibited threshold voltages about 300 mV

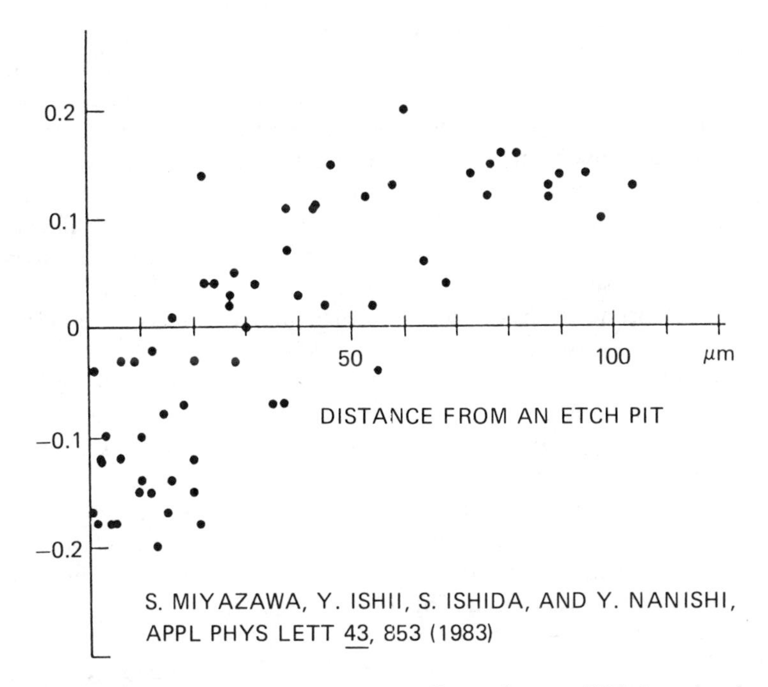

Fig. 6. Plots of FET threshold voltages versus distance between FET channel and nearest dislocation, measured randomly for 60 FETs along the ⟨110⟩ direction on a wafer. [From Miyazawa *et al.* (1983).]

higher than the FETs closer to a dislocation. The authors stressed that the variation in threshold voltage introduced by dislocations would be a serious limitation on GaAs integrated circuit design and fabrication. The low resolution studies of Honda *et al.* (1983) had suggested that uniform dislocation density would be required; now, Miyazawa *et al.* called for dislocation-free substrates, since, even in a substrate with uniform dislocation density, there would inevitably be a non-uniform distribution of dislocation-gate distances and, thus, an uncontrolled source of FET parameter dispersion that would limit integrated circuit performance. In this work, Miyazawa and his associates demonstrated the utility of closely-spaced FET arrays for mapping substrate-related effects on implant activation on a micro-scale. They also suggested that the influence of dislocations on nearby FETs was the result of regions a few tens of μm wide around dislocations being "denuded" of acceptor-like impurities or point defects. Such a depletion of acceptors would account for the higher activation of implanted donors and the corresponding, more negative threshold voltages observed close to a dislocation. Denuded zones had previously been invoked to account for the variation of photoluminescence intensity around dislocations (Heinke and Queisser, 1974). While it is plausible that recombination centers that influence the free carrier lifetime and the luminescence intensity might be gettered by dislocations, it does not seem likely that enough acceptor impurities could have been present initially to account for the changes around dislocations. Watanabe *et al.* (1986), noted that the concentration of C, the major acceptor known to be present, is as much as an order of magnitude smaller than the presumed decrease in acceptor concentration necessary to explain the increase in net implanted donor concentration near a dislocation.

In the meantime, a dislocation-free InAs-alloyed GaAs of dimensions suitable for integrated circuit fabrication was becoming available. Hobgood *et al.* (1984) made high-quality FETs on alloy material, with electrical parameters comparable to those of FETs fabricated on normal dislocated material. Winston *et al.*, (1984a, 1984b) prepared a high-density array of FETs on low-dislocation alloy material, and showed that the variation of threshold voltage across the wafer was about half that of arrays fabricated on dislocated substrates. Not only were the excursions of threshold voltage over the whole wafer lower for the alloy substrates, but the short-range uniformities were better as well. The short-range uniformities available from different substrates were assessed by finding the lowest standard deviation of threshold voltage for any group of 48 contiguous FETs occupying a 1.2 mm square area. This method is appropriate for comparing the effects of different substrates, though by picking the best local values of uniformity, it obviously is providing a lower bound, rather than a representative value, for standard deviations of FET properties. The lower bound on threshold voltage standard deviation selected in this way was 14 mV for the alloy substrate and 27 mV for the standard LEC substrate. However, in this work, there was no discernible effect of proximity to a dislocation on the threshold voltage of the FETs, with either an unalloyed dislocated substrate with the usual 10^4–10^5 cm^{-2} dislocation density, or with the almost dislocation-free InAs-alloyed GaAs substrates, in apparent direct contradiction to the Japanese report. Winston *et al.* did find that the local average threshold voltage of groups of 100 FETs occupying 1 mm $\times$ 3 mm areas was generally more negative in regions of higher local dislocation density, in agreement with earlier results on the correlation of FET properties with dislocation densities.

Other workers from the same laboratory as Miyazawa (Yamazaki, 1984) conducted FET array measurements on low-dislocation InAs-alloyed GaAs, confirming the improvement in FET uniformity with alloy substrates. They observed a standard deviation of threshold voltage for 625 FETs in a 5 mm square of 20–30 mV, compared with 50–60 mV values for ordinary LEC substrates. For the ordinary dislocated substrates, they confirmed

Miyazawa's earlier observation of a dislocation proximity effect on the threshold voltage of nearby gates, but with the low-dislocation alloy substrates they found no such effect, in agreement with Winston *et al.* (1984a, 1984b). Still another early report on the improvement of uniformity when InAs-alloyed GaAs substrates are used was made by Tada *et al.* (1984). They measured the average threshold voltage and local standard deviation for 379 2 mm square cells, each containing 100 FETs. The average threshold voltages on the low-dislocation alloy material were between −1.1 V and −1.3 V, with the standard deviations within each cell ranging from 20 to 80 mV. In contrast, with a conventional GaAs substrate wafer, the average threshold voltages were from −1.2 to −1.9 V, and the standard deviations from 50 to over 200 mV. For a lower energy implant, the average threshold voltage for 625 FETs in a 5 mm square region on an alloyed low-dislocation substrate was −22 mV with a standard deviation of 13 mV, as compared to 50–100 mV for a conventional substrate.

Before recounting other work on low-dislocation substrates, we should point out that an explanation has been offered for the differences between Miyazawa's and Winston's results on a dislocation proximity effect. Hyuga *et al.* (1986) noted that the post-implant anneals in the Japanese work were conducted with a silicon nitride cap, while the American workers used a silicon oxide cap. Hyuga *et al.* repeated both kinds of post-implant anneals in their own laboratory. They carried out measurements of sheet carrier concentration by a high-resolution method originated by Hyuga (1985), in which an array of van der Pauw Hall samples, with 40 μm square active regions and a spacing of 400 μm in each direction, is applied to an implanted and annealed substrate by photolithographic techniques. There was no effect of nearby dislocations on sheet carrier concentration with silicon oxide encapsulation, just as in the work of Winston *et al.* With a silicon nitride cap, however, the sheet carrier concentration was influenced by nearby dislocations, in agreement with Miyazawa. Hyuga *et al.* explained the influence of the post-implant anneal cap by reference to the commonly held idea that Ga diffuses out through the oxide cap during annealing, but not through the nitride cap. Therefore, the gallium vacancy concentration in the annealed region under the oxide is expected to be greater than under the nitride. Implanted Si, which can occupy either Ga sites as a donor or As sites as an acceptor, would have a greater tendency to occupy Ga sites after oxide annealing, which provides a greater concentration of gallium vacancies than nitride annealing. In the case of the nitride cap, the availability of sites for the Si dopant ion is apparently controlled by the nearby dislocations, which are known to have high concentrations of As-rich entities, such as EL2, in their vicinity. Hyuga *et al.* demonstrated that the characteristic photoluminescence of the silicon acceptor, Si_{As}, is, in fact, to be found after nitride annealing, as shown by Watanabe *et al.* (1986), but not after oxide annealing. Evidently, the availablity of gallium sites for silicon in the oxide-annealed substrate is so great that very little silicon goes into the As site to form acceptors. In nitride-annealed implants, it appears that the As-rich entities (EL2, As_{Ga}, As interstitials, As clusters, As precipitates) can modulate the availability of As sites for occupancy by Si (Watanabe *et al.*, 1986), reducing the Si_{As} acceptor concentration near dislocations. It may be the decrease of Si_{As} aceptors around dislocations that corresponds to the denuded zone previously mentioned, but in this interpretation, it is not acceptors originally present that are gettered out but, rather, the implanted amphoteric Si atoms that are kept out of acceptor sites by the excess As in the vicinity of the dislocation. Another piece of evidence in agreement with this interpretation is that the net donor activation of oxide-annealed Si implants, in which apparently none of the Si activates as Si_{As} acceptors, is considerably higher than the net donor activation after nitride annealing, in which enough Si enters acceptor sites to give rise to the corresponding Si_{As} photoluminescence line. The crucial role of excess As, as measured by the local concentration of EL2, in controlling the degree of activation of a Si implant, was pointed out by Dobrilla *et al.* (1985), who found that FET parameters correlated more closely with the EL2 density than with the dislocation distribution when these were appreciably different. Miyazawa and Wada (1986) also proposed an explanation of the effects of dislocations on implant activation and FET parameters in terms of the excess As around the dislocations and its effect on the availability of acceptor sites for implanted Si.

Hyuga (1985), using his high-resolution sheet carrier concentration method, and Miyazawa and Hyuga (1986), using FET arrays, further clarified the effect of dislocation proximity on donor activation and FET threshold voltage for nitride-annealed Si implants. Hyuga showed that regions more than 75 μm from a dislocation had implanted sheet carrier concentrations independent of the distance, but that regions closer to a dislocation exhibited higher sheet carrier concentrations; a dislocation within a 75 μm radius increased the carrier concentration by about 3×10^{15} cm^{-3}. The effect is additive, increasing with the number of dislocations within a 75 μm radius. This corresponds to a dislocation density effect; Hyuga noted that the effect saturates for about 10 dislocations within 75 μm, or a dislocation density about 5×10^{4} cm^{-2}. Thus, a wafer with dislocation density higher than that might be expressed to exhibit uniform sheet carrier concentration and FET threshold voltage. However, Hyuga noted that high dislocation densities in GaAs are usually associated with the presence of dislocation networks, which introduce a very large micro-scale fluctuation in the local dislocation density. He concluded that dislocation-free substrates were essential for high-performance integrated circuits in GaAs. Miyazawa and Hyuga found a scattering of threshold voltage for FETs within 30 μm of a dislocation that was independent of the dislocation-gate distance, but reaffirmed the existence of a proximity effect, with the sheet carrier concentration decreasing and FET threshold voltage increasing with distance from a dislocation in the range from 30 to about 50 μm under the conditions of their experiment, with a nitride post-implant anneal cap.

Hyuga *et al.* (1985a, 1985b) studied the implanted sheet carrier distribution in both dislocated and dislocation-free InAs-alloyed GaAs and in conventional dislocated material. They showed that it is the presence or absence of dislocations that influences the sheet carrier density, rather than the presence of In. Also, for comparable implants, the implanted sheet carrier density for a dislocation-free alloy substrate (and for an In-free and dislocation-free substrate as well) was lower than when a conventional dislocated GaAs substrate was used. Thus, the absence of dislocation leads to a lower activation efficiency of implanted Si atoms. [Winston *et al.* (1985) and Lee *et al.* (1984) reported higher activation efficiency on InAs-alloyed low-dislocation substrates than on conventional LEC GaAs; possibly the difference is the result of their use of oxide, rather than nitride, caps in post-implant anneals.] In Hyuga's experiments, the standard deviation of the sheet carrier concentration on the alloy substrate was only one fifth that of the conventional substrate. High temperature annealing of the dislocation-free alloy material led to a substrate wafer that yielded an even higher sheet carrier density after implantation and post-implant anneal than the conventional substrate, without degradation of the excellent uniformity across the wafer. Osaka *et al.* (1986) also emphasized that low activation efficiency for implanted Si in InAs-alloyed GaAs substrates can be improved by annealing without any degradation of electrical uniformity. The experiments of Hyuga *et al.* also included comparisons of dislocation-free, striated alloys and dislocation-free, striation-free alloys prepared by the fully encapsulated applied-magnetic-field method of Kohda *et al.* (1985). The implanted sheet carrier density was noticeably more uniform for the unstriated substrate, with a standard deviation half that of the striated material. The elimination of substrate striations appears to be effective in improving the uniformity of the implanted sheet carrier density. The striations that appear in high pressure liquid encapsulated Czochralski GaAs are the result of large fluctuations in temperature near the growth interface, and, no doubt, reflect variations in the stoichiometric ratio of As to Ga + In in the crystal, as well as in the concentrations of In and residual impurities. These investigations make it clear that the use of InAs-alloyed GaAs is not, by itself, a complete solution to the problem of implanted layer electrical uniformity. Any dislocations not eliminated by the alloying make their contribution to non-uniformity, just as in conventional material; post-growth annealing has a significant effect on implant activation even in dislocation-free material; and striations present in dislocation-free substrates evidently contribute to non-uniform activation of an implant. The utility of InAs-alloyed GaAs as an implantation substrate should be considered in the light of these issues, in comparison and conjunction with other methods for assuring more uniform implant activation.

Miyazawa *et al.* (1984) demonstrated that the use of ordinary LEC GaAs substrate wafers that had been annealed before implantation (800°C for 8–24 hr with a silicon nitride protective layer in flowing N_2, followed by removal of 50 μm by mechanochemical polishing) led to standard deviations of FET threshold voltage about half those of non-annealed substrates. In addition, they observed that, in the annealed substrate, the range of variation in threshold voltage with distance of the FET from a dislocation was half that in a non-annealed substrate, although the dislocation density was essentially unchanged by the annealing. Substrate wafers from ingots subjected to annealing (900–1000°C for more than 3 hr) also exhibit improved FET uniformity, according to Kasahara *et al.* (1985a, 1986). These workers noted that non-annealed substrates, ingot-annealed substrates, and InAs-alloyed substrates all showed the presence of a bimodal distribution of FET threshold voltages, with an "extraordinary" group of FETs exhibiting thresholds about 150 mV less negative than the main group. The improvement in threshold voltage uniformity for ingot-annealed substrates was caused primarily by a drastic decrease of the number of FETs in the extraordinary group, and the extraordinary group was still smaller and the uniformity better in dislocation-free InAs-alloyed GaAs. However, there did not appear to be a direct connection between observed etch pits and the occurrence of the extraordinary FETs.

It was suggested early in the study of FET behavior on InAs-alloyed GaAs substrates (Winston *et al.*, 1984b) that some of the improvement in uniformity was the result of an annealing effect during the additional time at high temperatures spent by an alloy crystal in the course of its growth at pull rates substantially lower than those for unalloyed crystal. The same kinds of migrations and reactions of lattice defects might be expected to occur as in a separate post-growth annealing step. McGuigan *et al.* (1986) conducted post-growth ingot annealing on all their alloy crystals (950°C for 18 hr), as well as decelerating the growth rate, so that the effects of alloying, reduced dislocation density, and post-growth annealing could not be evaluated separately. However, in comparing substrates of different dislocation densities and In contents used with different post-implant annealing techniques, they found that the tightest distributions of FET threshold voltages were obtained for low-dislocation alloy substrates processed with a phosphorus-doped silicon oxide glass cap. No correlation was observed between the proximity of a dislocation and FET threshold voltage for substrates from ingot-annealed crystals.

The spatial resolution of FET arrays for determining the range of dislocation effects depends, of course, on the spacing of the FETs. Matsuura *et al.* (1985) introduced an evaluation method using FETs with gates spaced on a 60 μm × 60 μm grid, which they used to compare conventional, ingot-annealed, and InAs-alloyed substrates. They used Si implantation and capless annealing in an atmosphere of AsH_3. The resolution of the array was sufficient to reveal a pattern of threshold voltage distribution in the conventional substrates that corresponded to the dislocation networks common in this material. The threshold voltages in the cell walls of the networks where the dislocations are clustered were 50–100 mV more negative than in the more or less dislocation-free areas enclosed by the cell walls. The effect of dislocation networks was not apparent in ingot-annealed substrates, which were quite uniform in threshold voltage, with the exception of half-millimeter circular areas of noticeably more negative threshold values. The arrays on alloy substrates were as uniform as those on ingot-annealed substrates, with standard deviations as small as 12 mV for 700 FETs in an area 0.84 mm × 3.0 mm. Some alloy substrates exhibited half-millimeter circular regions with threshold values 100–300 mV less negative than the rest of the wafer. A comparison of wafers from the same ingot, part of which had been ingot-annealed, showed that the FETs on the ingot-annealed wafer had an average threshold voltage almost half a volt more negative, corresponding to greater activation, with only half as large a standard deviation across the wafer. The authors concluded that dislocations affected implant activation not directly but, rather, through the distribution of crystal imperfections near dislocations.

Maluenda *et al.* (1986) and Schink *et al.* (1986) introduced a FET test pattern, including 30 FETs packed 10 μm apart along a 300 μm long row, providing the highest resolution reported to date. Their post-implant anneal was capless in an AsH_3 atmosphere. One advantage of this dense row pattern is that the effects of a single dislocation on several different FETs can be observed. These authors also conducted measurements on the current-voltage characteristic of the forward biased gate to detect any variation in the Schottky barrier height of the FETs that might be contributing to threshold voltage variation. They found that the barrier height variation is of second order, and that almost all the observed threshold voltage variation is ascribable to the activation of the implant. InAs-alloyed low-dislocation GaAs provided the best uniformity of threshold voltage, on both a microscopic and a full wafer scale, of any substrate examined.

It is evident that low-dislocation InAs-alloyed GaAs substrates lead to excellent uniformity of the activation of implanted Si with several different kinds of post-implant annealing caps or with capless annealing in an arsenic-rich atmosphere, and, correspondingly, to excellent FET threshold voltage uniformity. Miyazawa (1986) emphasized that it is the low dislocation content, rather than the In content, that is important for achieving uniformity. He pointed out that dislocated regions of alloy substrates have a large scattering of threshold voltage, and that the behavior of sheet carrier density around individual dislocations in the alloy substrates is similar to that in normal unalloyed dislocated substrates. He pointed to the excellent uniformities obtained with alloy substrates grown with full encapsulation of the growing crystal by boric oxide and with an applied vertical magnetic field as the best yet obtained. Post-growth annealing of unalloyed dislocated GaAs does not result in quite as good device uniformity, though Miyazawa noted that post-growth annealing has not yet been fully optimized.

VI. Discussion

The mechanism by which In reduces the dislocation density in GaAs grown by the liquid encapsulated Czochralski process is still not clear, since measurements on actual InAs-alloyed GaAs do not indicate a value of critical resolved shear stress high enough to prevent the formation and multiplication of dislocations under the thermoelastic stresses experienced by the crystal as it is pulled from the melt. On the other hand, there is considerable evidence to help us understand how dislocation-free or low-dislocation alloy substrates are effective in promoting the uniformity of activation of Si implants. Whereas a normal dislocated GaAs contains very inhomogeneous distributions of excess As in the form of precipitates along dislocations, and enhanced As_{Ga} concentrations near dislocations, the alloy, to the extent that it is dislocation-free, appears to possess a uniform stoichiometry. There may be As present in excess of the thermodynamically-stable amount after the alloy crystal, grown from an As-rich melt, has cooled down, but the absence of dislocations assures that the excess As has no nucleation sites around which to cluster. We believe, in agreement with Miyazawa (1986) and Miyazawa and Wada (1986), that it is, ultimately, the large deviation from stoichiometry in the vicinity of dislocations that is responsible for the nonuniformity of activation of implanted Si. InAs-alloyed GaAs leads to uniform implant activation because it has uniform stoichiometry on the spatial scale of interest for integrated circuits. Post-growth annealing of dislocated substrated material enhances uniformity by reducing the deviations in stoichiometry around dislocations. It presumably provides enough time at a high enough temperature for the lattice defects involved in the stoichiometry deviation to achieve a spatially uniform distribution independent of the dislocations.

Still another way to eliminate the spatial fluctuations of stoichiometry is to grow the crystal originally without so much excess As that there is a tendency for it to cluster at any available nucleation site as the crystal cools down. Recent crystal growth studies have shown that this can, in fact, be done. The usual high-pressure liquid encapsulated Czochralski process employs a large stoichiometric excess of As in the original charge to ensure that, even after losses of As to the cold walls of the furnace, there will still be enough As to guarantee an adequate concentration of the deep level EL2, related to As_{Ga}, for semi-insulating behavior. For example, the completely dislocation-free and striation-free material grown by Kohda *et al.* (1985), with In alloying, full encapsulation of the crystal by boric oxide, and an applied vertical magnetic field, was compounded so as to have an almost exactly stoichiometric melt at the outset of growth. The melt did not become depleted of As because of the thick boric oxide layer, and striations, which indicate local variations in crystal stoichiometry, were eliminated because the melt and growth interface were stabilized by the magnetic field. The observed uniformity of an implanted layer in this material (Hyuga *et al.*, 1985a) may be due, in part, to the overall near-equilibrium stoichiometry of the material, though it is difficult to disentangle the effects of stoichiometry and of the absence of dislocations. A more definite indication of the effectiveness of stoichiometry control for achieving uniform substrates, even when dislocations are present in the 10^4–10^5 cm^{-2} range, is provided by the work of Inada *et al.* (1986). They modified a conventional high-pressure liquid encapsulated Czochralski crystal puller to provide a separately-controllable reservoir for As injection, so that the melt composition could be maintained very close to the congruent point. Evidently, growth at the congruent composition provides enough As to assure semi-insulating behavior. However, since Si-implanted layers in this material were electronically highly uniform, and the standard deviations of FET threshold voltages were comparable to those obtained on dislocation-free InAs-alloyed GaAs, it is also evident that the crystals did not contain so much excess As that As clustered at the dislocations.

We may conclude that the use of InAs-alloyed low-dislocation GaAs substrates is an excellent expedient for reducing or eliminating the electrical nonuniformity of implant activation and FET characteristics that result from the clustering of excess As near dislocations. Competitive methods exist, including the annealing of dislocated ingots and the growth of dislocated ingots with control of the stoichiometry by As injection to limit the excess As in the first place. Whichever method or combination of methods ultimately prevails will probably depend upon economic considerations, since all appear to be capable of providing adequate FET threshold voltage uniformity, even for the most demanding applications. The full benefits of improved substrates depend upon corresponding improvements in device fabrication. As Miyazawa (1986) has commented, the quality of substrate material is "a complicated function of device type, device density, process parameters and other factors. *When process parameters are well controlled*, the bulk material then becomes key to the development of GaAs integrated circuits."

VII. Recent Developments

Additional work on InAs-alloyed GaAs has been reported since the manuscript of this chapter was originally completed. Many relevant papers appear in Semi-Insulating III–V Materials, the proceedings of the conference at Hakone in 1986, published by OHM-North Holland. Additional useful papers appear in the proceedings of the DRIP conferences at Montpelier in 1985 and Monterey in 1987, published by Elsevier as Defect Recognition and Image Processing in III–V Compounds. Here we will note some other recent publications.

1. Crystal Growth

Yamada *et al.* (1986) described their procedure for eliminating grown-in dislocations in the alloy material. These are the dislocations that propagate from the seed along the growth axis and produce a relatively high dislocation density in the center of wafers. They found that the dislocation-free seed containing In at the same concentration as the crystal to be grown is effective in preventing the appearance of axial dislocations. The beneficial effect of reducing lattice misfit between the seed and the growing crystal is interpreted in terms of the elimination of misfit dislocations that tend to form composite type axial dislocations. Kimura *et al.* (1987) have alloyed GaAs with both In and P, with the intention of reducing stress in the resulting crystal by compensating the effects of the In atom (larger than Ga) by the presence of the P atom (smaller than P). Because In concentrates in the melt during growth while P is depleted, the compensation for stress would be expected to be most effective at only one position along the growth axis, depending on the initial concentrations of In, introduced as the metal, and of P, introduced as InP or GaP. The concentration of In employed in these experiments would not have been completely effective in eliminating dislocations at the seed end of the crystal if it were the only alloying element; however, it appears that the initially high concentration of P complemented the effects of In, because the seed-end wafers were substantially dislocation-free. The use of both In and P also allows some degree of control of the lattice constant in the resulting crystal, so that dislocation-free wafers of specified lattice constant can be obtained. Kimura *et al.* (1988) have replenished the depleted P during growth to increase length of crystal exhibiting a desired lattice constant. McGuigan *et al.* (1986) presented a review of the work on the growth and properties of InAs-alloyed GaAs at Westinghouse, and Matsumura *et al.* (1986) showed that dislocation-free crystals can be obtained with In concentrations as low as 10^{19} cm^{-3} by reducing the thermal gradients and thus the thermal stresses near the growth interface. This is desirable to avoid striations and inclusions observed with larger In contents. Lopez Coronado *et al.* (1986) have demonstrated that vapor phase epitaxial growth of GaAs containing In from a Ga/In source leads to layers with a dislocation density of a few thousand per cm^2 when the In content in the mid-10^{19} cm^{-3} range, instead of the usual value of 10^5 cm^{-2}.

2. Crystallography and Structure

Tohno *et al.* (1986), using *in situ* X-ray topographic observations at high temperature, observed differences in the generation and propagation of dislocations between unalloyed and InAs-alloyed GaAs, with a one-directional predominance in the alloyed material. Guruswamy *et al.* (1986) and Bourret *et al.* (1987) reported experimental determinations of the critical resolved shear stress (CRSS) as a function of temperature in alloyed GaAs. However, the nature of the mechanism by which the addition of In (or other impurities) reduces dislocation formation in GaAs remains obscure.

Kidd *et al.* (1987), Suchet *et al.* (1987), and Nakamura *et al.* (1985) have reported on optical methods for observing precipitates in GaAs. A notable observation by Suchet *et al.* is that precipitates (presumed to be As) are never observed in the dislocation-free regions of InAs-alloyed GaAs.

3. Optical Properties

Kirillov *et al.* (1987) reported the use of room-temperature photoluminescence to determine In content in InAs-alloyed GaAs. Like the low-temperature methods, this technique is based on the shift of bandgap with In content. Since the room-temperature photoluminescence is caused by band-to-band transitions, there is only one peak. There is no necessity for identifying the exact transition involved, as in the relatively complex low-temperature spectrum, and there is of course no need to cool the sample.

Hunter (1987) has addressed the issue of the large luminescence contrast observed around dislocations in LEC GaAs. This phenomenon is especially prominent around the isolated dislocations in InAs-alloyed GaAs of low dislocation density. The contrast is caused by the inhomogeneous distribution of deep recombination centers that interact in various ways with dislocations. However, the entities responsible for these recombination centers have not been identified; they are certainly not the predominant deep level EL2.

4. InAs-Alloyed GaAs as an Ion Implantation Substrate

InAs-alloyed GaAs crystals have been shown to undergo exodiffusion of In by Krawczyk *et al.* (1986). Auger and SIMS analysis shows that In concentrates within about a hundred Angstroms of the surface after 15-minute anneals at 700°C in N_2/H_2. This conclusion is confirmed by the photoluminescence spectrum, which shows displacement toward higher energies corresponding to depletion of In below the surface in regions reached by the exciting light; there is also a deformation and displacement of the low-energy side of the peak corresponding to increase in In concentration (and lowering of bandgap) very close to the surface. The photoluminescence effects are observed at temperatures as low as 400°C. This phenomenon may be significant in the high-temperature processing of InAs-alloyed GaAs.

Inada *et al.* (1987) have studied the effects of annealing on the uniformity of EL2 concentration and resistivity in InAs-alloyed GaAs, and on the uniformity of the characteristics of FETs on the material as a substrate by Si ion implantation followed by capless annealing in arsine gas. They found great improvements in the uniformity of the EL2 concentration and the resistivity, but the FET threshold voltages did not improve in uniformity; in fact, they continued to be correlated with the original EL2 concentration as measured before annealing. This experiment serves as a warning that uniformity in one characteristic does not necessarily imply uniformity in others.

A particularly interesting study of amphoteric doping in Si-implanted InAs-alloyed and unalloyed GaAs was reported by Warwick *et al.* (1987). The implanted wafers were capped by pyrolytic chemical vapor deposition of either silicon oxide or silicon nitride, and annealed in either a furnace or an infrared rapid annealer. The donor activation of Si was inferred from the sheet carrier concentration and the presence of Si_{As} (an acceptor) from the relative intensity of the appropriate low-temperature photoluminescence line. Only in the case of InAs-alloyed GaAs substrate with a silicon oxide anneal cap was Si_{As} detected, and in this case the sheet carrier concentration was nonuniform, high in the regions with low Si_{As} and low where Si_{As} was high. Warwick *et al.* ascribe these effects to the presence of the large In atom on Ga sites, its influence on the Ga and As vacancy concentrations, and the further influence of the capping material on vacancy concentrations in the implanted region. The effects of dislocation density were presumably below the detection limit for the experimental conditions employed. The authors believe that the non-uniformities observed with InAs-alloyed GaAs substrates and silicon nitride capping were not caused by substrate non-uniformities but rather by an extreme sensitivity under their experimental conditions to furnace non-uniformities. Indeed, all other combinations of substrate and capping, in some cases on other quadrants of the same wafer, showed excellent uniformity. These results again serve as a warning that the activation of Si implants in GaAs substrates is not yet fully understood.

Johannessen *et al.* (1988) studied the influence of substrates, including an InAs-alloyed GaAs wafer, on implanted layer characteristics. Their Si implants were furnace-annealed under a silicon oxide cap. They did not find a consistent local correlation on single wafers between EL2 concentrations and Si donor activation, but comparing different wafers (the InAs-alloyed material, two unalloyed LEC GaAs samples, and a horizontal Bridgman wafer) there was excellent wafer-to-wafer correlation between average EL2 concentration and average FET threshold voltage (a measure of donor activation). The highest activation corresponded to the lowest EL2 concentration. The activation in the low-dislocation InAs-alloyed GaAs wafer was higher than in the unalloyed LEC wafers, in agreement with previous studies using silicon oxide capping reported by Winston *et al.* (1985).

REFERENCES

Avdonin, N. A., Vakhrameev, S. S., Mil'vidskii, M. G., Osvenskii, V. B., Sakharov, B. A., Smirnov, V. A., and Shchelkin, Yu. F. (1972). *Sov. Phys. Doklady* **16**, 772 (original in Russian, 1971, *Doklady Akademii Nauk SSSR* **200**, 316).

Barrett, D. L., McGuigan, S., Hobgood, H. M., Eldridge, G. W., and Thomas, R. N. (1984). *J. Crystal. Growth* **70**, 179.

Blakemore, J. S., and Dobrilla, P. (1985). *J. Appl. Phys.* **58**, 204.

Bol'sheva, Yu. N., Grigor'ev, Yu. A., Grishina, S. P., Mil'vidskii, M. G., Osvenskii, V. B., and Shifrin, S. S. (1982). *Sov. Phys. Crystallogr.* **27**, 433.

Bonnet, M., Visentin, N., Gouteraux, B., Lent, B., Duchemin, J. P. (1982). *IEEE GaAs IC Symp.*, p. 54.

Bourret, E. D., Tabache, M. G., and Elliot, A. G. (1987). *Appl. Phys. Lett.* **50**, 1373.

Burle-Durbec, N., Pichaud, B., Minari, F., Soyer, A., and Epelboin, Y. (1986). *J. Appl. Cryst.* **19**, 140.

Carlos, W. E., Bishop, S. G., McGuigan, S., and Thomas, R. N. (1985). *Bull. Am. Phys. Soc.* **30**, 417 (discussed in McGuigan, 1986).

Chin, A. K., Camlibel, I., Caruso, R., Young, M. S. S., and Von Neida, A. R. (1984). *J. Appl. Phys.* **57**, 2203.

Cornier, J. P., Duseaux, M., and Chevalier, J. P. (1985). *Inst. Phys. Conf. Ser.* **74.** (Int. Symp. GaAs and Related Cpds., Biarritz, 1984), 95.

Cullis, A. G., Augustus, P. D., and Stirland, D. J. (1980). *J. Appl. Phys.* **51**, 2556.

Dobrilla, P., and Blakemore, J. S. (1986). *J. Appl. Phys.* **60**, 169.

Dobrilla, P., Blakemore, J. S., McCamant, A. J., Gleason, K. R., and Koyama, R. Y. (1985). *Appl. Phys. Lett.* **47**, 602.

Duseaux, M., and Jacob, G. (1982), *Appl. Phys. Lett.* **40**, 790.

Duseaux, M., and Martin, S. (1984). "Semi-Insulating III–V Materials, Kah-nee-ta 1984," p. 118.

Duseaux, M., Schiller, C., Cornier, J. P., Chevalier, J. P., and Hallais, J. (1983). *J. de Physique* **C4**, 397.

Ehrenreich, H., and Hirth, J. P. (1985). *Appl. Phys. Lett.* **46**, 658.

Elliott, A. G., Wei, C., Farraro, R., Woolhouse, G., Scott, M., and Hiskes, R. (1984). *J. Cryst. Growth* **70**, 169.
Foulkes, E. J., Brozel, M. R., Grant, I., Singer, P., Waldock, B., and Ware, R. M. (1984). "Semi-Insulating III–V Materials, Kah-nee-ta 1984," p. 160.
Fujimoto, I., Kamata, N., Kobayashi, K., and Suzuki, T. (1986). *Inst. Phys. Conf. Ser.* **79** (Int. Symp. GaAs and Related Cpds., Karuizawa, 1985), 199.
Gant, H., Koenders, L., Bartels, F., and Mönch, W. (1983). *Appl. Phys. Lett.* **43,** 1032.
Guruswamy, S., Hirth, J. P., and Faber, K. T. (1986). *J. Appl. Phys.* **60,** 4136.
Heinke, W., and Queisser, H. J. (1974). *Phys. Rev. Lett.* **33,** 1082.
Hirayama, M., Togashi, M., Kato, N., Suzuki, M., Matsuoka, Y., and Kawasaki, Y. (1986). *IEEE Trans. Electron Dev.* **ED-33,** 104.
Hobgood, H. M., Thomas, R. N., Barrett, D. L., Eldridge, G. W., Sopira, M. M., and Driver, M. C. (1984). "Semi-Insulating III–V Materials, Kah-nee-ta 1984," p. 149.
Holmes, D. E., Kuwamoto, H., Kirkpatrick, C. G., and Chen, R. T. (1984). "Semi-Insulating III–V Materials, Kah-nee-ta 1984," p. 204.
Honda, T., Ishii, Y., Miyazawa, S., Yamazaki, H., and Nanishi, Y. (1983). *Jap. J. Appl. Phys.* **22,** L270.
Hunter, A. T. (1985). *Appl. Phys. Lett.* **47,** 715.
Hunter, A. T. (1987). Defect Recognition and Image Processing in III–V Compounds II, 137.
Hunter, A. T., Kimura, H., Baukus, J. P., Winston, H. V., and Marsh, O. J. (1984). *Appl. Phys. Lett.* **44,** 74.
Hunter, A. T., Kimura, H., Olsen, H. M., and Winston, H. V. (1986). *J. Electron. Mat.* **15,** 215.
Hyuga, F. (1985). *Jap. J. Appl. Phys.* **24,** L160.
Hyuga, F., Kohda, H., Nakanishi, H., Kobayashi, T., and Hoshikawa, K. (1985a). *Appl. Phys. Lett.* **47,** 620.
Hyuga, F., Watanabe, K., and Hoshikawa, K. (1985b). *IEEE GaAs IC Symp.*, p. 63.
Inada, T., Sato, T., Ishida, K., Fukuda, T., and Takahashi, S. (1986). *J. Electronic Mat.* **15,** 169.
Inoue, T., Nishine, S., Shibata, M., Matsutomo, T., Yoshitake, S., Sato, Y., Shimoda, T., and Fujita, K. (1986). *Inst. Phys. Conf. Ser.* **79** (Int. Symp. GaAs and Related Cpds., Karuizawa, 1985), 7.
Ikoma, T., and Mochizuki, Y. (1985). *Jap. J. Appl. Phys.* **24,** L935.
Inada, T., Fujii, T., and Fukuda, T. (1987), *J. Appl. Phys.* **61,** 5483.
Ishikawa, T., Kitano, T., and Matsui, J. (1985). *Jap. J. Appl. Phys.* **24,** L968.
Ishii, Y., Miyazawa, S., and Ishida, S. (1984a). *IEEE Trans. Electron Dev.* **ED-31,** 800.
Ishii, Y., Miyazawa, S., and Ishida, S. (1984). *IEEE Trans. Electron Dev.* **ED-31,** 1051.
Jacob, G. (1982). "Semi-Insulating III–V Materials, Evian 1982," p. 2.
Jacob, G., Duseaux, M., Farges, J. P., Van den Boom, M. M. B., and Roksnoer, P. J. (1983). *J. Cryst. Growth* **61,** 417.
Johannessen, J. S., Harris, J. S., Rensch, D. B., Winston, H. V., Hunter, A. T., Kocot, C., and Bivas, A. (1988). *Inst. Phys. Conf. Ser.* **91,** 113 (Int. Symp. GaAs and Related Cpds, Crete, 1987).
Jordan, A. S., Von Neida, A. R., and Caruso, R. (1980). *Bell Syst. Tech. J.* **59,** 593.
Jordan, A. S., Von Neida, A. R., and Caruso, R. (1984). *J. Cryst. Growth* **70,** 555.
Kasahara, J., Arai, M., and Watanabe, N. (1985a). *Electronics Lett.* **21,** 1040.
Kasahara, J., Sakurai, H., Suzuki, T., Arai, M., and Watanabe, N. (1985b). *IEEE GaAs IC Symp.,* p. 37.
Kasahara, J., Arai, M., and Watanabe, N. (1986). *Jap. J. Appl. Phys.* **25,** L85.
Kidd, P., Booker, G. R., and Stirland, D. J. (1987). *Appl. Phys. Lett.* **51,** 1331.

Kimura, H., Afable, C. B., Olsen, H. M., Hunter, A. T., Miller, K. T., and Winston, H. V. (1984a). *Extended Abstracts, 16th Conf. on Sol. State Devices and Materials, Kobe, 1984*, p. 59.

Kimura, H., Afable, C. B., Olsen, H. M., Hunter, A. T., and Winston, H. V. (1984b). *J. Cryst. Growth* **70**, 185.

Kimura, H. (1986). unpublished.

Kimura, H., Hunter, A. T., and Olsen, H. M. (1986). *Inst. Phys. Conf. Ser.* **79** (Int. Symp. GaAs and Related Cpds., Karuizawa, 1985), 1.

Kimura, H., Hunter, A. T., Cirlin, E.-H., and Olsen, H. M. (1987). *J. Cryst. Growth* **85**, 116.

Kimura, H., Olsen, H. M., and Afable, C. B. (1988). Unpublished.

Kirillov, D., Vichr, M., and Powell, R. A. (1987). *Appl. Phys. Lett.* **50**, 262.

Kirkpatrick, C. G., Chen, R. T., Holmes, D. E., Asbeck, P. M., Elliott, K. R., Fairman, R. D., and Oliver, J. R. (1984). *In* "Semiconductors and Semimetals" (R. K. Willardson and A. C. Beer, eds.) Vol. **20**, Chap. 3. Academic Press, Orlando, Florida.

Kitagawara, Y., Itoh, T., Noto, N., and Takenaka, T. (1986). *Appl. Phys. Lett.* **48**, 788.

Kitagawara, Y., Noto, N., Takahashi, T., and Takenaka, T. (1986). *Appl. Phys. Lett.* **48**, 1664.

Kitahara, K., Kodama, K., and Ozeki, M. (1985). *Jap. J. Appl. Phys.* **24**, 1503.

Kitano, T., Matsui, J., and Ishikawa, T. (1985). *Jap. J. Appl. Phys.* **24**, L948.

Kohda, H., Yamada, K., Nakanishi, H., Kobayashi, T., Osaka, J., and Hoshikawa, K. (1985). *J. Cryst. Growth* **71**, 813.

Krawczyk, S. K., Khoukh, A., Olier, R., Chabli, A., and Molva, E. (1986). *Appl. Phys. Lett.* **49**, 1776.

Kuriyama, K., Satoh, M., and Kim, C. (1986). *Appl. Phys. Lett.* **48**, 411.

Lee, R. E., Hunter, A. T., Bryan, R. P., Olsen, H. M., Winston, H. V., and Beaubien, R. S. (1984). *IEEE GaAs IC Symp.*, p. 45.

Lessoff, H. (1984). *Materials Letters* **3**, 251.

Lessoff, H., and Gorman, R. (1984). *J. Electron. Mat.* **13**, 733.

Lopez Coronado, M., Abril, E. J., and Aguilar, M. (1986). *Jap. J. Appl. Phys.* **25**, L899.

Makram-Ebeid, S., Gautard, D., Devillard, P., and Martin, G. M. (1982). *Appl. Phys. Lett.* **40**, 161.

Maluenda, J., Martin, G. M., Schink, H., and Packeiser, G. (1986). *Appl. Phys. Lett.* **48**, 715.

Martin, S., Duseaux, M., and Erman, M. (1985). *Inst. Phys. Conf. Ser.* **74** (Int. Symp. GaAs and Related Cpds., Biarritz, 1984), 53.

Matsui, M., and Yokayama, T. (1986). *Inst. of Phys. Conf. Ser.* **79** (Int. Symp. GaAs and Related Cpds., Karuizawa, 1985), 13.

Matsumoto, K., Morishita, H., Sasaki, M., Nishine, S., Yokogawa, M., Sekinobu, M., Tada, K., and Akai, S. (1984). "Semi-Insulating III–V Materials, Kah-nee-ta 1984," p. 175.

Matsumura, T., Obokata, T., and Fukuda, T. (1985). *J. Appl. Phys.* **57**, 1182.

Matsumura, T., Sato, F, Shimura, A., Kitano, T., and Matsui, J. (1986). *J. Cryst. Growth* **78**, 533.

Matsuura, H., Nakamura, H., Sano, Y., Egawa, T., Ishida, T., and Kaminishi, K. (1985). *IEEE GaAs IC Symp.*, p. 67.

McGuigan, S., Thomas, R. N., Barrett, D. L., Eldridge, G. W., Messham, R. L., and Swanson, B. W. (1986). *J. Cryst. Growth* **76**, 217.

McGuigan, S., Thomas, R. N., Barrett, D. L., Hobgood, H. M., and Swanson, B. W. (1986). *Appl. Phys. Lett.* **48**, 1377.

McGuigan, S., Thomas, R. N., Hobgood, H. M., Eldridge, G. W., and Swanson, B. W. (1986). "Semi-Insulating III–V Materials, Hakone 1986," p. 29.

Mikkelsen, Jr., J. C., and Boyce, J. B. (1983). *Phys. Rev.* **B28**, 7130.

Mil'vidskii, M. G., and Bochkarev, E. P. (1978). *J. Cryst. Growth* **44**, 61.

Mil'vidsky, M. G., Osvensky, V. B., and Shifrin, S. S. (1981). *J. Cryst. Growth.* **52,** 396.
Mircea, A., Mitonneau, A., Hallais, J., and Jaros, M. (1977). *Phys. Rev.* **B16,** 3665.
Mitchel, W. C., and Yu, P. W. (1985). *J. Appl. Phys.* **57,** 623.
Miyairi, H., Inada, T., Obokata, T., Nakajima, M., Katsumata, T., and Fukuda, T. (1985). *Jap. J. Appl. Phys.* **24,** L729.
Miyazawa, S. (1986). "Semi-Insulating III–V Materials, Hakone 1986," p. 10.
Miyazawa, S., Ishii, Y., Ishida, S., and Nanishi, Y. (1983). *Appl. Phys. Lett.* **43,** 853.
Miyazawa, S., Honda, T., Ishii, Y., and Ishida, S. (1984). *Appl. Phys. Lett.* **44,** 410.
Miyazawa, S., and Hyuga, F. (1986). *IEEE Trans. Electron Dev.* **ED-33,** 27.
Miyazawa, S., and Wada, K. (1986). *Appl. Phys. Lett.* **48,** 905.
Nakamura, H., Tsubouchil, L., Mikoshiba, N., and Fukuda, T. (1985). *Jap. J. Appl. Phys.* **24,** L876.
Nanishi, Y., Ishida, S., Honda, T., Yamazaki, H., and Miyazawa, S. (1982). *Jap. J. Appl. Phys.* **21,** L335.
Nanishi, Y., Ishida, S., and Miyazawa, S. (1983). *Jap. J. Appl. Phys.* **22,** L54.
Noto, N., Kitagawara, Y., Takahasi, T., and Takenaka, T. (1986). *Jap. J. Appl. Phys.* **25,** L394.
Ono, H., Watanabe, H., Kamejima, T., and Matsui, J. (1986). *J. Cryst. Growth* **74,** 446.
Osaka, J., and Hoshikawa, K. (1984). "Semi-Insulating III–V Materials, Kah-nee-ta 1984," p. 126.
Osaka, J., Okamoto, H., Hyuga, F., Watanabe, K., and Yamada, K. (1986). "Semi-Insulating III–V Materials, Hakone 1966," p. 279.
Ozawa, S., Miyairi, H., Nakajima, M., and Fukuda, T. (1986). *Inst. Phys. Conf. Ser.* **79** (Int. Symp. on GaAs and Related Cpds., Karuizawa, 1985), 25.
Pichaud, B., Burle-Durbec, N., Minari, F., and Duseaux, M. (1985). *J. Cryst. Growth* **71,** 648.
Rumsby, D., Grant, I., Brozel, M. R., Foulkes, E. J., and Ware, R. M. (1984). "Semi-Insulating III–V Materials, Kah-nee-ta 1984," p. 165.
Rumsby, D., Ware, R. M., Smith, B., Tyjberg, M., Brozel, M. R., and Foulkes, E. J. (1983). *IEEE GaAs IC Symp.,* p. 34.
Schink, H., Packeiser, G., Maluenda, J., and Martin, G. M. (1986). *Jap. J. Appl. Phys.* **25,** L369.
Scott, M. P., Laderman, S. S., and Elliott, A. G. (1985). *Appl. Phys. Lett.* **47,** 1280.
Solov'eva, E. V., Rytova, N. S., Mil'vidskii, M. G., and Ganina, N. V. (1981). *Sov. Phys. Semicond.* **15,** 2141.
Solov'eva, E. V., Mil'vidskii, M. G., Osvenskii, V. B., Bol'sheva, Yu. N., Grigor'ev, Yu. A., and Tsyganov, V. P. (1982). *Sov. Phys. Semicond.* **16,** 366.
Stirland, D. J., Brozel, M. R., and Grant, I. (1985). *Appl. Phys. Lett.* **46,** 1066.
Stolte, C. A. (1984). *In* "Semiconductors and Semimetals" (R. K. Willardson and A. C. Beer, eds.) Vol. **20,** Chap. 2. Academic Press, Orlando, Florida.
Suchet, P., Duseaux, M., Gillardin, G., LeBris, J., and Martin, G. M. (1987). *J. Appl. Phys.* **62,** 3700.
Tabache, M. G., Bourret, E. D., and Elliott, A. G. (1986). *Appl. Phys. Lett.* **49,** 289.
Tada, K., Murai, S., Akai, S., Takashi, S. (1984). *IEEE GaAs IC Symp.,* p. 49.
Takebe, T., Murai, S., Tada, K., and Akai, S. (1986). *Inst. Phys. Conf. Ser.* **79** (Int. Symp. GaAs and Related Cpds., Karuizawa, 1985), 283.
Theis, W. M., Bajaj, K. K., Litton, C. W., and Spitzer, W. G. (1982). *Appl. Phys. Lett.* **41,** 70.
Thomas, R. N., Hobgood, H. M., Eldridge, G. W., Barrett, D. L., Braggins, T. T., Ta, L. B., and Wang, S. K. (1984). *In* "Semiconductors and Semimetals" (R. K. Willardson and A. C. Beer, eds.), Vol. **20,** Chap. 1. Academic Press, Orlando, Florida.
Tohno, S., Shinoyama, S., Katsui, A., and Takaoka, H. (1986). *Appl. Phys. Lett.* **49,** 1204.

Warwick, C. A., Ono, H., Kuzuhara, M., and Matsui, J. (1987). *Jap. J. Appl. Phys.* **26,** L1398.

Watanabe, K., Hyuga, F., Nakanishi, H., and Hoshikawa, K. (1986). *Inst. Phys. Conf. Ser.* **79** (Int. Symp. GaAs and Related Cpds., Karuizawa, 1985), 277.

Winston, H. V., Hunter, A. T., Olsen, H. M., Bryan, R. P., and Lee, R. E. (1984a). "Semi-Insulating III–V Materials, Kah-nee-Ta 1984," p. 402.

Winston, H. V., Hunter, A. T., Olsen, H. M., Bryan, R. P., and Lee, R. E. (1984b). *Appl. Phys. Lett.* **45,** 447.

Winston, H. V., Hunter, A. T., Kimura, H., Olsen, H. M., Bryan, R. P., Lee, R. E., and Marsh, O. J. (1985). *Inst. Phys. Conf. Ser.* **74** (Int. Symp. GaAs and Related Cpds., Biarritz, 1984), 497.

Yamada, K., Kohda, H., Nakanishi, H., and Hoshikawa, K. (1986). *J. Cryst. Growth* **78,** 36.

Yamazaki, H., Honda, T., Ishida, S., and Kawasaki, Y. (1984). *Appl. Phys. Lett.* **45,** 1109.

Yokogawa, M., Nishine, S., Sasaki, M., Matsumoto, K., Fujita, K., and Akai, S. (1984). *Jap. J. Appl. Phys.* **23,** L339.

Yonenaga, I., Sumino, K., and Yamada, K. (1986). *Appl. Phys. Lett.* **48,** 326.

Zuleeg, R. *Final Report*, Contract Nos. MDA-903-82-C-0445 and MDA-903-84-C-0386. Dept. of Army, Defense Supply Service, Washington, DC.

CHAPTER 3

Deep Levels in III–V Compound Semiconductors Grown by Molecular Beam Epitaxy

Pallab K. Bhattacharya and Sunanda Dhar[†]

DEPARTMENT OF ELECTRICAL ENGINEERING AND COMPUTER SCIENCE
THE UNIVERSITY OF MICHIGAN
ANN ARBOR, MICHIGAN

† Permanent address: The Institute of Radiophysics and Electronics, University of Calcutta, Calcutta, India.

ISBN 0-12-752126-7

List of Symbols

A	the diode area
C	the capacitance transient
C_0	the quiescent reverse bias capacitance
C	light velocity
$c_{n,\mathrm{p}}$	the capture rates of electron and hole traps
E_T	trap activation energy
$e_{n,\mathrm{p}}$, e_T	thermal emission rates of electrons and holes
$e^{\mathrm{O}}_{n,p}$	the optical emission rates of electron and hole traps
F_i	flux density of ith specie in thermal beam
f_T	the steady state occupation function
G	the Gibbs free energy
g	degeneracy factor of trap level
H	the enthalpy of ionization
h	the Plank constant
$\hbar$	$h/2\pi$
I	unit matrix
I_L	the steady state diode leakage current
k	the Boltzmann constant
m_0	the electron rest mass
m^*	the effective mass
N_A	total number of ionized acceptors
N_D	total number of ionized donors
$N_{\mathrm{c,v}}$	effective density of states in the conduction and valance bands
N_T	the total concentration of trap level
n_T	the density of traps occupied by electrons
n, p	the free carrier densities of electrons and holes
R	epitaxial growth rate
S	the entropy of ionization
T	the absolute temperature
V	defect perturbative potential
$\langle v_{n,p} \rangle$	average thermal velocity of electrons and holes
W	depletion layer width
α_i	sticking coefficient of ith specie on growing surface
ε_0	the permittivity in vacuum
ν	the extrinsic light frequency
φ	the incident photon energy
$\sigma_{n,p}$	capture cross section of electron and hole traps
σ_∞	capture cross-section at $T = 0\,\mathrm{K}$
σ^{O}	the photoionization cross section
$\tau_{n,p}$	the emission constant of electron and hole traps

I. Introduction

The rapid progress currently being made in the areas of fiber-optical communication, high-speed signal processing and supercomputing makes stringent demands for high-performance electronic, opto-electronic and optical devices. Techniques to synthesize well-tailored materials to suit such device applications, therefore assume roles of considerable importance. It is accepted that device fabrication and system performance are ultimately limited by the fundamental properties of the materials from which the devices are made. For many applications, it is also evident that semiconductor heterostructures and artificial synthetic modulated structures are needed. Their electronic and optical properties, which, in many cases, are affected by quantum-sized effects, are, in a sense, tunable to a great degree of precision. In such structures, the properties of the heterointerface, in addition to those of the bulk layers, become critical.

In the present review, we have therefore chosen to describe the present status regarding deep level traps in III–V semiconductors and heterostructures grown by molecular beam epitaxy (MBE). Both aspects of the review, molecular beam epitaxy and deep levels, are of paramount importance at the present time. Advances in molecular beam epitaxy have ushered in a new era of device technology based on precise thickness, composition control and the ability to grow modulated structures whose periods are typically less than the electron mean free path. The importance of deep states in semiconductors and their eventual role in limiting device performance cannot be overemphasized. In what follows, we will make a brief review of the two aspects of this chapter and then proceed to a more detailed description in the subsequent sections.

The importance of epitaxy in semiconductor device fabrication is a direct consequence of two critical needs: for thin, defect-free single crystal films with precisely defined geometrical, electrical and optical properties, and for heterojunction structures free of interfacial impurities and defects. The traditional techniques of liquid phase epitaxy, which is dependent on thermodynamic phases equilibria, and vapor phase epitaxy, which achieves growth by chemical reactions in the gas phase on a heated substrate, can certainly meet certain subsets of the above requirements, but not all of them. The newer technique, molecular beam epitaxy, achieves epitaxial growth by the reaction of one or more thermal atomic or molecular beams of the constituent elements with a crystalline substrate surface held at a suitable elevated temperature under ultrahigh vacuum (UHV).

The technique, as it stands today, has come a long way since the pioneering studies of Davey and Pankey (1968), Arthur (1968), Cho and Arthur (1975), Ilegems (1975, 1977), Robinson and Ilegems (1978) and Dingle, Gossard and coworkers (Gossard, 1982). Work by Esaki, Chang and coworkers (1973, 1975), Ploog (1980), Joyce, Foxon and coworkers (1975, 1977) and Holloway, Walpole and coworkers (1976, 1979) have led to a better understanding of growth kinetics and to an improved quality of materials and devices. Essentially confined to research and development until about 1976, MBE (molecular beam epitaxy) has now emerged as a viable growth technique for the realization of stringent device requirements. The uniqueness of the technique lies mainly in the tremendous precision in controlling layer dopings and thicknesses and in achieving near-perfect heterointerfaces and surface morphologies.

To cite just a few milestones, MBE-grown GaAs/AlGaAs lasers using bulk epitaxial layers have achieved performance levels comparable to or better than LPE-grown lasers (Tsang, 1985), and modulation-doped field effect transistors (MODFETs) have demonstrated extremely encouraging high-frequency and low noise performance (Chao *et al.*, 1985; Gupta *et al.*, 1985). In addition, unique quantum effects and efficient light generation and detection based on these effects are being demonstrated with multilayered materials grown by MBE (Capasso, 1985; Juang *et al.*, 1985, 1986; Tsang, 1981a, 1982). The stringent requirements of interface abruptness, layer thickness and composition control are now being routinely achieved by MBE. For the first time, growth of polar III–V semiconductors on non-polar Si substrates has been achieved with a great degree of reliability, and several electronic and opto-electronic devices with III–V materials grown on Si substrates have been demonstrated (Morkoç *et al.*, 1985; Sakai *et al.*, 1985; Windhorn and Metze, 1985).

In evaluating the quality of semiconducting materials for device applications, it has been found that centers with deep energy levels in the forbidden energy gap play an important role. Deep levels essentially act as carrier recombination or trapping centers and adversely affect the performance of almost all electronic and opto-electronic devices. Deep level effects have been the subject of study for over thirty years. The two material properties most commonly affected by deep levels are radiative efficiency and minority carrier lifetime, It can be shown very simply that for two levels in the forbidden energy gap of a semiconductor, separated in energy by more than 0.1 eV, the lifetime will be determined by the deeper level, even if the concentration is two orders of magnitude less than that of the shallow one. Excellent reviews on deep levels have been made by Grimmeiss (1977), Neumark and Kosai (1983) and Milnes (1973). It is therefore felt that a lengthy review is unnecessary here. However, the essential properties of deep levels and techniques being used by us and other workers to characterize deep levels will be described in this and subsequent sections.

Deep levels can essentially be described by the Shockley–Read–Hall (1952) recombination statistics. However, a classical deep level trap is one whose capture cross-section for one type of carrier is many times larger than that for the other type. The trap, therefore, normally interacts with one of the band-edges during emission and capture processes. To distinguish it from shallow level impurities, deep levels are usually non-hydrogenic with ionization energies several times kT. They are, therefore, usually non-radiative. Deep levels can have single ground states or multiple states, and the level can go from one state to the other under electrical or optical excitation.

Finally, it should be noted that we will restrict this chapter to the wider bandgap III–V semiconductors being grown by molecular beam epitaxy. With respect to Fig. 1., which illustrates the variation of bandgap with lattice-constant in common III–V compounds and their ternary and quaternary derivatives, we will review the properties of deep levels in GaAs, AlGaAs, InGaAs and InAlAs lattice-matched to InP, and heterostructures made from suitable combinations of these semiconductors. Traps in GaAs–InGaAs strained-layer superlattices will also be discussed. It is to be realized that, with the exception of MBE GaAs, not much work has been reported on deep level traps in the compounds mentioned above. A complete study of traps should include their electrical and optical characterization and their identification. The latter includes the elucidation of their physio-chemical origin. Molecular beam epitaxy is a complex growth process and traps in MBE-grown materials can arise from native defects, impurities or complexes involving both. A study of traps in MBE-grown semiconductors is, therefore, both fascinating and difficult.

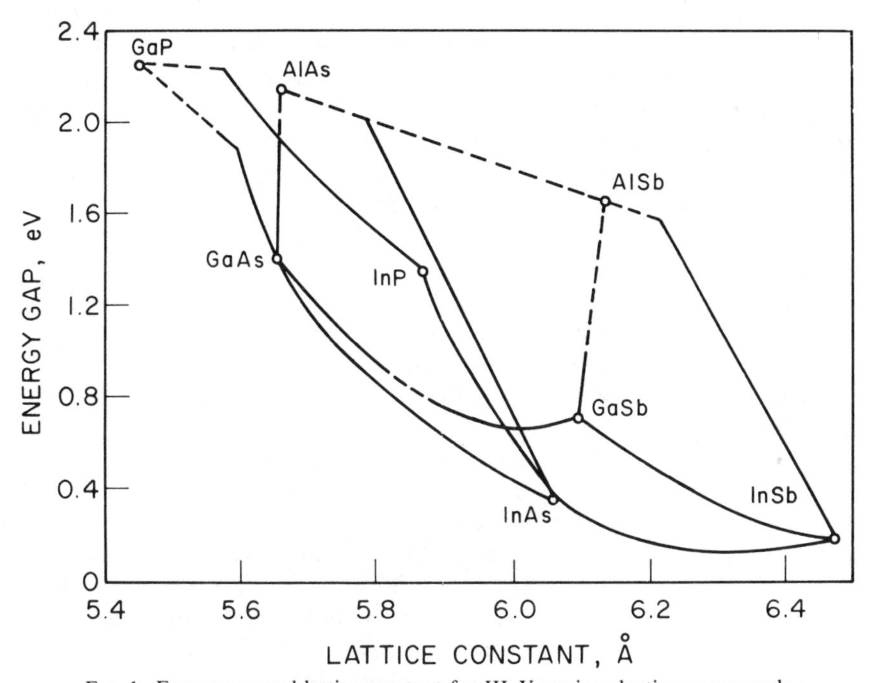

FIG. 1. Energy gap and lattice constant for III–V semiconducting compounds.

II. Molecular Beam Epitaxial Growth

1. Introduction

In contrast with liquid phase epitaxial (LPE) and vapor phase epitaxial (VPE) growth of semiconducting crystals under quasi-equilibrium conditions, growth by MBE is accomplished under non-equilibrium conditions and is principally governed by surface kinetic processes. Since the initial demonstration of the principle and process, MBE has come a long way and has emerged as a technique with unprecedented control over growth parameters. This has resulted from extensive research and is leading to a better understanding of the growth processes and to great advances in system design and reliability. Excellent reviews on the subject of MBE have been made by several authors (Cho and Arthur, 1975; Ploog, 1980; Tsang, 1985; Foxon and Joyce, 1975, 1977; Wood, 1982; Esaki and Tsu, 1970; Ilegems, 1985). We will therefore review the subject very briefly and dwell on more recent issues concerning the understanding of hetero-interface growth and deep level traps.

2. The Basic Growth Process

Molecular beam epitaxy is a controlled thermal evaporation process under ultra-high vacuum conditions. The process is schematically shown in Fig. 2 for the growth of GaAs with the possibility of dopant incorporation. The cells are designed in such a way that realistic fluxes for crystal growth at the substrate can be realized, while the Knudsen effusion condition is maintained (i.e., the cell aperture is smaller than the mean free path of the vaporized effusing species within the cell). The individual cells are provided with externally-controlled mechanical shutters whose movement times are less than the time taken to grow a monolayer. Therefore, very abrupt composition and doping profiles are possible. Interfaces that are one monolayer abrupt can be obtained fairly easily.

The process of crystalline growth by MBE involves the adsorption of the constituent atoms or molecules, their dissociation and surface migration and, finally, incorporation, resulting in growth. From pulsed molecular beam experiments, Arthur (1968, 1974) observed that, below a substrate temperature of 480°C, Ga had a unity sticking coefficient on (100) GaAs. Above this temperature, the coefficient is less than unity. The adsorption and incorporation of As_2 or As_4 molecules is more complex. It was found that, in general, As sticks only when a Ga adatom plane is already established. Joyce and coworkers (1977, 1978) have described the kinetic processes leading to the growth of GaAs from Ga and As_2 or As_4 molecules. The cation is in atomic form when it reaches the heated surface. It attaches randomly to a

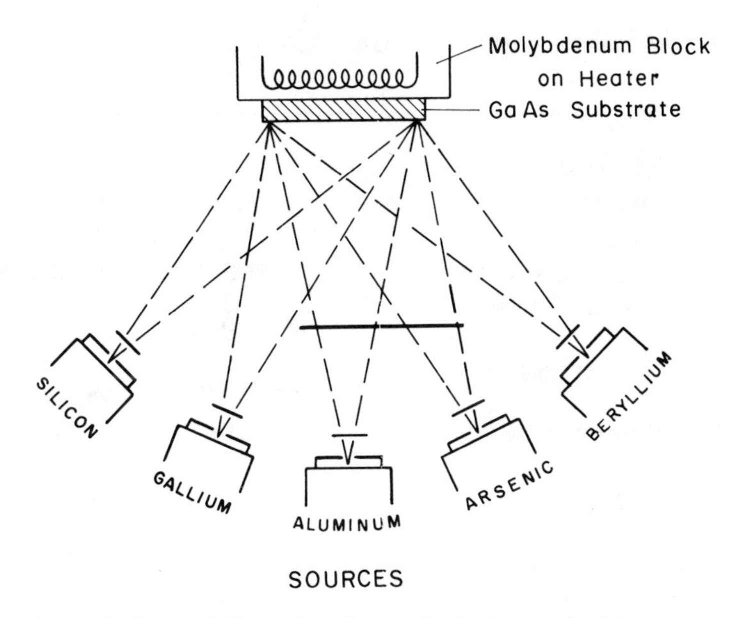

FIG. 2. Schematic illustration of the molecular beam epitaxial process.

surface site and undergoes several kinetically-controlled steps before it is finally incorporated. The As_2 or As_4 atom is first physisorbed into a mobile, weakly-bound precursor state. As this state moves on the surface, some loss occurs due to re-evaporation, and the rest are finally incorporated in paired Ga lattice sites by dissociative chemisorption.

The processes involved in cation and anion incorporation are schematically illustrated in Fig. 3. The important fact that emerged from the early kinetic studies is that stoichiometric GaAs can be grown over a wide range of substrate temperatures by maintaining an excessive overpressure of As_2 or As_4 over the Ga beam pressure. Substrate temperatures during growth are usually close to, and slightly higher than, the congruent evaporation temperature of the growing compound. Under normal MBE growth conditions, where the incorporation rate of the cations is nearly 100%, the cation surface migration needs to be high. Otherwise, MBE growth will occur by a three-dimensional island mode, rather than a step-growth mode, which is necessary for producing high-quality and abrupt interfaces.

3. SYSTEM DESCRIPTION

A typical, present-day MBE growth facility consists of the UHV growth chamber into which the growth substrate is introduced through one or two sample-exchange loadlocks. The base pressure in the growth chamber is

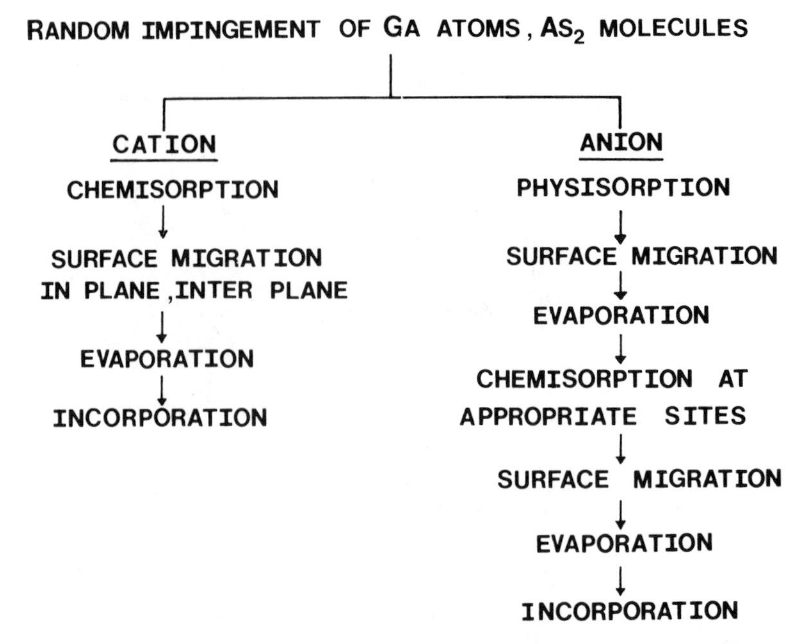

FIG. 3. Physico-chemical processes associated with the growth of GaAs byMBE.

usually $\sim 10^{-11}$ torr, while the other chambers are at $\sim 10^{-10}$ torr. The schematic of the typical growth chamber is shown in Fig. 4. The effusion cells are made of pyrolytic BN. To obtain the dimers from the tetramer species of the Gr V elements, external cracking cells are incorporated. Cracking is enhanced in the presence of a loosely-packed catalytic agent. Charge interlocks with auxiliary pumping are sometimes used for the more rapidly depleting Gr V species. The growth chamber and effusion cells are provided with liquid N_2 cryoshrouds, which are kept cold during growth. One of the systems being used in the authors' laboratory is shown in Fig. 5.

Most growth systems are equipped with *in situ* surface diagnostic and analytical capabilities in the growth and auxiliary chambers. The most common facilities in the growth chambers are a quadrupole mass spectrometer (or residual gas analyzer), which gives important information regarding the ambient in the growth chamber at all times, and a reflection high-energy electron diffraction (RHEED) system, which gives an insight to the growth mechanism. Electrons from a high-energy (~ 10 keV) electron gun strike the substrate or the growing layer surface at glancing incidence, and the diffraction pattern is monitored on a fluorescent screen.

In addition to the static RHEED pattern, a study of the oscillations in intensity of various beams during growth give information regarding the

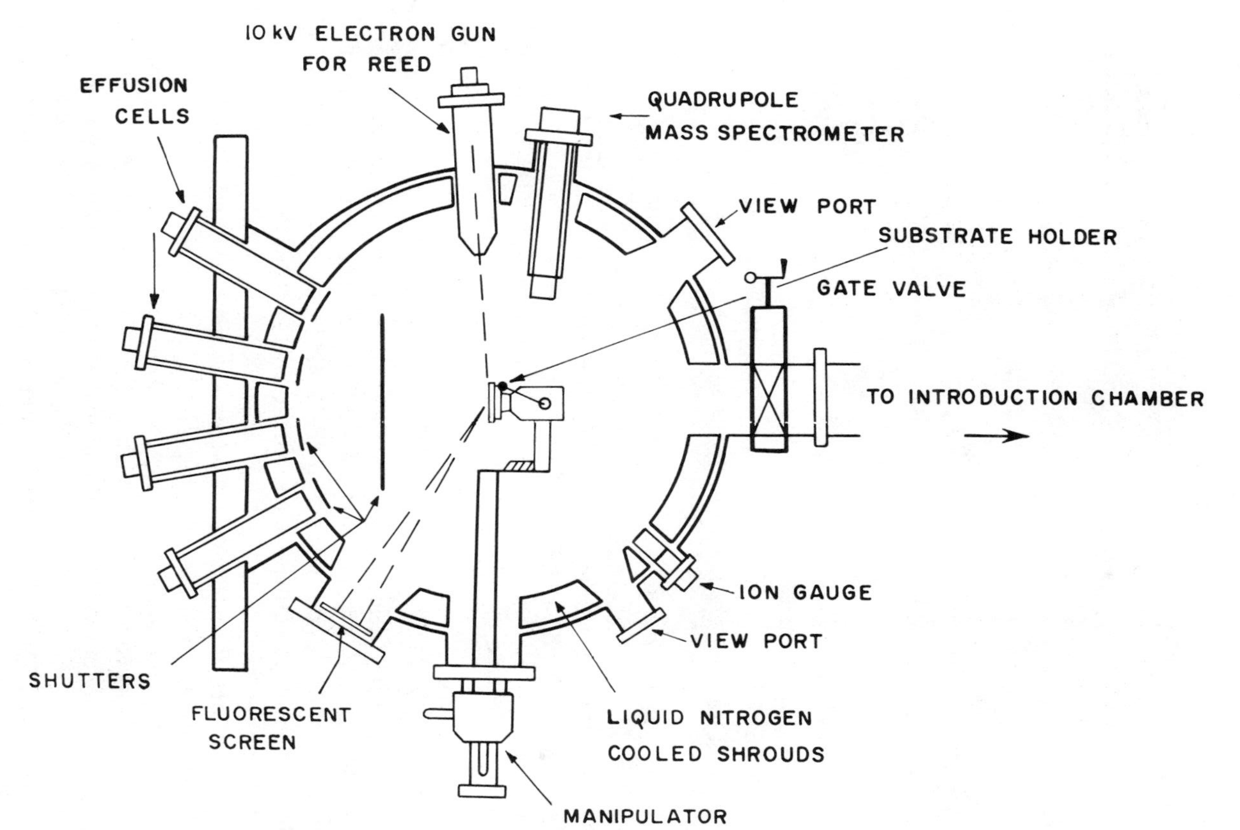

FIG. 4. Schematic cross-section of a typical MBE growth chamber equipped with *in situ* diagnostic tools.

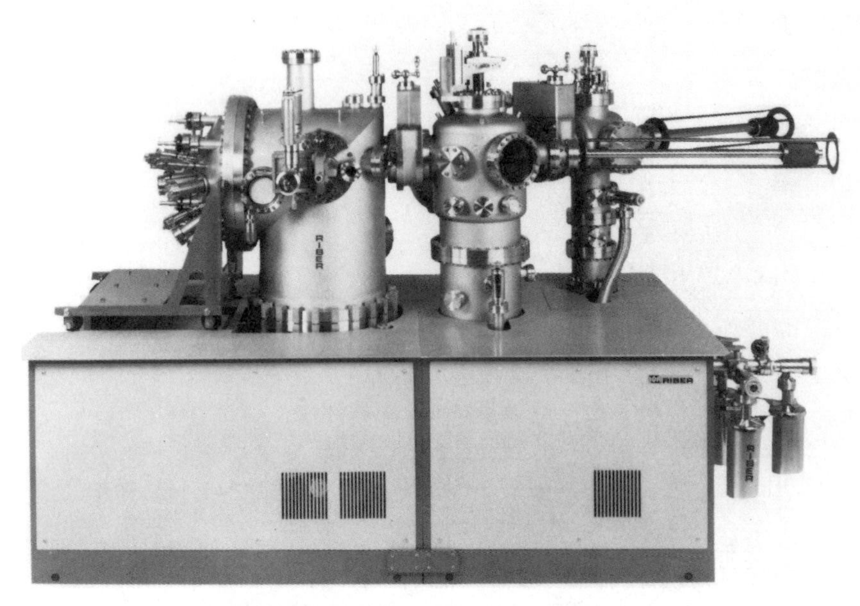

FIG. 5. A RIBER 2300 molecular beam epitaxy system.

mode of growth (Neave and Joyce, 1983). Such RHEED oscillation measurements are also being done by other workers (Ghaisas and Madhukar, 1985; Van Hove *et al.*, 1983). It would be important to pursue such studies during heterostructure and multilayered growth.

4. GROWTH OF TERNARY COMPOUNDS

In the present article, we will be principally concerned with the ternary alloys $Al_xGa_{1-x}As$, $In_xGa_{1-x}As$ and $In_xAl_{1-x}As$. The first alloy system is almost perfectly lattice-matched to GaAs for $0 \leq x \leq 1$. Though work is being done by us and other workers (Bhattacharya *et al.*, 1986; Quillec *et al.*, 1984; Laidig *et al.*, 1984; Biefeld *et al.*, 1983) on strained $In_xGa_{1-x}As$ on GaAs substrates (the lattice mismatch between GaAs and InAs is ~7%), the compositions $In_{0.53}Ga_{0.47}As$ and $In_{0.52}Al_{0.48}As$, lattice-matched to InP (see Fig. 1), are of tremendous technological importance for optical communication and high-speed device applications. In this case, a slight deviation from the stated solid solution compositions will create a large mismatch. This is illustated in Fig. 6, where the data of Cheng *et al.* (1981a) for $In_{0.53}Ga_{0.47}As$ are reproduced. A similar situation holds for $In_{0.52}Al_{0.48}As$. In spite of these difficulties, these alloys are being grown by several groups of workers (Miller and McFee, 1978; Ohno *et al.*, 1981; Cheng *et al.*, 1981b, 1982;

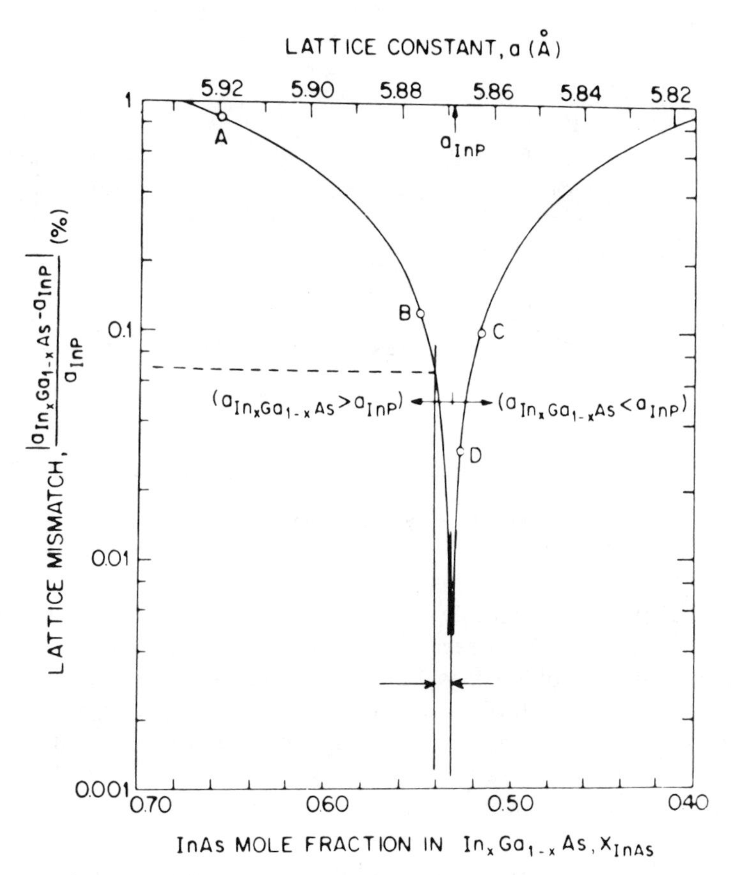

FIG. 6. Lattice mismatch between the $In_xGa_{1-x}As$ epitaxial layer and the InP substrate as a function of the InAs mole fraction in the ternary epitaxial layer. Lattice mismatch is expressed in absolute value to accommodate both positive and negative values on the same coordinate. Note that a variation of 1% in InAs mole fraction from $x = 0.53$ will cause a lattice mismatch of 7×10^{-4}. [From Cheng *et al.* (1981a).]

Kawamura *et al.*, 1982; Massies *et al.*, 1982; Mizutani and Hirose, 1985; Seo *et al.*, 1985) with quality almost comparable to LPE material.

The growth rate R of III–V compounds is entirely controlled by the flux densities of the Gr III beams F_i:

$$R = \sum_{i=1}^{n} \alpha_i F_i, \tag{1}$$

where n is the number of the different Gr III elements and α_i are their respective sticking coefficients. At normal growth temperatures, $\alpha_i \approx 1$, but decreases for elevated growth temperatures.

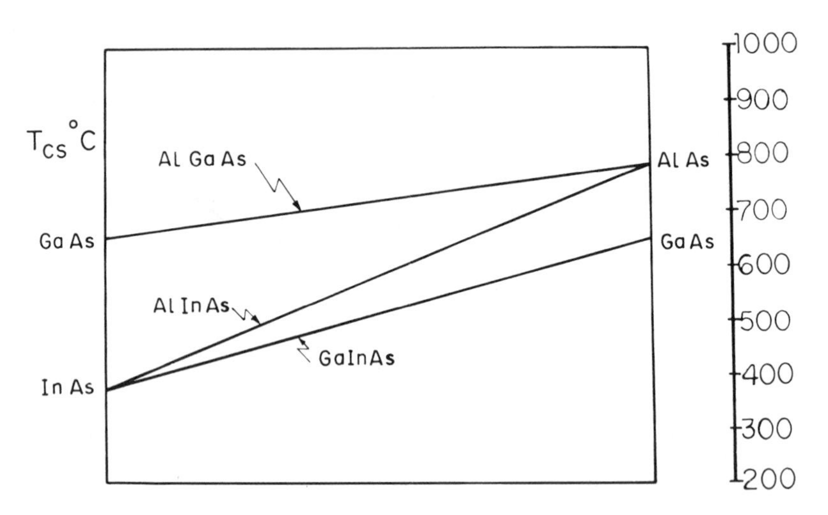

FIG. 7. Vegard approximations for congruent sublimation temperatures for GaInAs, AlInAs and AlGaAs. [From Wood (1982).]

From a study conducted by Wood (1982), it is apparent that the best growth temperatures for the ternary compounds are based on Vegard's rule for averaging the congruent sublimation temperatures of the constituent binaries. The data for $In_xGa_{1-x}As$, $In_xAl_{1-x}As$ and $Al_xGa_{1-x}As$ are shown in Fig. 7. Thus, $In_{0.53}Ga_{0.47}As$ is usually grown in the temperature range 480–520°C, and $Al_xGa_{1-x}As$/GaAs is grown in the range 600–700°C, depending on the application. At low growth temperatures, $Al_xGa_{1-x}As$ has high-resistivity and its luminescence efficiency is very poor. The luminescent properties steadily improve with increase of growth temperature (Swaminathan and Tsang, 1981). More recently, it has been demonstrated that $Al_xGa_{1-x}As$ of good optical quality can be grown in the temperature range 620–660°C (Erickson *et al.*, 1983). It was established that deep trap concentrations decrease with the increase of substrate temperature, as evidenced by the data of McAfee, Tsang and coworkers (Tsang, 1985), reproduced in Figs. 8 and 9.

5. Effects of Growth Ambient

System preparation before growth has changed dramatically over the years. It is evident that all active components and the cryoshrouds need to be baked very extensively before initiating growth of high quality layers. By monitoring the ambient with the mass-spectrometer, it can be observed that concentrations of CO, CO_2, etc. are substantially reduced. Typical baking schedules followed in the authors' laboratory have been mentioned in detail

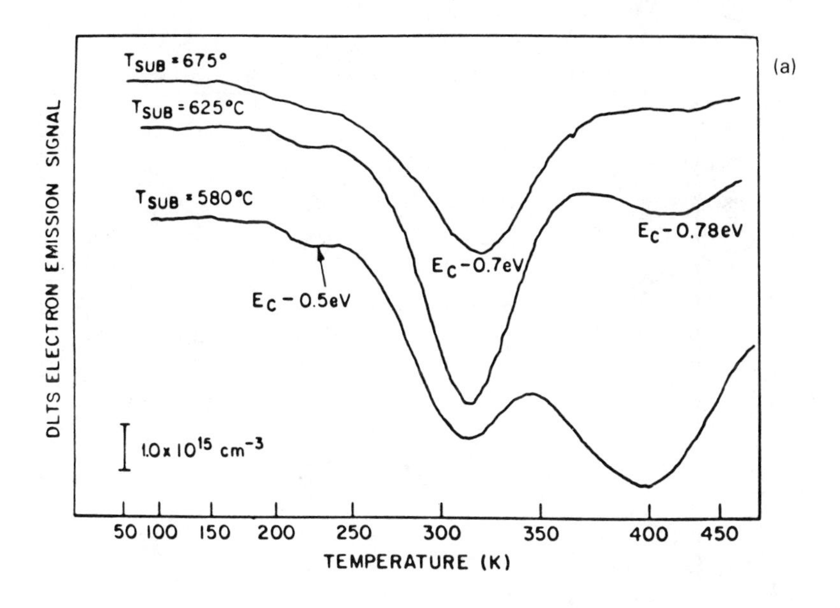

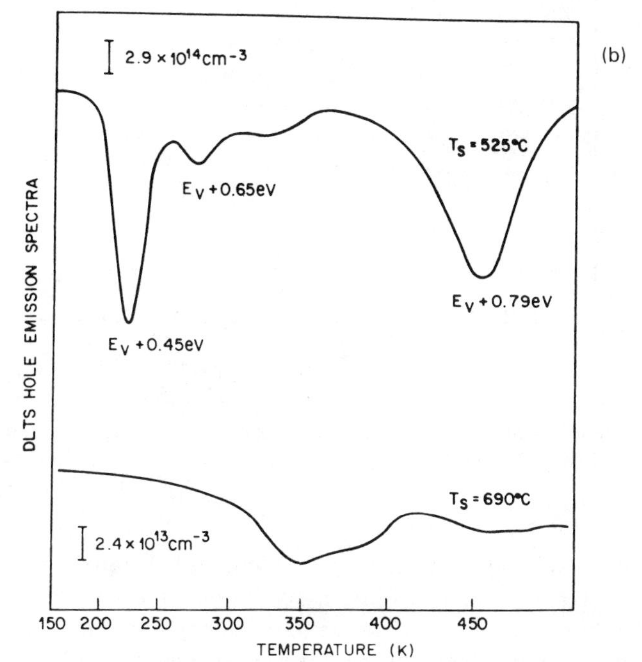

FIG. 8. DLTS spectra (50 sec^{-1} rate window) of (a) electron traps in MBE Sn-doped $Al_{0.25}Ga_{0.75}As$, and of (b) Be-doped $Al_{0.25}Ga_{0.75}As$ for various substrate growth temperatures. [From Tsang (1985).]

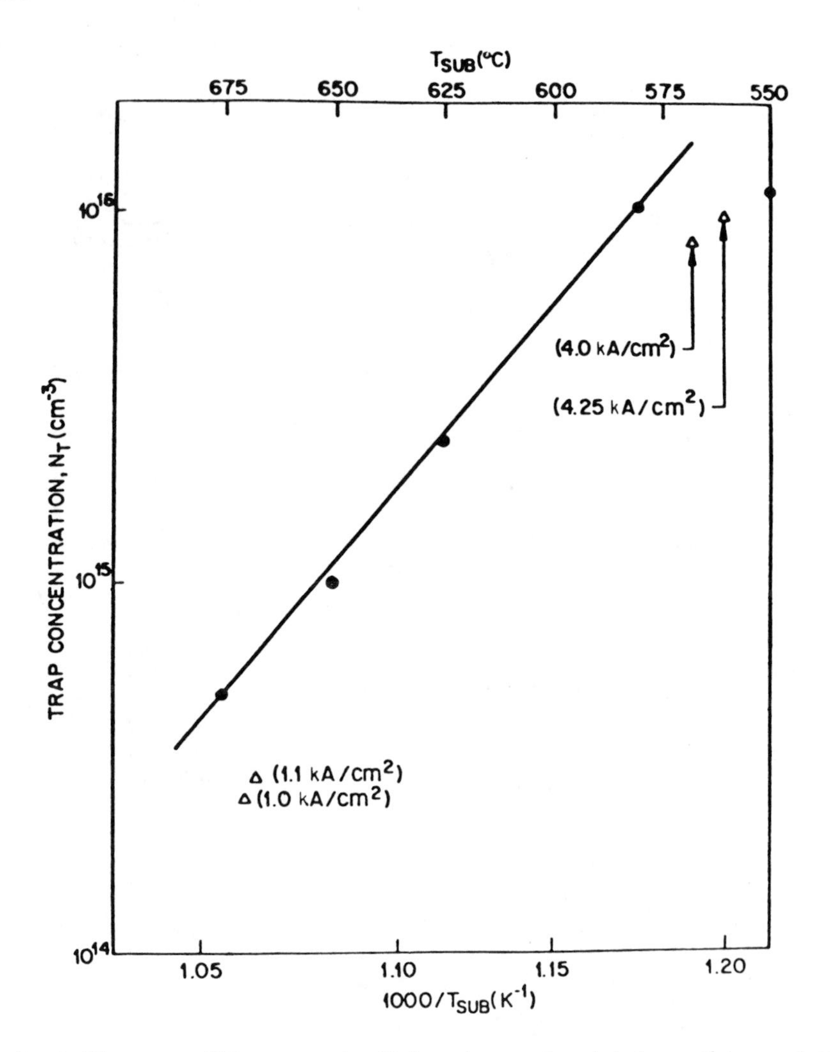

FIG. 9. Electron trap (E3) concentration N_T dependence on the substrate growth temperature in (●) MBE-grown n-$Al_{0.25}Ga_{0.75}As$ Schottky barriers and (△) MBE DH lasers. The threshold current densities of the lasers are indicated in parentheses. [From Tsang (1985).]

in an earlier publication (Juang *et al.*, 1985). It is, therefore, improbable that any appreciable amount of impurity could be incorporated in the growing layer. Carbon is the dominant acceptor impurity found in MBE compounds and is probably produced by the reaction of background CO with the Ga adatoms (Stringfellow *et al.*, 1981). It is also suspected that O from volatile Ga_2O may be incorporated (Kirchner *et al.*, 1981a), but the electrical and optical properties of O in MBE III–V compounds have not been clearly established. At the present time, it is believed that the most likely source of ubiquitous impurities in the films are the effusing species.

6. Electrical Doping

The common *n*-type dopants in III–V compounds are Si, Sn, Te and Se, while the *p*-type dopants are Cd, Zn, Be, Mg and Mn. Ge is usually amphoteric. Due to various reasons, as tabulated by Ploog (1980), most of the elements have been eliminated, and Si and Be have emerged as the best *n*- and *p*-dopants, respectively. However, there still exist the problems of dopant diffusion with the growth front in both cases.

7. Growth of Strained Layers and Strained Layer Superlattices

InGaAs–GaAs and GaAsP–GaAs (GaP) strained layer superlattices are currently being investigated for several active and passive device applications (Bhattacharya *et al.*, 1986). More freedom of bandgap tailoring is made possible with strained layer superlattices (SLS). We will briefly describe the growth of InGaAs–GaAs SLS. Assuming unity sticking coefficients for In and Ga at the growth temperature ~520–560°C, the In concentration in the $In_xGa_{1-x}As$ layers is given by

$$x = \frac{F_{In}}{F_{In} + F_{Ga}} = 1 - \frac{n_1 t_2}{n_2 t_1}, \tag{2}$$

where t_1 and t_2 are the growth times of the GaAs and InGaAs layers, respectively, and n_1 and n_2 are the number of molecular layers of GaAs and InGaAs in the superlattice. A typical SLS structure is schematically shown in Fig. 10. In such a heterostructure, the lattice mismatch is accommodated by alternate expansion and compression in the plane of the layer, resulting in a two-dimensional lattice strain. An obvious and important question in relation to device applications is the structural integrity of the SLS and the effects of doping, implantation and diffusion on them. We have investigated the behavior of deep levels in as-grown, annealed and implanted-and-annealed SLS structures. The results will be described and discussed in Section VI.

8. Growth of Heterostructures and Multiquantum Wells

An increasing number of present-day electronic and opto-electronic devices make use of lattice matched heterostructures and multilayered materials. Interesting carrier dynamics in such structures have given rise to

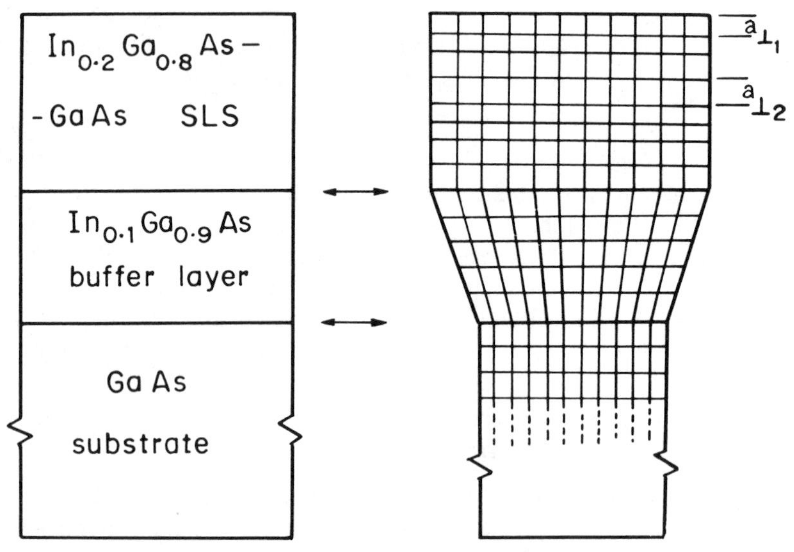

FIG. 10. Schematic of a strained layer superlattice and the alternate layers within the superlattice in biaxial compression and tension. $a_{\perp_1}$ and $a_{\perp_2}$ are the lattice constants of alternate layers in the direction normal to the plane of the layers.

a variety of interesting and novel concepts that are being applied to detectors, lasers, modulators and heterostructure electronic devices. Most of these devices have stringent requirements regarding interface quality. Though near-perfect lattice matching at the interface can be maintained during MBE growth, the microscopic nature of the region may be far from perfect. In fact, *interface roughness* can extend to several monolayers, which can be of the order of the thickness of active regions of devices. The problem can, therefore, have very severe implications, since it gives rise to additional carrier scattering at the interface and lower luminescence efficiency of the structure.

The magnitude of the interface roughness is related to the III–V bond strengths and cation migration rates under particular growth conditions. It is, therefore, likely that the two layers forming a heterostructure may have different ideal growth conditions, and, therefore, depending on the sequence of growth, the cation adatoms may "pile up" in the layer-by-layer growth mode. Thus, the interface, instead of being abrupt to one monolayer, can extend to several monolayers. Ideal and non-ideal interface structures are schematically shown in Fig. 11.

The problem of interface roughness has been theoretically studied by Singh *et al.* (Singh and Bajaj, 1985a,b,c), and these authors have suggested a number of techniques to improve the interface, such as different growth

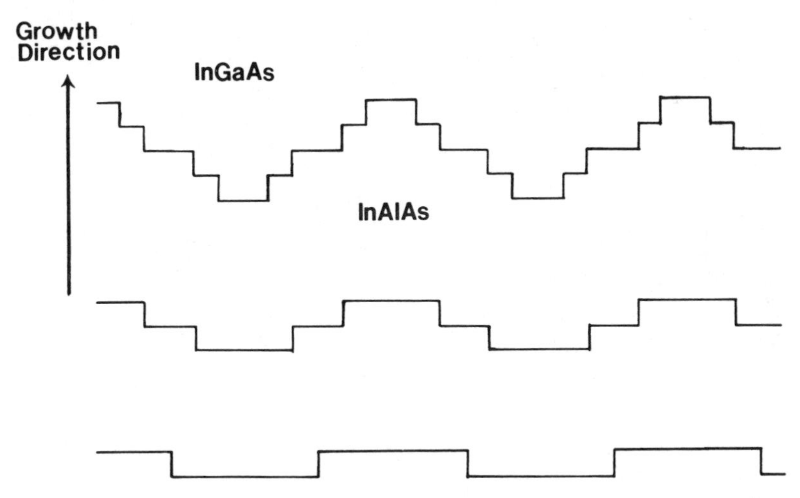

FIG. 11. The InGaAs/InAlAs interface structure, indicating various degrees of surface roughness, as obtained during MBE growth. The step height corresponds to a monolayer.

temperatures for the two layers, growth interruption, resonant laser excitation and adding a few monolayers of a Gr III adatom with high surface migration rate just before the interface. Interrupted growth is being implemented by other authors (Hayakawa *et al.*, 1985a), and we have investigated growth interruption at the InGaAs/InAlAs normal and inverted interface in detail. The probe used is the exciton transition intensity in the low temperature photoluminescence spectra (Juang *et al.*, 1986). Photoluminescence (PL) studies were carried out on 120 Å InGaAs/InAlAs single quantum well structures grown by molecular beam epitaxy. Three types of samples were grown, with the growth being interrupted before interface formation. The interruption times were zero, two and three minutes. The corresponding linewidths of the main excitonic transition associated with the quantum well were found to be 20, 16 and 8 meV, respectively. Analysis of the 8 meV linewidth, which is close to the intrinsic limit for this system, suggests that the inverted InAlAs/InGaAs interface can be described by two dimensional steps that have a height of two monolayers and a lateral extent of 100 Å. The normal interface (InAlAs on top of InGaAs) is more ideal, and roughness of the order of one monolayer can be achieved.

The photoluminescence spectrum for a 120 Å single-quantum well grown with interruption at the interfaces is shown in Fig. 12. The decrease of PL linewidth with interruption is depicted in Fig. 13. However, as shown in Fig. 13, the improvement in interface quality is accompanied by a decrease in the PL intensity. We believe this is due to impurity accumulation and the creation of non-radiative defects at the interface due to interruption. This aspect is

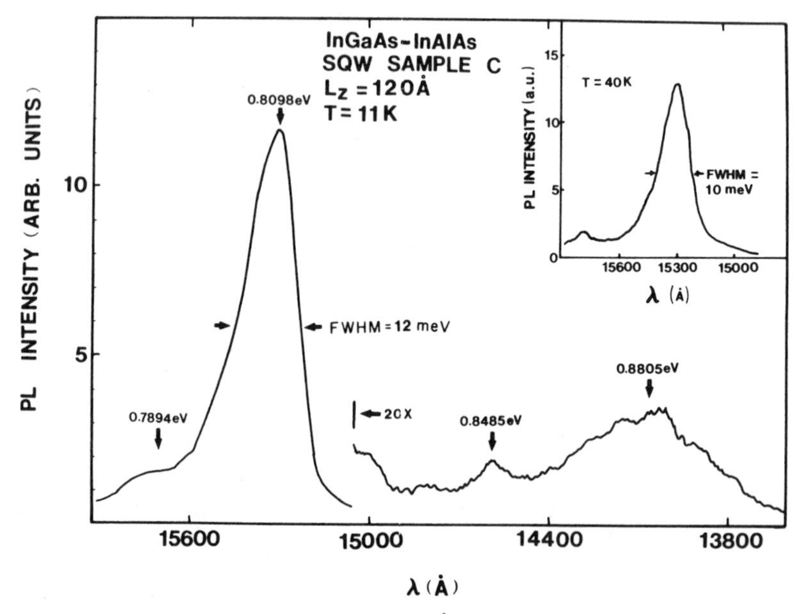

FIG. 12. Measured photoluminescence from 120 Å $In_{0.53}Ga_{0.47}As$ single quantum well with $In_{0.52}Al_{0.48}As$ barriers. The main peak and the shoulders in the lower energy side are attributed to bound exciton transitions. Inset shows the phototoluminescence spectrum at 40 K. The measured linewidth (FWHM) is 10 meV.

being further investigated by us. Improvement of the interface quality will also lead to higher carrier mobilities parallel to the layers, such as in modulation-doped structures. It is also important to investigate experimentally other growth techniques that might improve the interface, as suggested by theoretical models.

9. MBE Growth Models

The MBE growth process is an extremely complex one and is a non-equilibrium thermodynamic problem. Therefore, theoretical studies involving analytical treatment and simulation prove to be very useful in proving an insight to MBE growth and complementing experiments. They can also provide guidelines for better growth, as mentioned earlier. Singh and Bajaj (1985d) have subdivided the problems in three key areas: (a) energetics, (b) incorporation process, and (c) growth kinetics. Early work by Arthur (1974) and Foxon and Joyce (1977) has led to a good understanding of the epitaxial process and layer growth. More recent work by Singh *et al.* (1985d), using sophisticated computer simulation models and the Monte Carlo technique, has led to an understanding of the detailed atomistic processes underlying

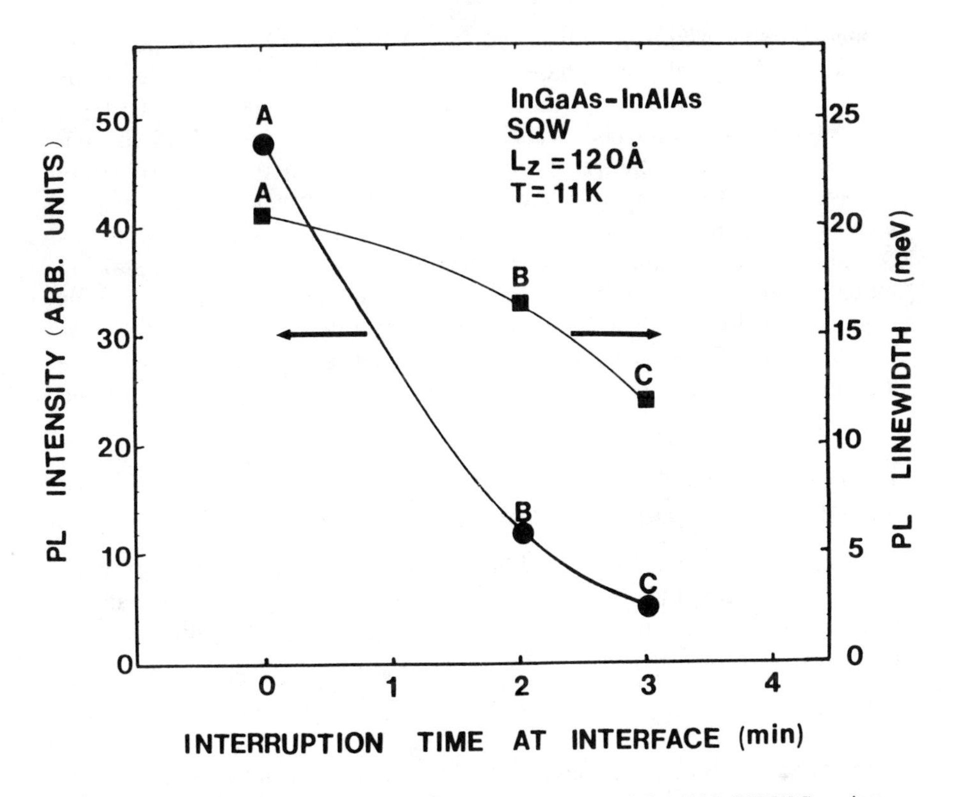

FIG. 13. Dependence of measured luminescence intensity and linewidth (FWHM) on interruption interval during MBE growth of the heterointerfaces in a single quantum well.

MBE growth. From these studies, the importance of flux rates, cation migration and growth rates and their relation to each other can be understood. In conclusion, it is important to realize that all these factors have to be understood and implemented in order to grow high-quality, defect-free materials.

10. METAL–ORGANIC MOLECULAR BEAM EPITAXY

This growth technique, also called chemical beam epitaxy (CBE), has recently demonstrated potential for the growth of InP- and GaAs-based compounds (Panish, 1980; Tsang, 1984). In this technique, all the sources are gaseous Gr III and Gr V alkyls. The In and Ga are derived by the pyrolysis of either trimethylindium or triethylindium and either trimethylgallium or triethylgallium at the heated substrate surface, respectively. The As_2 and P_2 are obtained by themal decomposition of triethylphosphine and trimethylarsine in contact with heated Ta or Mo at 950–1200°C, respectively. Unlike conventional vapor phase epitaxy, in which the chemicals reach the substrate surface by diffusing through a stagnant carrier gas boundary layer above the substrate, the chemicals in CBE are admitted into a high vacuum growth chamber and impinged directly, line-of-sight onto the heated substrate surface in the form of molecular beams. It is apparent that this growth technique will be important for the growth of P-containing compounds on large-area substrates, and for hetero-epitaxy on Si. Gaseous starting materials allow semi-infinite source volumes. The quality of III–V materials being grown by this technique has not yet attained the level of MBE or metalorganic chemical vapor deposited (MOCVD) materials. A large amount of work in perfecting this important growth technique remains to be done.

11. Transport and Photoluminescence in MBE-Grown Semiconductors

A recent review by Tsang (1985) gives detailed data obtained from high-quality MBE-grown III–V compounds. We will, therefore, briefly describe some important results. Furthermore, we will restrict this discussion to GaAs, $Al_xGa_{1-x}As$, $In_{0.53}Ga_{0.47}As$ and $In_{0.52}Al_{0.48}As$.

a. GaAs

Very high levels of purity have been obtained in this compound by several groups of authors (Morkoc and Cho, 1979; Calawa, 1981). Unintentionally doped samples are generally *p*-type with $N_A - N_D \sim (0.2\text{–}1.0) \times 10^{14}\ \mathrm{cm^{-3}}$ at 300 K. In lightly-doped (Si and Sn) *n*-type layers, $N_D - N_A \sim 1.0 \times 10^{14}\ \mathrm{cm^{-3}}$ have been achieved (Morkoc and Cho, 1979; Hwang *et al.*, 1983; Heiblum *et al.*, 1983) with mobilities ranging from 100,000 to 144,000 $\mathrm{cm^2\ V^{-1}\ s^{-1}}$ at 77 K. Typical variations of mobility and carrier concentration in a lightly Si-doped layer are depicted in Fig. 14.

Low-temperature photoluminescence spectra of high-purity and lightly doped MBE GaAs have the same sharp features as in LPE and in VPE materials. The excitonic spectra are characterized by transitions with linewidths (full width at half maximum) ~0.1 meV (Skromme *et al.*, 1985). The excitonic spectra in MBE GaAs has additional transitions in the 1.504–1.512 and 1.466–1.482 eV ranges (Ploog, 1980). More recently, Skromme *et al.* (1985) have ascribed these lines to complexes involving C and native defects. These authors have also investigated the donor impurities in MBE GaAs by photothermal ionization spectroscopy. The main residual donor species are S and Si, with the S possibly originating from As source material.

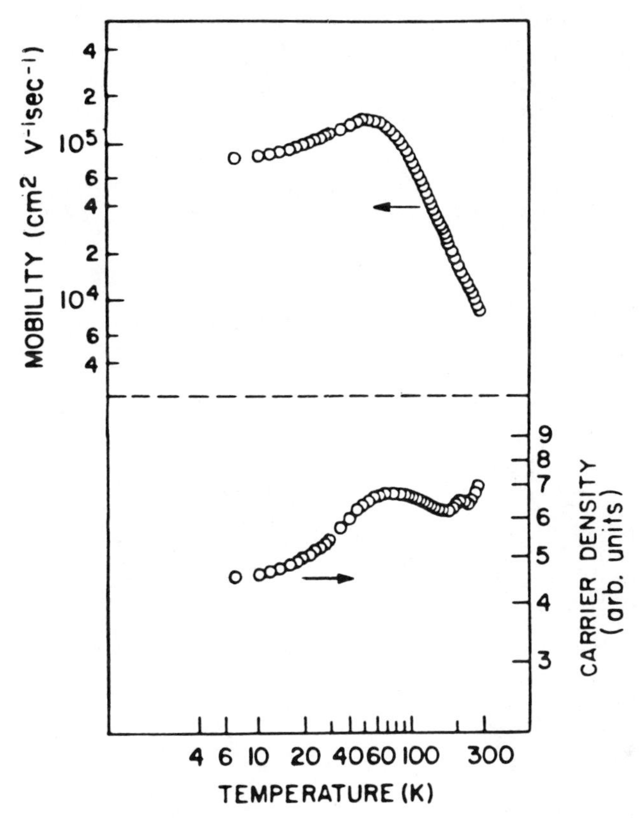

FIG. 14. The temperature dependence of carrier concentration and electron mobility of a typical GaAs sample measured by Hall effect. [From Hwang *et al.* (1983).]

b. $Al_xGa_{1-x}As$

Undoped $Al_xGa_{1-x}As$ crystals grown at temperatures lower than 650°C are usually of high-resistivity, and, therefore, their transport properties cannot be evaluated at room temperature. At growth temperatures of 700°C and above, the layers are *p*-type (Heiblum *et al.*, 1983) with $(N_A - N_D) \sim 5 \times 10^{14}\ cm^{-3}$. Studies on *n*-doped AlGaAs (Morkoc *et al.*, 1980; Ishibashi *et al.*, 1982) indicate that higher electron mobilities can be obtained by growing at higher temperatures.

Low-temperature photoluminescence in high-quality MBE $Al_xGa_{1-x}As$ has been measured (Wicks *et al.*, 1981; Heiblum *et al.*, 1983; Hayakawa *et al.*, 1985b; Reynolds *et al.*, 1986), and the excitonic transition linewidths are broader than those in GaAs, due to the effects of alloying. Singh and Bajaj (1985b) have recently calculated the linewidths as a function of alloy

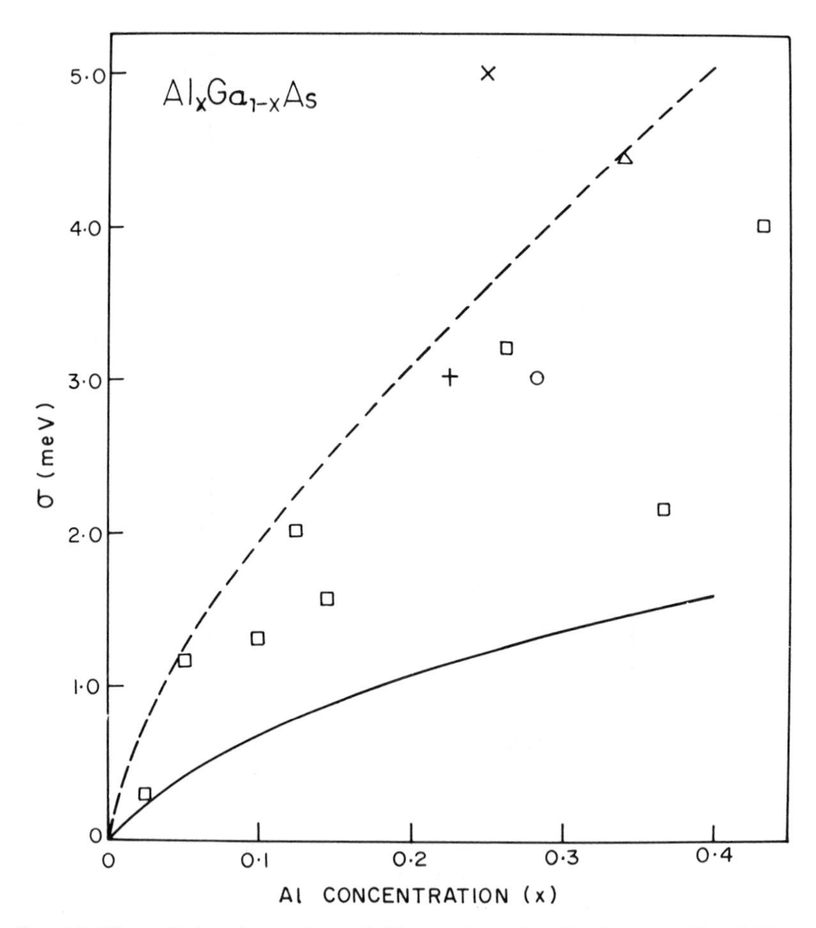

FIG. 15. Theoretical and experimental (data points) photoluminescence linewidths as a function of alloy composition. The solid and dashed curves are from quantum-mechanical and classical calculations, respectively. [From Singh and Bajaj (1985b).]

composition by using a quantum mechanical approach. Their theoretical data and some experimental data are shown in Fig. 15 for comparison. It is clear that the best possible quality of $Al_xGa_{1-x}As$ has yet to be attained.

c. $In_{0.53}Ga_{0.47}As/InP$

This ternary compound lattice matched to InP is being grown for electronic and opto-electronic device applications. 300 and 77 K mobilities of 10,000 and 45,000 $cm^2/V\,s$ with $n_{300K} \sim 5 \times 10^{15}\ cm^{-3}$ can be obtained routinely, as mentioned earlier. It appears that impurities in the source In and outdiffusion from the substrate are the two main causes for high levels of background doping. Improvements are observed by pre-baking the source In (Seo *et al.*, 1985) and the InP substrate (Brown *et al.*, 1985). Mizutani and Hirose (1985) have recently grown material with μ_{300K} and μ_{77K} equal to 13,000 and 55,000 cm^2/V s, respectively, which are comparable to those obtained in high-quality LPE layers. These authors believe that initial flux transients create a few monolayers of a severely mismatched region after initiation of growth of the ternary. By removing these effects, high-quality materials could be grown. Doping studies of this alloy has been done by Cheng *et al.* (1981c).

The low-temperature photoluminescence spectrum of undoped $In_{0.53}Ga_{0.47}As$ is dominated by a single peak, which is probably composed of several bound exciton transitions. A typical spectrum is shown in Fig. 16. The line width of this transition is approximately 3.8 meV and is comparable to that measured in LPE material and to values calculated theoretically, assuming a perfectly random alloy. The data indicate that clustering effects are almost non-existent in high-quality MBE InGaAs. Furthermore, PL intensities are critically dependent on lattice matching (Tsang, 1985).

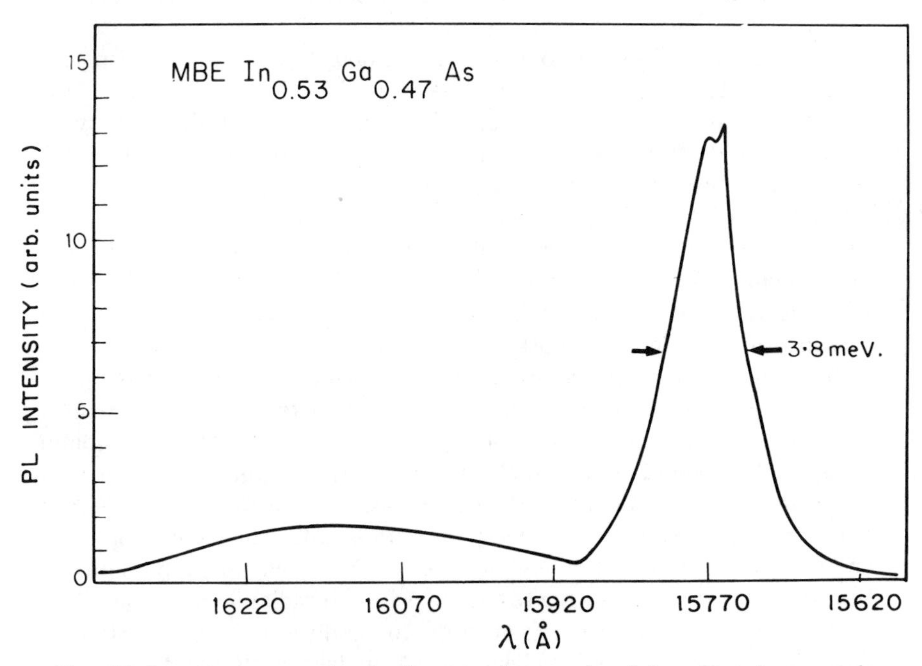

FIG. 16. Low-temperature photoluminescence from undoped $In_{0.53}Ga_{0.47}As$ grown by molecular beam epitaxy. The main peak arises from bound exciton transition, while the low-intensity broad peak at lower energies originates from silicon and zinc impurities.

d. $In_{0.52}Al_{0.48}As/InP$

Very little work has been done in studying the properties of undoped or lightly doped $In_{0.52}Al_{0.48}As$. Growth and doping (*n*-type) of this compound has been investigated by some authors (Massies *et al.*, 1983; Wakefield *et al.*, 1984; Welch, 1985). PL linewidths are ~20 meV, which suggests that clustering effects may be present during growth. This alloy forms the barrier regions in InGaAs/InAlAs superlattices and the doping layer of modulation-doped structures. Hence, more work is necessary to optimize its growth and measure the properties of this compound.

III. The Physics of Deep Levels

12. Defects in Semiconductors

The term imperfection or defect is generally used to describe any deviation from an orderly array in a crystal lattice. Imperfections within the semiconductor can disrupt the perfect periodicity of the crystal lattice and, as a result, can introduce energy levels deep in the forbidden gap, much as donor or acceptor impurities do. Three kinds of defects can be described. An atom missing from a correct site gives rise to a Schottky defect. An extra atom in an interstitial site gives rise to an interstitial defect, and an atom displaced to an interstitial site creating a nearby vacancy gives rise to a Frenkel defect. Defects such as impurity-vacancy complexes may also occur during growth, and antisites are formed in compound semiconductors. In the latter defect, one atom occupies the position of the other in the lattice.

Considering the charged states and the electronic configurations, three types of defects may be distinguished. Impurities with a Coulombic potential which include all interstitials and any substitutional impurities of valence different from that of the atom they replace; isoelectronic impurities in which the substitutional impurity must have the same number of valence electrons as the normal host atom; and vacancies. Two principal mechanisms control the recombination processes observed in semiconductors. The first is a direct recombination of electrons and holes, accompanied by photon and phonon emission. The second mechanism for recombination requires the presence of localized energy states in the forbidden gap of the semiconductor. The two processes are shown in Fig. 17. Deep level traps fall in the second category. The energy of the recombining carriers can be taken up by phonons or photons in one or more of the following ways: (i) radiative recombination, (ii) Auger mechanism, (iii) exciton formation and (iv) phonon emission (or emission and absorption). Excellent reviews on deep levels and their characterization have been made by Grimmeiss (1977), Lang and coworkers (1977, 1979), Neumark and Kosai (1983) and Bourgoin and Lanoo (1983).

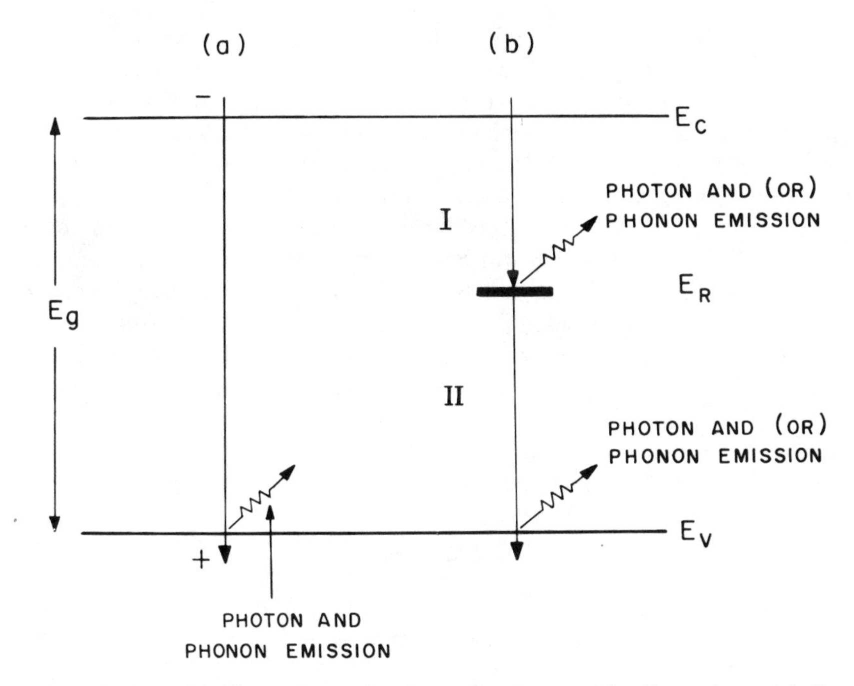

FIG. 17. Recombination and trapping in semiconductors. The figure shows (a) direct radiative recombination, and (b) indirect, two-step recombination in the presence of a recombination center E_R.

The calculation of the electronic structure induced by ideal and non-ideal point-defects is a matter of great interest and importance. In essence, these defects introduce energy levels located deep within the bandgap. The corresponding eigenfunctions are localized, compared to those for shallow levels, and, hence, the effective mass theory is, generally, not valid. One of the most general methods of calculation from which the properties of defects can be determined is the Green's function method, which can, in principle, yield exact results for a given defect perturbative potential, V. The bound energy levels associated with the localized states are obtained from a solution of the Schrödinger equation, provided the condition $\det|IG_0V| = 0$ holds, where G_0 is the crystalline Green's function for the site replaced by the defect, and I is the unit matrix.

13. Thermal Ionization and Capture Rates

A deep level in a semiconductor may act either as a trap or as a recombination center, depending on the temperature and doping conditions. If a carrier is captured in a center and, after living a mean lifetime in the captured state, is ejected thermally to the band from which it came, the center may be regarded as a trap. If, however, a carrier of opposite polarity is also trapped before thermal emission can occur, the deep level is a recombination center. Which role a center will play depends primarily on the relative cross section for capture of electrons and holes. The capture cross section $\sigma_{n,p}$ for electrons (or holes) is a measure of how close to the center the carrier has to come to get captured. Usually, a center acts as an electron trap if $\sigma_n \gg \sigma_p$ and as a hole trap if $\sigma_p \gg \sigma_n$. Note that the charged state or the donor- or acceptor-like nature of the centers has not been specified. This is a more complex behavior and is partly decided by the absolute values of the capture cross sections. We will therefore avoid a lengthy discussion of the properties of deep levels, except for a brief review of the important features relevant to data, presented later.

From the principle of detailed balance, the thermal emission rates for electrons or holes from their respective trapping centers are given by:

$$e_n = \sigma_n \langle v_n \rangle g N_c \exp\left(-\frac{E_T}{kT}\right) \tag{3a}$$

$$e_p = \sigma_p \langle v_p \rangle g^{-1} N_v \exp\left(-\frac{E_T}{kT}\right). \tag{3b}$$

Here, $\sigma_{n,p}$ are the capture cross sections of the traps; $\langle v_{n,p} \rangle$ are the average thermal velocities of electrons and holes; and g is the degeneracy factor for the level. The quantity E_T in these equations (Fig. 18) is, in fact, the free

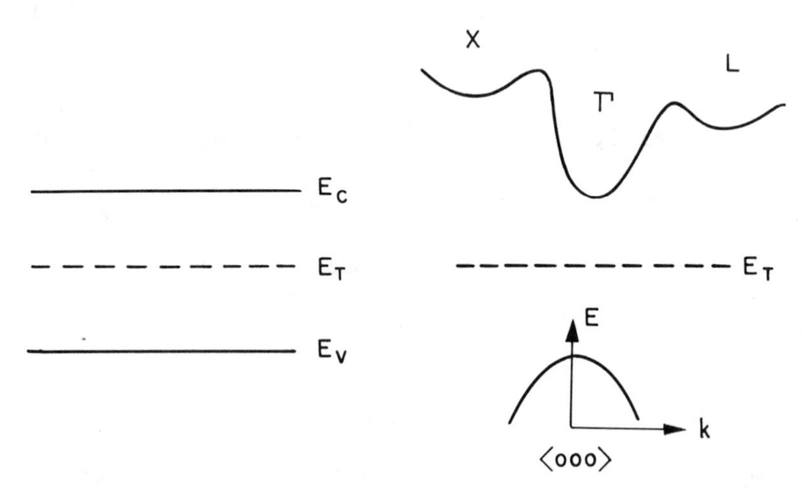

FIG. 18. Energy position of a deep level trap with respect to the conduction and valence bands in a direct bandgap semiconductor.

energy of ionization or the standard Gibbs free energy of the ionization reaction, denoted by G. It relates to the ionization of an electron to the conduction band, or of a hole to the valance band. The enthalpy of ionization, then, is given by $H = G + ST$, where S is the entropy. Equation (3a) can therefore be expressed as

$$e_n = \sigma_n \langle v_n \rangle g N_c \exp\left(\frac{S}{k}\right) \exp\left(-\frac{H}{kT}\right), \tag{4}$$

and Eq. (3b) can be expressed similarly. Therefore, the activation energy determined from an Arrhenius plot is actually the enthalpy H for $T > 0$. The change in the force constants between the defect and its neighbors, caused by the modification of bonding with the host lattice, due to the ionization process, gives rise to the entropy term (Bourgoin and Lannoo, 1983). It is also evident from Eq. (4) that, unless S and g can be accurately known, the capture cross section determined from the emission rate prefactor is inclusive of these parameters and is, therefore, erroneous.

Large cross sections ($>10^{-15}$ cm^2) are usually associated with attractive centers. Cross-sections in the range 10^{-17} to 10^{-15} cm^2 are usually associated with neutral centers, and very small cross sections, of the order of 10^{-22} cm^2, are associated with repulsive centers. For repulsive centers that possess no excited states, multiphonon emission is the only capture process available other than optical and Auger processes, and very small cross sections are expected. Lang and Henry (1975) have, however, shown that multiphonon emission can also account for rapid non-radiative capture processes involving capture cross sections as large as 10^{-14} or 10^{-13} cm^2. In the theory of multiphonon capture, it is assumed that the deep center is strongly coupled to the lattice, and, for large vibrations of the latter, the deep level can cross into the band and capture a carrier. Jaros (1978) has proposed an Auger-type recombination mechanism that can explain the capture cross sections in the range 10^{-13} to 10^{-16} cm^2 that exhibit only weak temperature dependence, and that are independent of the carrier cross section. The model proposed by Jaros considers a helium-type defect, e.g., a center binding deeply *two* electrons. The recombination process consists of a free hole moving from the valence band with thermal velocity and recombining with one of the electrons of the center. The excess energy arising out of this annihilation is then carried away by the other electron, which will be ejected into the conduction band where its energy is dissipated by collisions with phonons. This aspect of trap behavior is discussed at some length, since the traps with anomalous capture properties discussed above are found to be present in MBE-grown semiconductors, as will be evident in later sections.

14. Optical Ionization Properties

In addition to the thermal emission and capture properties discussed above, it is important to determine the optical emission and capture behavior of the center. This information, together with the thermal ionization properties, gives an insight to the coupling of the defect to the host lattice. The optical emission rate e^{O} is related to the photoionization cross section σ^{O} by

$$e^{\mathrm{O}} = \sigma^{\mathrm{O}}\phi, \tag{5}$$

where ϕ is the incident photon flux density in $\mathrm{cm}^{-2}\,\mathrm{s}^{-1}$. Measurements of e^{O}, σ^{O} and the optical ionization energy $\mathcal{E}^{\mathrm{O}}$ are discussed in the next section in some detail, but the physical principles are briefly reviewed here.

An estimate of the optical ionization energies is obtained by comparing measured spectral variations of the photoionization cross section with theoretically estimated values. One of the simplest and most useful forms of $\sigma^{\mathrm{O}}(h\nu)$ has been calculated by Grimmeiss and Ledebo (1975), by assuming excitation to a single parabolic band. According to these authors,

$$\sigma^{\mathrm{O}}(h\nu) = Rg\sqrt{\mathcal{E}^{\mathrm{O}}}\,\frac{\sqrt{m_{\mathrm{T}}m^*}}{m_{\mathrm{H}}^2}\,\frac{(h\nu - \mathcal{E}^{\mathrm{O}})^{3/2}}{h\nu\{h\nu + \mathcal{E}^{\mathrm{O}}[(m_{\mathrm{T}}/m^*) - 1]\}^2}, \tag{6}$$

where

$$R = \frac{16\pi}{3}\,\frac{q^2\hbar}{\varepsilon_0 cn}\left(\frac{E_{\mathrm{eff}}}{E_0}\right)^2. \tag{7}$$

Here, the defect potential is assumed to be of the form of a delta function; m_{T} is the mass of the bound particle; and m_{H} is defined by a perturbation operator.

For a single discrete trap level of total concentration N_{T}, the rate equation governing its occupation is

$$\frac{dn}{dt} = (c_n + e_p + e_p^{\mathrm{O}})(N_{\mathrm{T}} - n_{\mathrm{T}}) - (c_p + e_n + e_n^{\mathrm{O}})n_{\mathrm{T}}, \tag{8}$$

where n_{T} is the density of the traps occupied by electrons, and the capture rates $c_{n,p}$ are given by

$$c_{n,p} = (n, p)\sigma_{n,p}\langle v\rangle. \tag{9}$$

It is assumed, then, $e_n \gg e_p$ for electron traps, and vice-versa for hole traps. The steady state occupation function is given by

$$f_{\mathrm{T}} = \frac{c_n + e_p + e_p^{\mathrm{O}}}{c_n + c_p + e_n + e_p + e_n^{\mathrm{O}} + e_p^{\mathrm{O}}}. \tag{10}$$

The various rates in the equation are small or large, depending on the experimental conditions and on the measurement techniques.

The activation energies derived from the Arrhenius Eq. (3) assume that there are no field-enhanced emission effects. In actual practice, measurements are made on biased Schottky barriers or junction diodes, and the defect potential may be considerably perturbed by this field. The effect, known as the Poole–Frenkel effect, asymmetrically lowers the defect potential, whereby carrier emission rates are enhanced. The lowered potential effectively reduces the measured ionization energy (Bourgoin and Lannoo, 1983). This effect should be minimized during measurement of trap thermal ionization properties. Other competing processes that could play a role are pure and phonon-assisted tunneling (Martin *et al.*, 1981).

IV. Measurement Techniques

15. General Review

Deep levels, whose concentrations are usually two or three orders smaller than the shallow impurity level concentrations, are hard to investigate. This is especially true because the conventional sensitive methods for characterizing shallow levels, such as luminescence, photothermal ionization and absorption cannot be readily applied to the characterization of deep centers, as most of them are usually non-radiative in nature. However, the emergence of the transient capacitance technique in the last decade has made it possible to gain better insight into most of the important characteristics of deep levels, e.g., their position with respect to the band edges, thermal emission and capture properties, photoionization characteristics and so on. An excellent review of the application of space charge regions in a semiconductor junction device to the study of deep levels was made by Sah and coworkers (1970), and new techniques for this purpose have been subsequently proposed (Sah, 1976). Various aspects of the capacitance transient techniques have been discussed by Miller *et al.* (1977). Bleicher and Lange (1973) and Eron (1985).

In the capacitance transient technique, the change in depletion region capacitance of a semiconductor junction diode due to a change in trap occupancy is monitored to derive the trap emission rates. The experiment has to be performed at several temperatures to determine the trap activation energy, and it lacks the capability of isolating traps with overlapping emission rates. Several spectroscopic methods, based on capacitance change due to carrier emission and capture, have been proposed to enable a quick indentification of the traps present in a material. In the thermally-stimulated capacitance (TSCAP) technique (Sah and Walker, 1973), the capacitance change at a junction due to carrier emission from traps, which are initially filled at 77 K, is plotted against temperature. The maxima or minima in this profile correspond to majority or minority carrier trap emissions, respectively. The admittance spectroscopy technique proposed by Losee (1972) depends on the measurement of junction admittance as a function of temperature. This technique is, however, limited to majority carrier traps. Another important technique is Isothermal Capacitance Transient Spectroscopy (ICTS) (Okushi and Tokumaru, 1981a). Fourier analysis techniques have also been used to resolve traps with overlapping time constants in a simple capacitance transient technique (Ikeda and Takaoka, 1982), as in the ICTS technique (Ishikawa *et al.*, 1985). A notable and very widely used modification of the transient capacitance technique is Deep Level Transient Spectroscopy (DLTS), proposed by Lang (1974). The spectroscopic nature of this technique and the ease with which it can be performed have made it very useful for deep level studies. A large number of variations and modifications, both in operating principle and in associated instrumentation, of the basic technique have been reported by several authors. A comprehensive review, covering most of these modifications, was made by Neumark and Kosai (1983) in an earlier volume of this series. In addition to the techniques discussed so far, there is another class of experimental methods that characterize the optical ionization properties of deep traps. Photocapacitance measurements (Sah *et al.*, 1969, 1971; Kotina *et al.*, 1969; Furukawa, 1967; Furukawa *et al.*, 1967; Fabre, 1970; Grimmeiss and Olofson, 1969; Kukimoto *et al.*, 1973) are generally made for this purpose. Other commonly used techniques for determining the photoionization properties of deep levels are photo-DLTS (Mooney, 1983), photo-FET measurements (Tegude and Heime, 1984) and transient photocurrent measurements (Grimmeiss, 1974). Because of their importance, we shall discuss the DLTS and photocapacitance techniques in somewhat greater detail.

16. DLTS and Associated Techniques

The original DLTS technique and some variation of it requiring special test structures have found widespread use in the study of MBE-grown materials. This method is superior to earlier TSCAP and admittance spectroscopy techniques in terms of improved flexibility towards measurement requirements and higher sensitivity. The principle of DLTS measurements is based on the physics of thermal emission, capture by traps and the associated junction capacitance variations. It depends on the repetitive filling and emptying of traps by use of positive and negative bias applied to a junction. The processes of carrier emission and capture in deep level traps are illustrated in Fig. 19, and can be briefly described as follows: Under quiescent reverse bias conditions, electron traps in a part of the depletion region of a Schottky

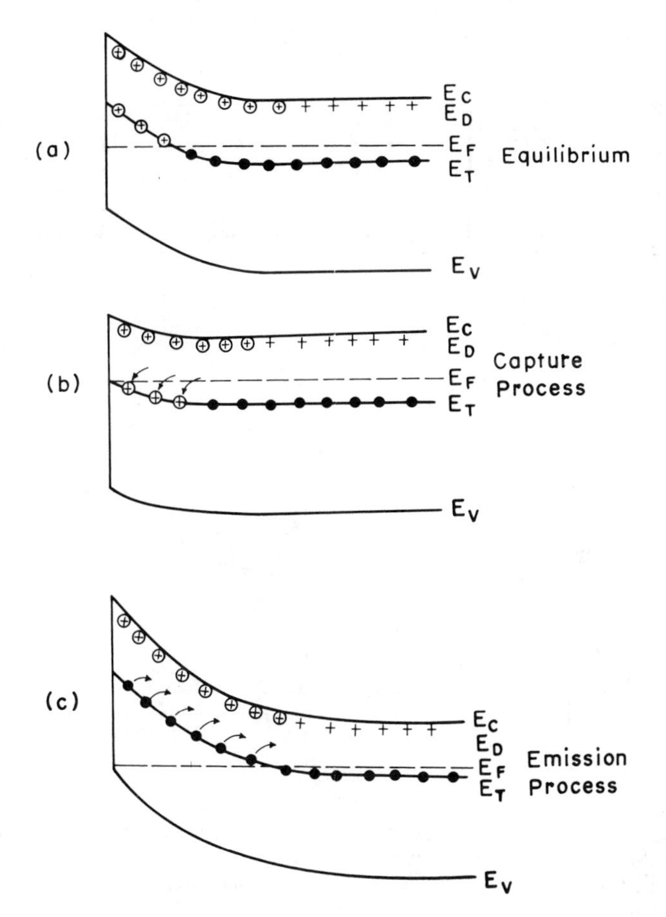

FIG. 19. Illustration of carrier emission and capture at a deep level electron trap located within the depletion region of a junction diode with applied (a) zero bias, (b) forward bias, and (c) reverse bias.

diode are unoccupied. By the application of a forward or zero bias pulse, the traps in this region can be partly or fully filled. On switching back to the quiescent reverse bias, the traps emit electrons to the conduction band with a characteristic time constant, resulting in a transient change of the depletion layer width and diode capacitance. The latter parameter is recorded and analyzed. We omit an elaborate discussion of these aspects, as they have been explained in many books and publications. In Fig. 20(a), we show the pulse cycling used in DLTS measurements, and the corresponding change in junction capacitance is depicted in Fig. 20(b). The major difference between the basic capacitance transient technique and DLTS lies in the use of a

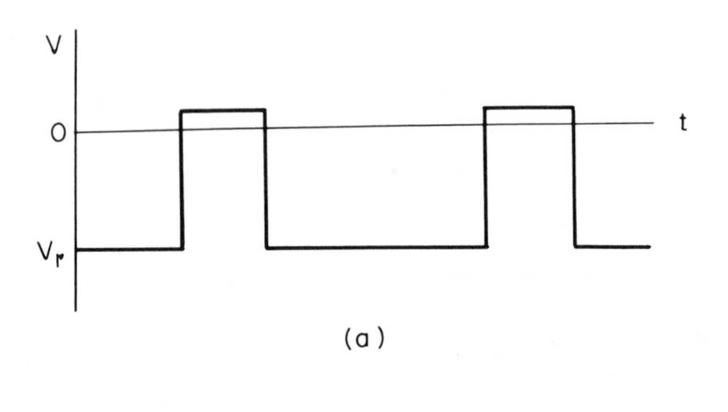

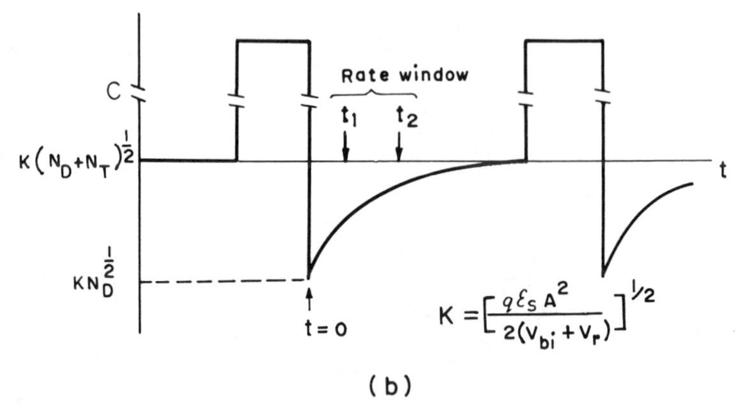

FIG. 20. (a) Pulse cycling used in DLTS measurements, and (b) the corresponding capacitance change at the semiconductor diode junction containing a deep level of density N_T. The test sample is assumed to be n-type with a shallow donor concentration N_D.

"rate window" in the latter, which enables one to "view" the traps in a spectroscopic manner. The choice of the rate window is critical, as it allows a compromise between the detector resolution and the signal to noise ratio (Thomas, 1985). A typical DLTS system that can be implemented with either a Boxcar averager or a lock-in amplifier for signal processing is shown in Fig. 21. Such a practical system suffers from many shortcomings (Day *et al.*, 1979) because of non-fulfilment of the conditions set forth by Lang (1974) and are related either to the device or to technical limitations. One of the basic problems associated with DLTS or any other technique depending on capacitance transients is that the ideal condition of exponential transients is rarely achieved under most experimental conditions. Non-exponential transients arise from large trap densities, electric field effects, non-uniform doping, non-abrupt junctions and depletion-layer edge effects. Okushi and

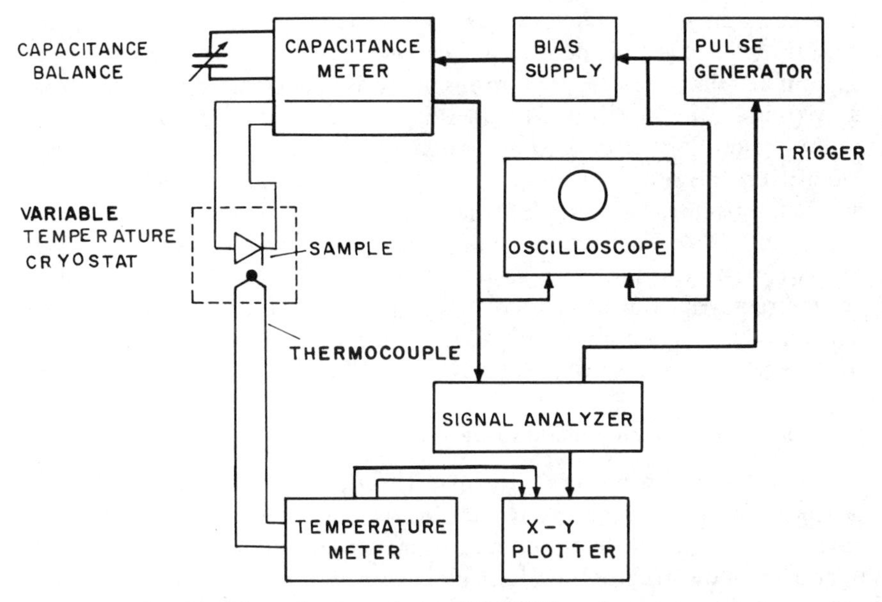

FIG. 21. Schematic of a typical 1 MHz DLTS measurement system.

Tokumaru (1981b) attempted to address the problem of large trap densities by redefining the normalized DLTS signal as

$$S(\mathrm{T}) = \frac{\Delta C^2(t_2) - \Delta C^2(t_1)}{C_0^2}, \tag{11}$$

where, obviously, the assumption $N_T/N_D \gg 1$ is not needed. Bhattacharya (1978) showed that the non-exponential transient can be approximately resolved into two components with a "short" and a "long" time constant, and that the latter was closer to the true value. Mathematical techniques like methods of moments and fast Fourier transforms have been used to analyze non-exponential transients yielding quite reliable data (Kirchner *et al.*, 1981b). Non-exponential transients arising out of the spatial dependence of capture cross section in the depletion region tail have been discussed by Zylbersztejn (1978), and remedies for the same have been proposed. The same phenomenon has also been used to profile accurately deep level densities (Gombia *et al.*, 1985).

The pulse cycling used in the DLTS measurements gives rise to a current transient through the diode, which can be expressed as

$$i(t) = \frac{qWAN_T}{2\tau}\exp\left(-\frac{t}{\tau}\right) + I_L, \tag{12}$$

where I_L is the steady state diode leakage current for the applied quiescent bias, W is the change in depletion-layer width, A is the diode area, and τ is the emission time constant. This forms the basis of current transient spectroscopy (Borsuk and Swanson, 1980), which, under certain conditions, gives higher resolution compared to capacitance spectroscopy techniques. Modified forms of this technique, utilizing drain current transients of large gate field-effect transistors (FET), have been used by several research groups (Chi *et al.*, 1984; Valois and Robinson, 1983; Takikawa and Ozeki, 1985) to characterize traps in MBE-grown $Al_xGa_{1-x}As$–GaAs modulation-doped heterostructures. For characterizing traps in MBE-grown multilayered structures, several variations of the DLTS technique have been used, and we shall discuss them in subsequent sections.

17. Photocapacitance Technique

Very little has been reported in the literature on the optical ionization properties of deep level traps in MBE-grown materials. We shall, however, have a brief discussion on the photocapacitance technique, which has proved to be a very powerful tool for determining the photoionization properties of commonly observed traps in III–V semiconductors.

The basic principle involved in this measurement technique consists of photo-exiting carriers from traps in the depletion region of a reverse-biased Schottky junction with extrinsic light of frequency ν. The sample is held at sufficiently low temperatures that thermal emission and capture processes are insignificant. The photo-excited carriers give rise to a capacitance transient corresponding to a carrier emission rate given by

$$e^{O}(h\nu) = \sigma^{O}(h\nu)\phi(h\nu). \tag{13}$$

For each incident photon energy $h\nu$, the photoionization cross section σ^{O} can be computed by using (13). Analysis of the photoionization cross section data, as described in Section III, yields most of the useful optical parameters of deep level traps.

V. Commonly Observed Deep Levels

18. GaAs

The first systematic study of electron traps in MBE-grown GaAs doped with Si, Sn and Ge was done by Lang and coworkers (1976), using Deep Level Transient Spectroscopy. They observed a total of nine electron traps labeled M0 through M8. All the samples studied by them except one were grown under As-stabilized conditions. The DLTS spectra for the electron traps for samples grown in four different MBE systems with three different dopants,

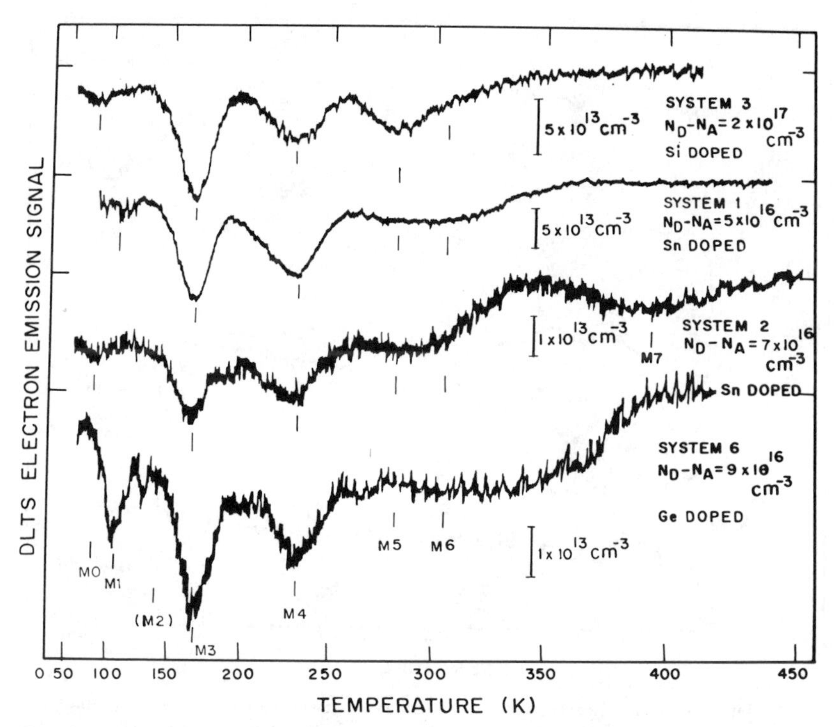

FIG. 22. DLTS spectra of electron traps (negative peaks) in MBE *n*-GaAs grown in four different systems with three different dopants (Si, Sn and Ge) under As-rich growth conditions. The rate window is 51 sec^{-1}. The trap concentration scale is indicated for each trace. [From Lang *et al.* (1976).]

as reported by these authors, are reproduced in Fig. 22. It should be noted that Fig. 22 presents DLTS data for samples grown under As-rich conditions where the traps M2 and M8 are noticeably absent. Traps M1, M3 and M4 are believed to be the dominant levels in MBE GaAs grown under As-stabilized conditions. It is also clear from Fig. 22 that the general behavior of the traps is more or less independent of the dopant species used and of the growth system. However, there is a marked difference between the DLTS spectra of samples grown under As-stabilized conditions and of samples grown under Ga-stabilized conditions. As seen from Fig. 23, the dominant traps in GaAs grown under Ga-stabilized conditions are M2 and M8. This observation is also supported by photoluminescence studies on undoped GaAs grown under these two conditions (Cho and Hayashi, 1971; Ilegems and Dingle, 1975). Activation energies and capture cross-sections of the traps giving well-resolved DLTS peaks were deduced from the temperature variation of emission rates (Fig. 24). For other traps that were just barely identified in

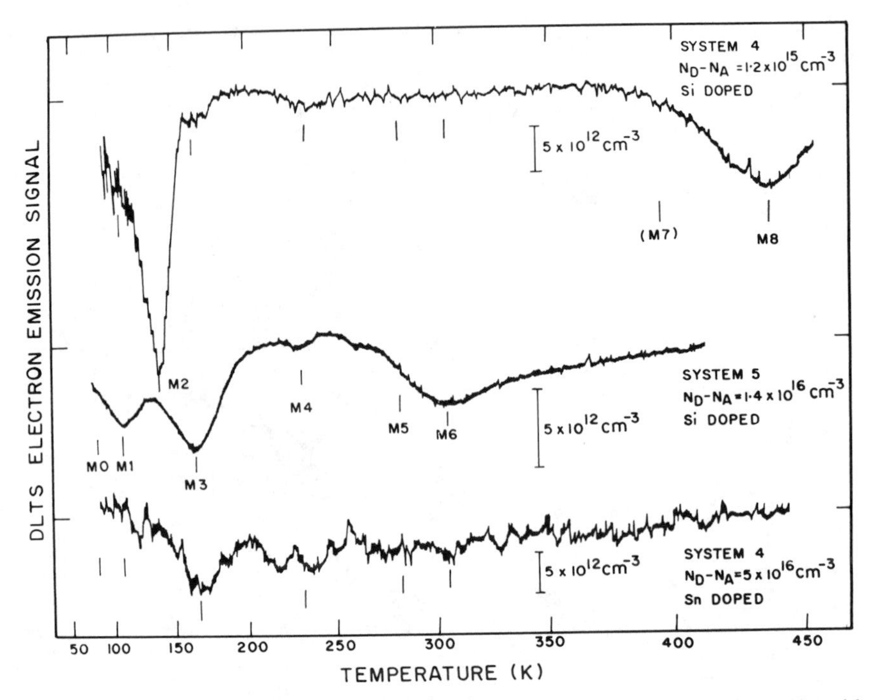

FIG. 23. DLTS spectra of electron traps in MBE n-GaAs showing two As-rich samples with lowest trap concentrations (lower two traces) and a Ga-rich sample (upper trace). The rate window is 51 sec^{-1}. The trap concentration scale is indicated for each trace [From Lang *et al.* (1976).]

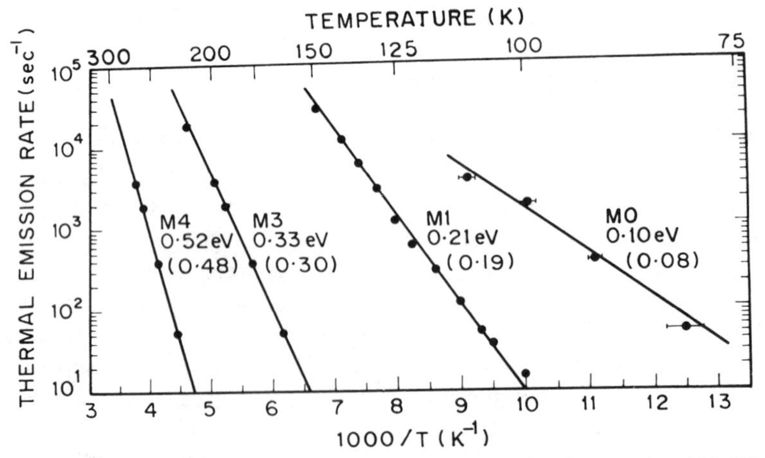

FIG. 24. Thermal emission rates versus inverse temperature for electron traps M0, M1, M3 and M4. The activation energies shown by each curve are ΔE_{meas} and, in parenthesis, $\Delta E = \Delta E_{meas} - 2kT$. [From Lang *et al.* (1976).]

TABLE I

MEASURED PROPERTIES OF ELECTRON TRAPS IN MBE n-GaAs. [From Lang *et al.* (1976).]

	MBE electron trap parameters			
Electron trap	E_{meas} [a] (eV)	$E_T = E_{meas} - 2kT$ [b] (eV)	$\sigma_\infty \langle v \rangle N_c / g$ [c] (sec^{-1})	$E_T \simeq 23.7kT \pm 10\%$ [d] (eV)
M0	0.10	0.08	1.5×10^7	—
M1	0.21	0.19	6.1×10^{10}	—
M2	—	—	—	0.29
M3	0.33	0.30	1.3×10^{11}	—
M4	0.52	0.48	4.0×10^{12}	—
M5	—	—	—	0.58
M6	—	—	—	0.62
M7	—	—	—	0.81
M8	—	—	—	0.85

[a] E_{meas} is the slope of the log of electron thermal emission rate versus inverse temperature.
[b] E_T is the emission activation energy corrected for the T^2 dependence of the prefactor.
[c] $\sigma_\infty \langle v \rangle N_c$ is the prefactor of the thermal emission rate at the temperature of measurement.
[d] The E_T values in the right-hand column are approximations derived from the DLTS peak position.

the DLTS spectra, the activation energy E_T was calculated using the empirical relationship $E_T = 23.7kT$. These values are presented in Table I. The concentrations of the traps are usually low, in the 10^{13} cm^{-3} range. Lang *et al.* (1976) concluded that the traps observed in MBE-grown GaAs are not simple native defects but are either related to chemical impurities in the growth system or complexes involving impurities and native defects. Neave *et al.* (1980) studied electron traps in MBE GaAs grown with As_2 and As_4 fluxes and found that the dominant traps M1, M3 and M4 have much higher densities in samples grown with As_4 than in those grown with As_2. They also showed that it was possible to grow GaAs at lower temperatures, using the As_2 source only so that the density of the deep levels is not significant. Regarding the variation of the density of deep states with two different As atomic species, these authors speculated that the density might be a direct consequence of the greater arsenic surface population obtainable with As_2 than with As_4. As-rich surfaces will have fewer number of vacancies, leading to a possible reduction in the trap concentrations.

The characteristics of traps in GaAs grown at different substrate temperatures were studied by Stall *et al.* (1980), who found that by lowering the growth temperature below 475°C, new traps involving Ga-vacancy complexes are created. At lower growth temperatures, the concentrations of

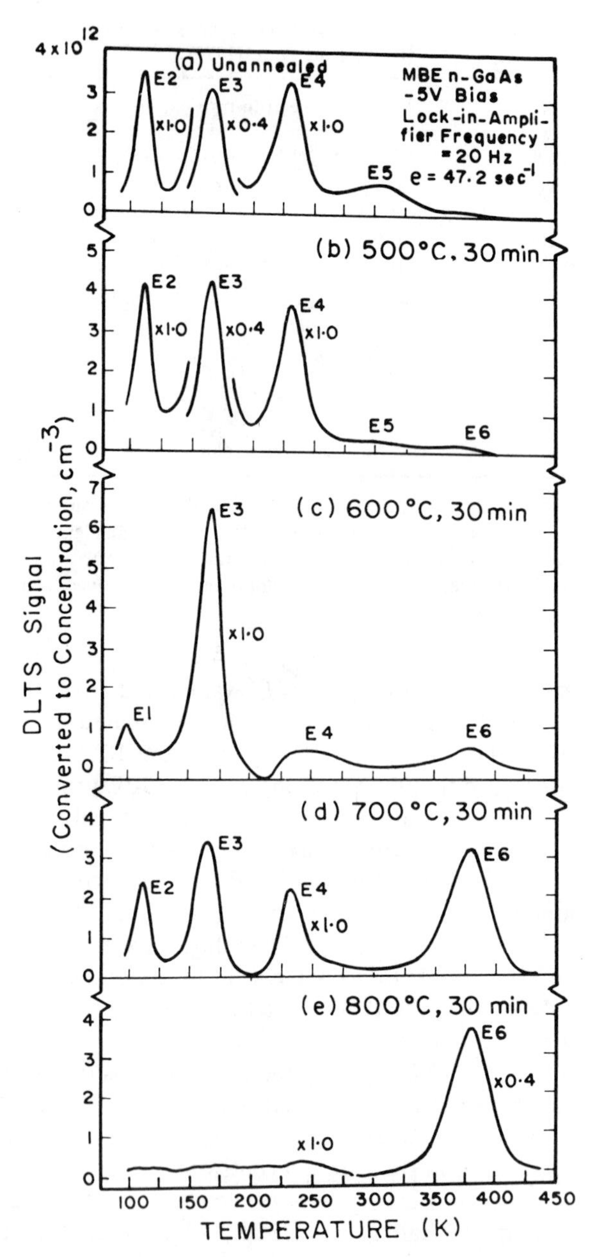

FIG. 25. DLTS spectra for MBE n-GaAs as a function of half-hour isochronal annealing. The samples were encapsulated on the top surface with 1200 Å of Si_3N_4 [From Day *et al.* (1981).]

M1, M3 and M4 increased, and at 460° and 430°C, the electron trap EB3, usually observed in electron irradiated material (Martin *et al.*, 1977), is detected along with a hole trap labeled HB6 in the literature (Mitonneau *et al.*, 1977). At a growth temperature of 300°C, electron trap EB6, identified as a simple Ga-vacancy (Lang *et al.*, 1977), is created together with trap EL3, observed in VPE materials (Martin *et al.*, 1977). At a growth temperature of 550°C, hole traps created by impurity diffusion from the substrate were observed. Hole traps HL9, HB4 (due to copper outdiffusion from the substrate), HB1 and Cr(A) (both due to Cr outdiffusion from the substrate) were observed. Day *et al.* (1981) performed high temperature annealing experiments on unintentionally doped *n*-type GaAs protected by a Si_3N_4 cap, and they observed the emergence of the well-known EL2 trap (Martin *et al.*, 1977) for anneals above 600°C. Figure 25 shows the gradual emergence of the E6 peak (EL2 level) after a series of anneals at gradually elevated temperatures. The result was explained in terms of outdiffusion of Ga into Si_3N_4, resulting in Ga-vacancies. It is also interesting to note that the other five traps observed in the same DLTS spectra, which are identical to the traps detected by Lang *et al.* (1976), disappear upon high-temperature annealing, which suggests that they are simple defects or complexes involving point defects. The activation energies and capture cross sections for the traps were measured by Day *et al.* (1981), and are summarized in Table II. Traps E2, E3 and E4 are identical to the traps M1, M3 and M4 observed by Lang *et al.* (1976). The appearance of trap E1 [Fig. 25(c)] is probably due to annealing and diffusion effects, and its presence masks the peak due to trap E2.

Deep levels in Be-doped GaAs grown by MBE were studied in detail by Bhattacharya *et al.* (1982). Figure 26 shows the Arrhenius plots of the electron and hole traps detected by these researchers in undoped, Be-doped and Sn-doped samples, and the trap characteristics are listed in Table III.

TABLE II

CHARACTERISTICS OF ELECTRON TRAPS OBSERVED IN MBE *n*-GaAs. [From Day *et al.* (1981).]

Trap	Activation energy E_T (eV)	Capture cross-section σ_∞ (cm^2)	Temperature range (K)
E1	0.16	5.0×10^{-15}	100–120
E2(M1)	0.18	9.7×10^{-16}	105–130
E3(M3)	0.32	2.6×10^{-14}	160–190
E4(M4)	0.51	5.1×10^{-13}	225–260
E5	0.76	2.3×10^{-13}	340–390
E6	0.83	1.8×10^{-13}	370–420

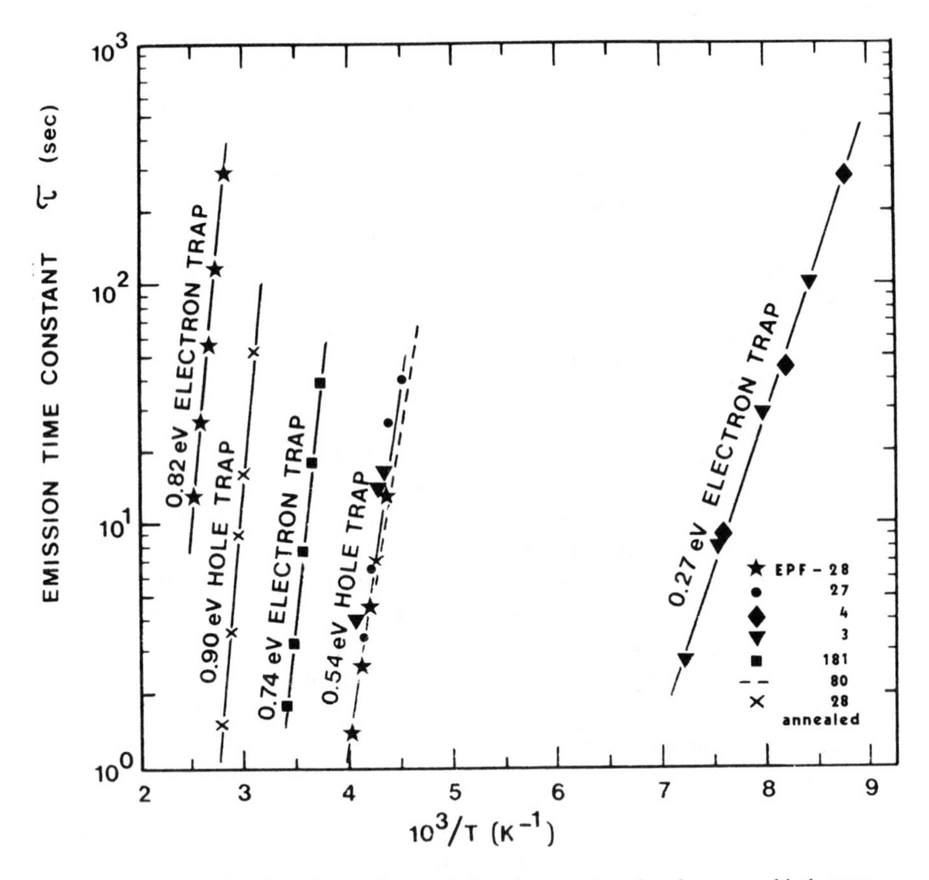

FIG. 26. Temperature dependence of the emission-time constant for electron and hole traps detected in undoped, Be-doped and Sn-doped MBE GaAs.

Besides the occurrence of the usual "M" levels, a hole trap with activation energy of 0.54 eV was consistently observed in all samples. Following Martin *et al.* (1977), this trap was attributed by Bhattacharya *et al.* (1982) to Fe impurities. Annealing enhanced the concentration of this trap further, and another deep hole trap, with an activation energy of 0.90 eV, possibly related to As-vacancies, appeared. Bhattacharya *et al.* (1982) also reported a very shallow electron trap having $E_T \simeq 0.10$ eV, with an unusually small capture cross-section ($1.2 \times 10^{-22}\ \text{cm}^{-2}$). The origin of this level was not clearly understood.

Amano *et al.* (1984) used DLTS and photocapacitance techniques on undoped GaAs to show that the density of electron traps are minimized for an As/Ga ratio of 3 : 7 under As-stabilized growth conditions, and a similar

TABLE III

CHARACTERISTICS OF DEEP-LEVEL TRAPS IN AS-GROWN AND HEAT-TREATED MBE GaAs [From Bhattacharya *et al.* (1982).]

MBE sample	Trap	Activation energy E_T (eV)	Thermal capture cross section[a] σ_∞ (cm^2)	Concentration N_T ($10^{13}\ cm^{-3}$)
GaAs : Be	Electron trap	0.82	1.6×10^{-16}	2.6
GaAs : Be	Hole trap	0.54	3.1×10^{-16}	0.8–16.0
GaAs : Sn	Electron trap	0.27	5.9×10^{-14}	0.5–0.9
GaAs : Sn	Hole trap	0.54	3.1×10^{-16}	0.9
Undoped GaAs (*n*-type)	Electron trap	0.74	1.6×10^{-13}	21.0
(As above)	Hole trap	0.54	3.0×10^{-16}	1.4
Heat-treated GaAs : Be	Hole trap	0.90	3.1×10^{-13}	3.3
(As above)	Hole trap	0.54	3.0×10^{-16}	17.0

[a] Obtained from the emission prefactor.

trend was observed for the background carrier concentration. Figure 27(a) and (b) illustrate the data. The electron traps with activation energies of 0.29, 0.35, 0.48 and 0.58 eV in Fig. 27(a) correspond to the trap levels labeled M2, M3, M4 and M5, respectively, by Lang *et al.* (1976). M2 was observed only in sample grown with an As/Ga ratio of 3. M3 was observed only in samples grown with an As/Ga ratio of 5. M4 was observed, except in samples grown with an As/Ga ratio of 3.7. M5 was observed in all samples, and, in the sample grown with an As/Ga ratio of 5, the concentration of M5 is greater than $3 \times 10^{13}\ cm^{-3}$. As-stabilized conditions during growth are maintained for As/Ga greater than 3.1, and Ga-stabilized conditions are maintained below this flux ratio. The horizontal line and arrow at As/Ga = 3.7 indicate the trap detection limit for the background carrier concentration measured for this flux ratio. Three additional traps detected by PHCAP measurements, shown in Fig. 27(b), show the same behavior with respect to the As/Ga ratio. Lang *et al.* (1976) have concluded that most of these traps are related to chemical impurities. Two unintentional impurities identified by Kop'ev *et al.* (1984) are C and Mn. An increase in growth temperatures reduced the density of both, whereas an increase in As pressure reduced the density of C but increased that of Mn.

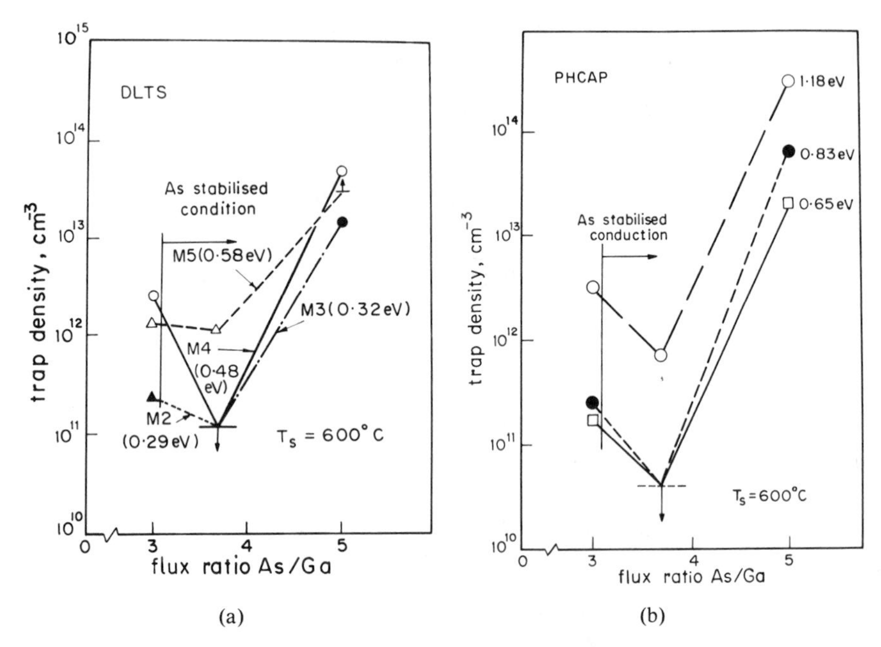

FIG. 27. As/Ga flux ratio dependence of densities for various electron traps measured by (a) DLTS, and (b) photocapacitance (PHCAP) for undoped GaAs layers grown at 600°C. [From Amano *et al.* (1984).]

Blood and Harris (1984) have made an extensive investigation of electron and hole traps in MBE GaAs, using transient capacitance techniques. In *p*-type materials, they found levels attributed to Fe and Cu. In addition, they detected the dominant M1, M3 and M4 levels and computed their characteristics at low electric fields. These characteristics were compared with previously published results. This comparison is shown in Fig. 28. These authors concluded that M1 and M4 are impurity-defect complexes, possibly involving As-vacancies. The trap M2 appears at growth temperatures above 600°C, and its concentration increases with increasing growth temperature and a decreasing As/Ga ratio. Therefore it is supposed that trap M2 is related to defect complexes involving As-vacancies. The general assignment of the "M" levels to impurity related defect complexes is, however, contradicted by Skromme *et al.* (1985). These authors have recently performed extensive constant-capacitance DLTS, capacitance-voltage, photoluminescence and photo-thermal ionization spectroscopy measurements on high purity and lightly Si-doped MBE GaAs. It was observed that trap concentrations in these materials are apparently not dependent on background or intentional doping levels, and are relatively constant. It was therefore suggested that the

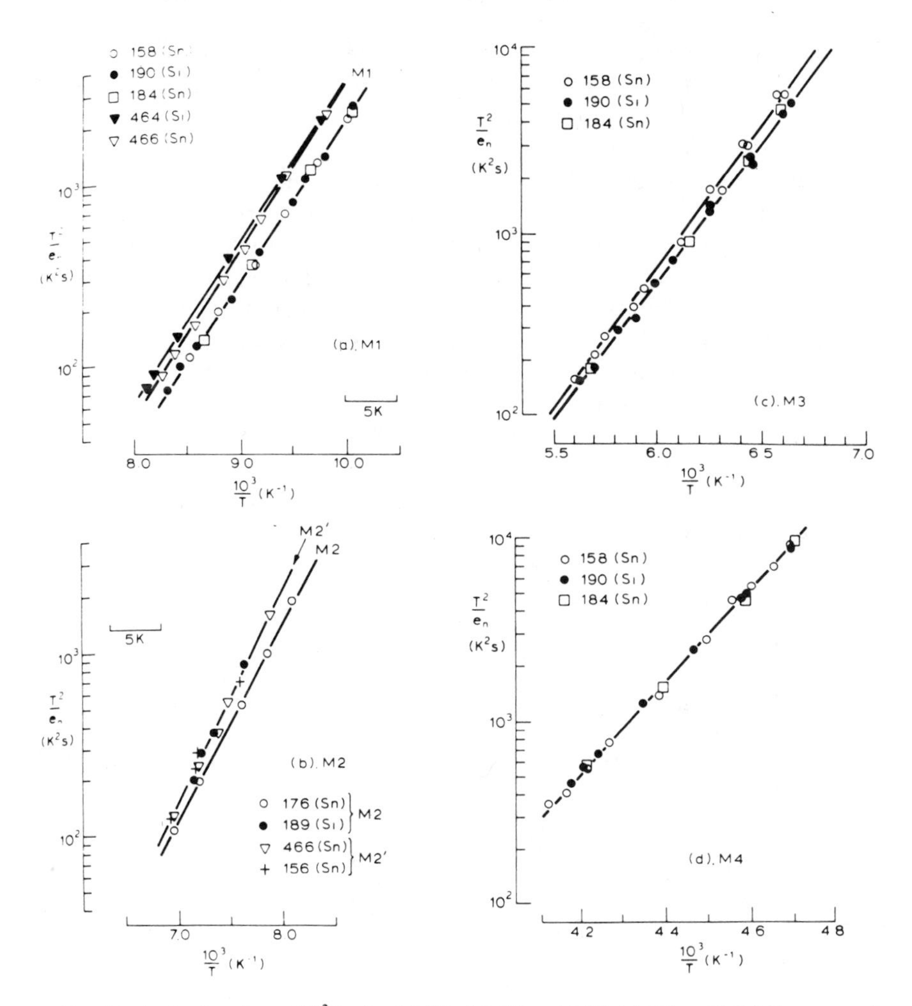

FIG. 28. Arrhenius plots of T^2/e_n for (a) $M1$, (b) $M2$ and $M2'$, (c) $M3$ and (d) $M4$, measured at low electric fields, and shown as experimental points with lines [From Blood and Harris (1984).]

observed deep levels in MBE GaAs are not impurity-related. In addition, the same group (De Jule *et al.*, 1985) reported the presence of a very shallow electron trap with a thermal activation energy of 30 meV in these materials. The origin of this level has not yet been established.

We have recently investigated the deep levels present in MBE-grown GaAs isoelectronically doped with In. Isoelectronic doping of GaAs with In has previously been used for LEC-grown bulk GaAs (Jacob, 1982; Mitchel and Yu, 1985) to reduce the defect densities, and, more recently, for LPE-grown layers (Narozny and Beneking, 1985). A series of GaAs layers were grown at 570°C, with the In content varying in the range 0.2–1.2%. Typical DLTS data for a sample with 0.9% In are shown in Fig. 29. All the "M" levels observed in MBE GaAs (Lang *et al.*, 1976) are present in this material, with concentrations $\sim 10^{13}\ cm^{-3}$. The striking observation is that the concentrations of all levels decrease by varying amounts as the In content in the material is increased. This is illustrated in Fig. 30 for the four dominant electron traps. It is very likely that In atoms, with a surface migration rate 20 times higher than that of Ga at the growth temperature used, reduce Ga vacancies in MBE GaAs. Our data strongly suggest that the electron traps in MBE GaAs are related to such defects—a conclusion that is also indirectly evident from the observations of Skromme *et al.* (1985). For higher In content materials, we also observe a sharp DLTS peak at around 77 K. We believe this peak is due to the 0.16 eV center observed by us in MBE-grown $In_{0.53}Ga_{0.47}As$ layers over InP, and this study indicates that the center is probably related to In or impurities therein.

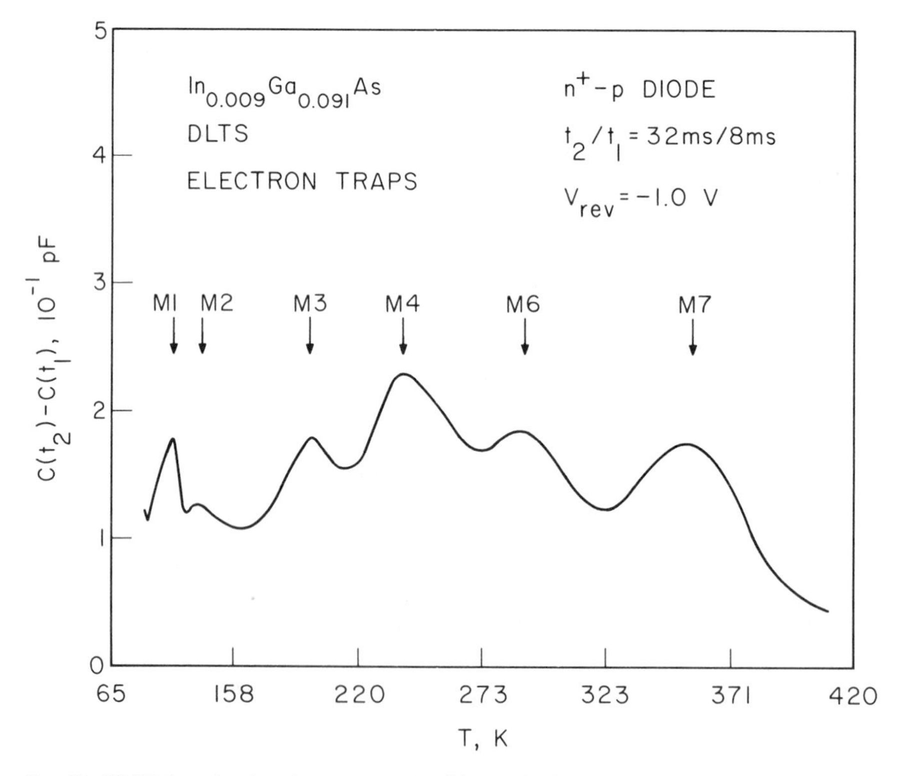

FIG. 29. DLTS data showing electron traps (positive peaks) identified in In-doped MBE GaAs.

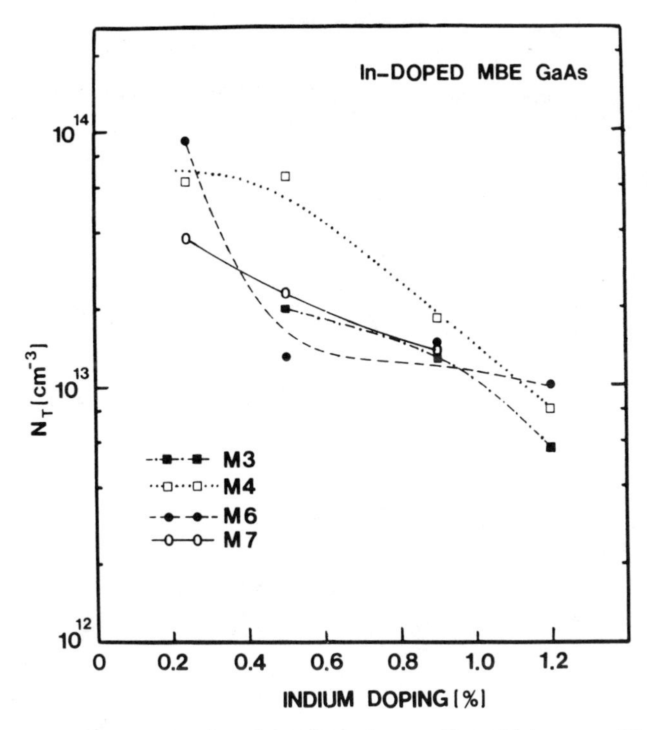

FIG. 30. Decrease in concentration of dominant electron traps with increase of In content in MBE GaAs.

19. $Al_xGa_{1-x}As$

a. Common Traps

The $Al_xGa_{1-x}As$ alloy system is particularly important for its application in high-speed electronic and optical devices (Mimura *et al.*, 1980, 1981; Lee *et al.*, 1983; Solomon and Morkoc, 1984; Cho and Casey, 1974; Cho *et al.*, 1976; Tsang and Ditzenberger, 1981; Tsang, 1981b; Tsang *et al.*, 1983; Capasso *et al.*, 1981, 1982). An excellent review of the properties of this material has recently been made by Adachi (1985). A study of deep levels in $Al_xGa_{1-x}As$ and of exploring ways to reduce their concentration is often found to be necessary, as some of the deep levels present in MBE $Al_xGa_{1-x}As$ are known to adversely affect heterojunction device performance (Störmer *et al.*, 1979; Valois *et al.*, 1983; Fischer *et al.*, 1984).

Deep level traps in MBE $Al_xGa_{1-x}As$ were first reported by Hikosaka *et al.*, (1982), who investigated Si doped material with $0 \leq x \leq 0.4$, grown at a substrate temperature of 580–680°C under As-stabilzed conditions. Six electron traps, labeled E1 through E6, were detected, as shown in Fig. 31.

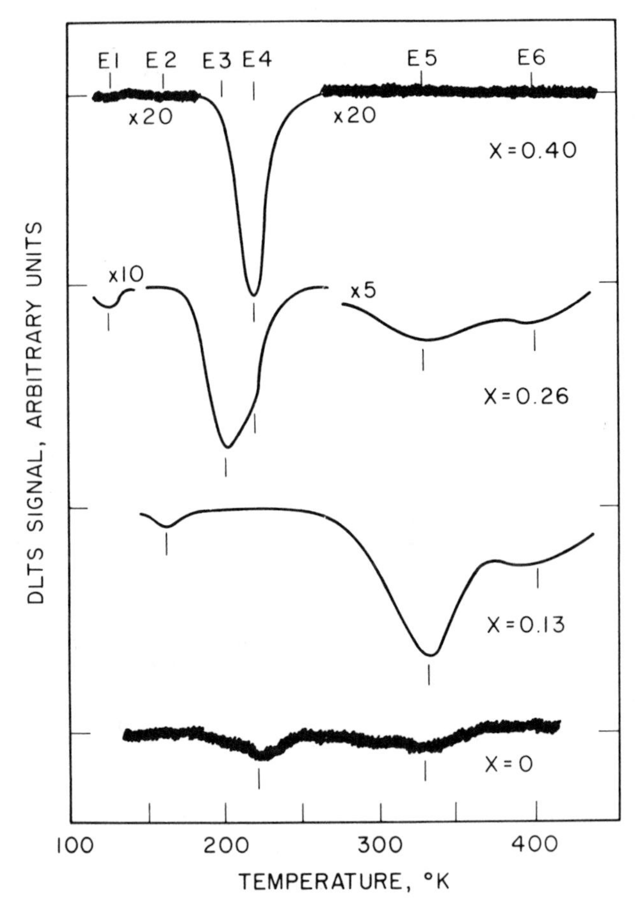

FIG. 31. Typical DLTS spectra of MBE-grown $Al_xGa_{1-x}As$ with $x = 0, 0.13, 0.26$ and 0.40. The rate window is 231 sec^{-1}. [From Hikosaka *et al.* (1982).]

The temperature dependence of thermal emission rates and the measured characteristics for these traps are depicted in Fig. 32 and Table IV, respectively. The density of trap E6 is found to decrease with increasing growth temperatures (McAfee *et al.*, 1981) as is evident from Fig. 8(a). This is accompanied by an increase in PL intensity in the same samples, and, therefore, it was concluded that trap E6 contributes to non-radiative recombination (McAfee *et al.*, 1981; Zhou *et al.*, 1982). Naritsuka *et al.* (1984) further observed that the concentration of E6 also decreases with a decreasing V/III flux ratio (Fig. 33), from which observation they concluded that the trap is related to a Ga vacancy or to a Ga vacancy complex. Recent data from the same group of authors (Naritsuka *et al.*, 1985) indicate more

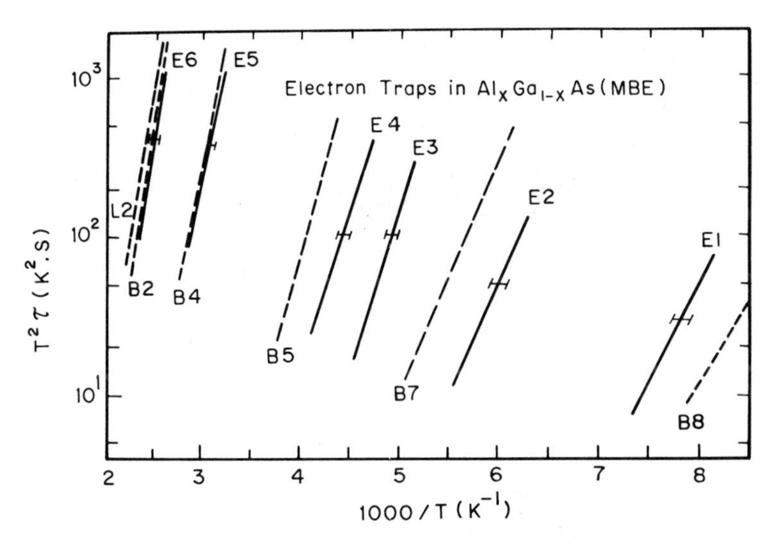

FIG. 32. Thermal emission time constant versus inverse temperature for the traps found in $Al_xGa_{1-x}As$. The horizontal error bars show the sample-to-sample scattering in levels. [From Hikosaka *et al.* (1982).]

specifically that trap E6 is due to a complex associated with a Ga vacancy and an oxygen atom. Most of the traps in MBE AlGaAs are found to have a dependence on alloy composition, as is evident from Fig. 34, and on high temperature annealing (Hikosaka *et al.*, 1982). In contrast with GaAs (Neave *et al.*, 1980), a dependence of trap densities on As specie has not been observed for higher growth temperatures (Mooney *et al.*, 1985a). Chen *et al.* (1985), however, observed a reduction by a factor of two in trap concentrations by using As_2 instead of As_4 at growth temperatures of 610°C or 640°C. A similar reduction was observed at higher growth temperatures only if the

TABLE IV

ELECTRONIC PROPERTIES OF DEEP LEVELS IN $Al_xGa_{1-x}As$. [From Hikosaka *et al.* (1981).]

Level	Activation energy (eV)	Capture cross section (cm^2)
E1	0.26 ± 0.03	2×10^{-12}
E2	0.28	2×10^{-14}
E3	0.42	9×10^{-13}
E4	0.40	3×10^{-14}
E5	0.60	2×10^{-14}
E6	0.78	5×10^{-14}

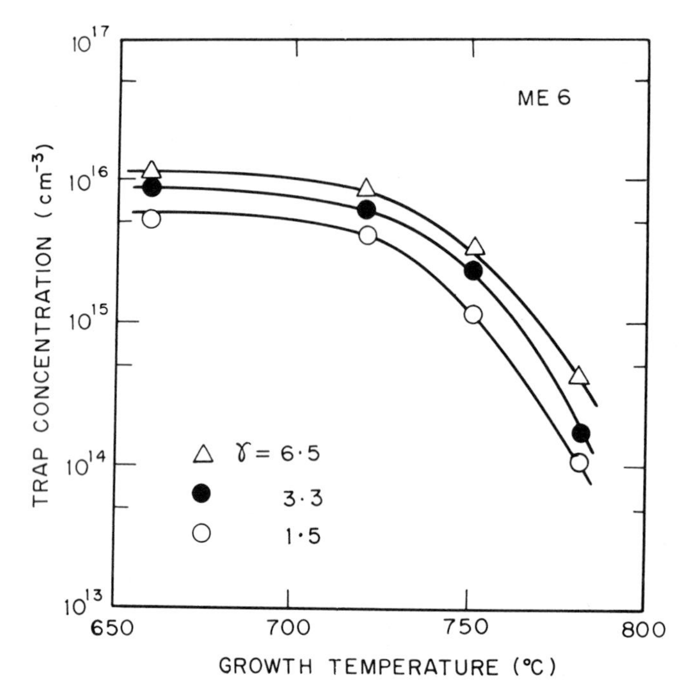

FIG. 33. Growth temperature dependence of the ME6 trap concentration in Si-doped $Al_{0.2}Ga_{0.8}As$ grown at group V/III beam flux ratios $\gamma = 1.5$, 3.3 and 6.5. [From Naritsuka *et al.* (1984).]

V/III flux ratio was increased. From such data, it was concluded that the trap levels E5 and E6 are related to Ga vacancies, and the trap E1 is linked to an As vacancy. Akimoto *et al.* have recently reported a deep trap with $E_T = 0.76$ eV in MBE $Al_{0.3}Ga_{0.7}As$ grown at a substrate temperature of 500°C. The concentration of this trap increases with increasing V/III ratio and alloy composition x, but remains constant with Si-doping. The photoluminescence intensity of the samples was found to be inversely proportional to the trap concentration. It was concluded that Al–O complexes are responsible for the formation of this deep trap. Data reported by Naritsuka *et al.* (1985) indicate that traps in MBE AlGaAs are strongly dependent on volatile impurities in the As source used. This is, however, not true for the 0.40 eV level, termed E3 by Hikosaka *et al.* (1986). This trap was identified with the D–X center (Lang *et al.*, 1979), which is, in fact, the dominant center in the ternary alloys. This center plays an important role in the performance characteristics of several $Al_xGa_{1-x}As$–GaAs heterojunction devices. It is, therefore, of interest to discuss some of the extraordinary carrier-dynamics associated with this trap in detail.

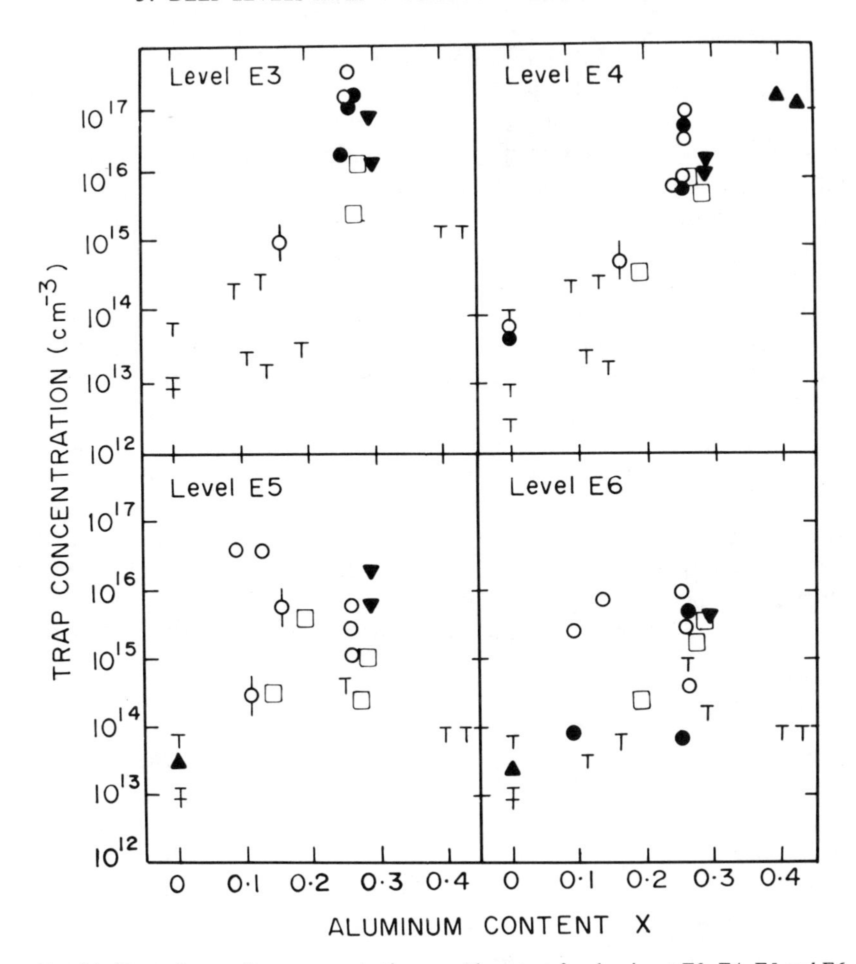

FIG. 34. Dependence of trap concentrations on Al content for dominant E3, E4, E5 and E6 levels. The symbols represent the following: (○) as-grown 580–600°C; (△) as-grown 610–620°C; (▽) as-grown 640°C; (□) as-grown 680°C; (○▲▼) annealed 800°C; (T) detection limit. [From Hikosaka *et al.* (1982).]

b. The D–X Center

This deep donor in $Al_xGa_{1-x}As$ ($0.25 \leq x \leq 1.0$) has been investigated and characterized by temperature- and pressure-dependent Hall, transient capacitance, photocapacitance, Deep Level Transient Spectroscopy and low temperature photoluminescence measurements. Most of the early work was done with LPE crystals for their application to lasers; more recently, work is being done with MOCVD and MBE crystals for their application to lasers and heterostructure electronic devices.

Typical values of the equilibrium donor binding energy E_0, the thermal and optical ionization energies E_T and E^O, and the thermal capture barrier energy E_B in an alloy with $x \approx 0.3$ are illustrated in Fig. 35. This is a modified configuration coordinate diagram constructed to fit the data from $Al_xGa_{1-x}As$ with $x \approx 0.25$–0.30. These and related data led Lang *et al.* (1979) to postulate that the microscopic nature of the center consists of a donor atom coupled to an As vacancy—Hence, the label D–X. This model is supported by the observed chemical shifts in the ionization energies caused by changing the donor atom species (Lang and Logan, 1979; Kumagai *et al.*, 1984), and from the symmetry of the defects determined from the absorption and scattering of ballistic phonons (Narayanamurti *et al.*, 1979). However, from pressure-dependent DLTS data on Sn- and Si-doped GaAs and $Al_xGa_{1-x}As$, Mizuta *et al.* (1985) have concluded that the D–X center is a simple substitutional donor. This point is still being debated. In fact, recent data from DLTS measurements on MOCVD $Al_xGa_{1-x}As$ by Bhattacharya *et al.* (1984a) suggest that Al or Al–O complexes could be a part of the microscopic model of the D–X center. It was also shown conclusively by Bhattacharya *et al.* (1979) that the deep centers are made up of L-valley wave functions. This has been subsequently confirmed (Saxena, 1980a; Henning *et al.*, 1984).

As x is increasd from zero in the alloys, an effective mass-like donor is first identified. At $x \approx 0.24$ the deep levels emerge, which, at $x > 0.3$, become the dominant contributor of electrons to the conduction band. The deep level

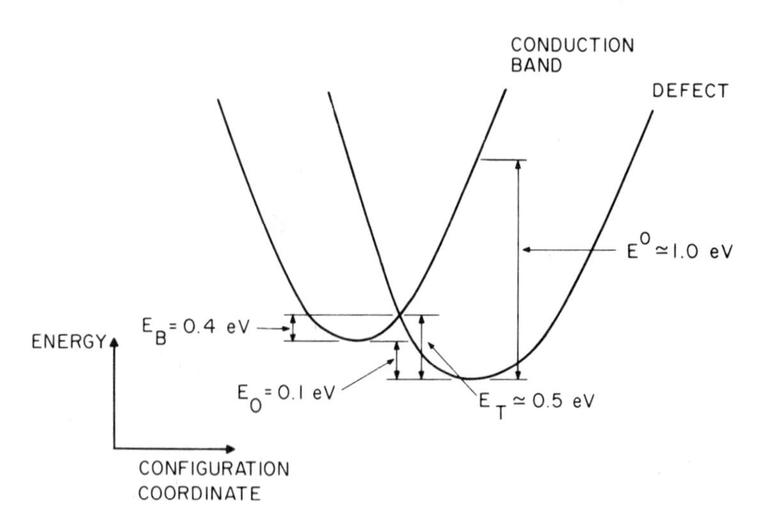

FIG. 35. Configuration coordinate diagram to fit the typical D–X center parameters from $Al_xGa_{1-x}As$ with $x = 0.25$–0.30.

concentration subsequently decreases in high Al-content materials, since the L-donor state is no more the lowest. The thermal ionization energy of the deep levels also reaches a maximum at $x \approx 0.45$, due to the proximity and superposition of Bloch waves from the Γ, L and X minima, and then decreases with increasing x.

The large relaxation of the defect gives rise to an unusually large (~1 eV) Franck–Condon shift, which is evident in Fig. 35. Furthermore, the D–X center in $Al_xGa_{1-x}As$ couples strongly with transverse-acoustic (TA) phonons at the edge of the Brillouin zone, in contrast with most well-known non-radiative deep levels, which couple with longitudinal-optical (LO) phonons (Lang, 1980). The measured capture cross-section of the defect is several orders smaller than most well-known deep centers. These properties give rise to the phenomenon of persistent photoconductivity in the alloys (Lang and Logan, 1977). Such photoconductivity can only be quenched by providing thermal energy to the lattice.

c. *D–X Center in MBE* $Al_xGa_{1-x}As$

This electron trap, which also behaves as a dominant deep donor in the alloys, has been studied by Hall-effect measurements in LPE (Saxena, 1980b, 1981), MOCVD (Bhattacharya *et al.*, 1984b) and MBE (Ishikawa *et al.*, 1982; Ishibashi *et al.*, 1982; Thorn *et al.*, 1982) crystals. Watanabe *et al.* (1984) used the DLTS technique to demonstrate the coexistence of this deep donor with a shallow donor level in Si-doped $Al_xGa_{1-x}As$ for a certain range of x, which explained some observed anomalies in Hall data (Watanabe and Maeda, 1984).

Usual DLTS data of Si-doped $Al_xGa_{1-x}As$ with $x > 0.2$ show either a single peak or two closely-spaced peaks, both believed to originate from the D–X center (Zhou *et al.*, 1982; Subramaniam *et al.*, 1985a; Dhar *et al.*, 1986a). Such multiple D–X centers have also been observed in MOCVD $Al_xGa_{1-x}As$ by us and by other groups (Sakamoto *et al.*, 1985). Theoretical curve-fitting of DLTS data indicate the existence of three levels related to the D–X center (Ohno *et al.*, 1985). The data reported by Zhou *et al.* (1982) are reproduced in Fig. 36. This figure shows two overlapping peaks, which were analyzed theoretically to yield the capture and emission properties (Figs. 37 and 38) of the two associated levels separately. Our own data (Dhar *et al.*, 1986a) in Figs. 39(a) and (b) illustrate that, by proper choice of doping, it is possible to get well-resolved peaks due to the two levels separately, and to characterize them experimentally. This is more apparent from Fig. 40, which shows the dependence of the concentration of the two levels on Si-doping. Experimental data on the emission and capture properties of these two levels (Fig. 41) clearly show that their carrier dynamics are distinctly separated.

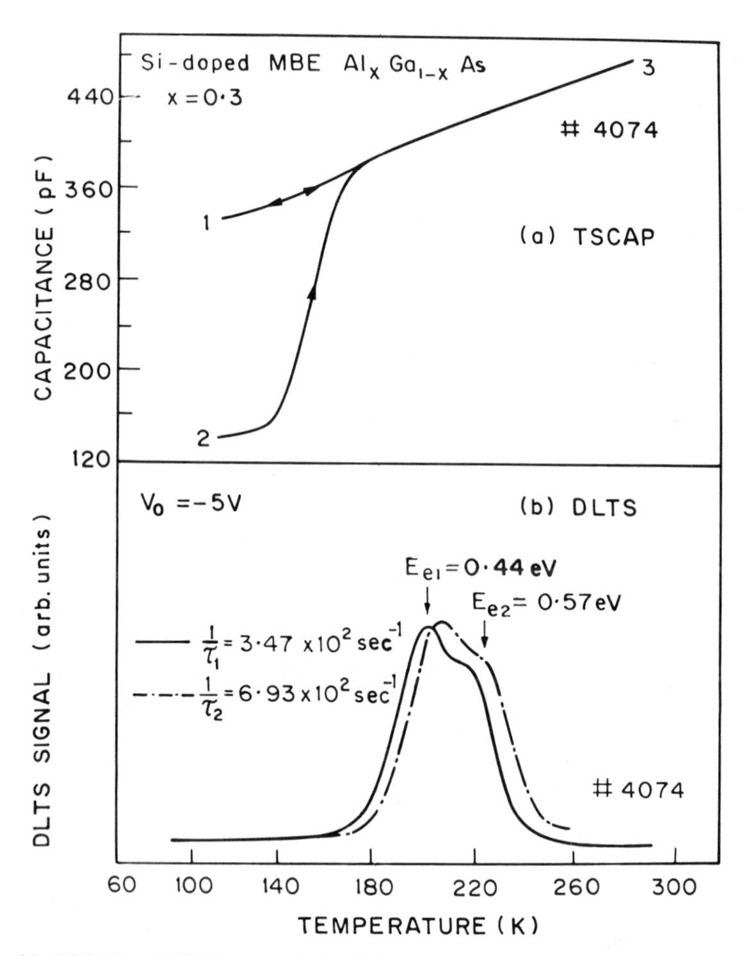

FIG. 36. TSCAP and DLTS curves for a Si-doped MBE $Al_xGa_{1-x}As$ sample with $x = 0.3$. The TSCAP heating rate was about 4 K/min. The DLTS rate windows are $3.47 \times 10^2\ sec^{-1}$ (solid line) and $6.93 \times 10^2\ sec^{-1}$ (dashed line). The bias voltage is -5 V. E_{e1} and E_{e2} are the activation energies for the electron emission rates from two D–X centers. [From Zhou *et al.* (1982).]

This is more evident from their photoionization behavior (Fig. 42). The levels also exhibit a slight difference in depth concentration profiles, as seen in Fig. 43. The increase in trap concentration beyond $0.26\,\mu m$ in the unannealed samples may be due to a change in the stoichiometry during growth. High-temperature annealing seems to have no significant effect on the densities of the two levels, indicating that they are not simple point defects, as recently suggested (Mizuta *et al.*, 1985).

Indirect evidence of the complex nature of the D–X center has also been obtained (Dhar *et al.*, 1986a) from field-dependent transient capacitance measurements on layers grown with different orientations. Figure 44 shows transient capacitance signals obtained from Schottky barriers formed on layers of (100), (111)A and (111)B orientations. It is clear that a strong discrepancy in the emission rates between the (100) and (111) directions, as observed for the EL2 center in GaAs (Mircea and Mitonneau, 1979), is not evident here. This result may be expected for a complex defect—for example, an impurity coupled to a vacancy. This conclusion is consistent with the original D–X center model proposed by Lang *et al.* (1979). However, the strong arguments presented recently in favor of a substitutional donor model (Mizuta *et al.*, 1985; Tachikawa *et al.*, 1985) for the D–X center cannot be fully disregarded. More work is necessary to elucidate the true nature and origin of the D–X center.

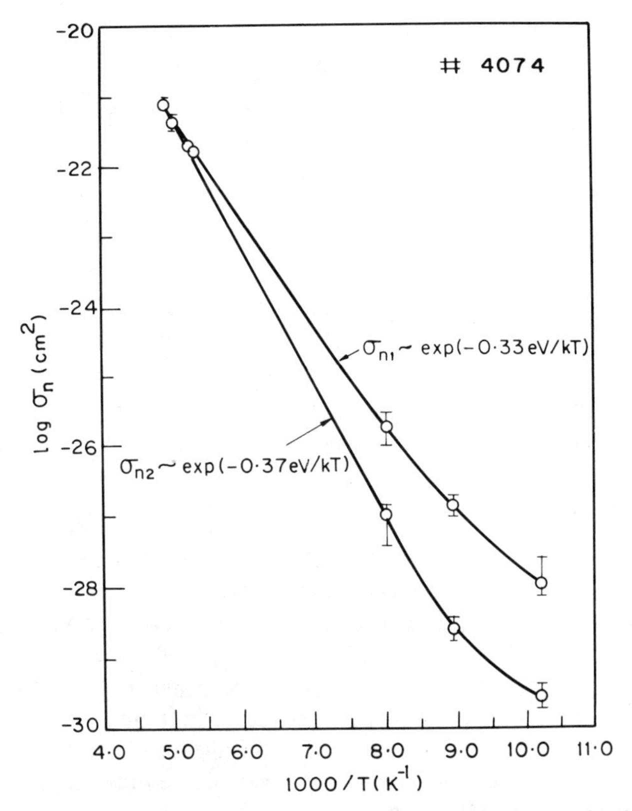

FIG. 37. The temperature dependence of electron capture cross sections σ_n for the two D–X centers in Si-doped MBE $Al_{0.3}Ga_{0.7}As$. [From Zhou *et al.* (1982).]

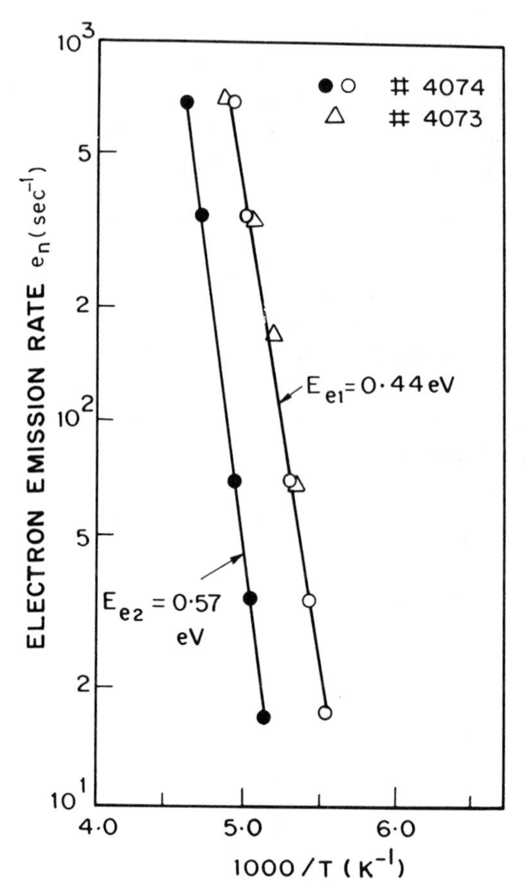

FIG. 38. The temperature dependence of the electron emission rate e_n for the two D–X centers in Si-doped MBE $Al_xGa_{1-x}As$. [From Zhou *et al.* (1982).]

20. $In_{0.53}Ga_{0.47}As$

Single $In_{0.53}Ga_{0.47}As$ layers, lattice-matched to InP substrates and $In_{0.54}Ga_{0.47}As$–$In_{0.52}Ga_{0.48}As$ heterostructures, are currently being grown by MBE for many electronic and optoelectronic device applications (Wicks *et al.*, 1982; Pearsall *et al.*, 1983; Capasso *et al.*, 1983; Mizutani and Hirose, 1985). Deep levels in $In_{0.53}Ga_{0.47}As$ layers grown on InP by LPE have been identified and discussed by Forrest and Kim (1982). Similar studies on lattice-mismatched, VPE-grown materials have also been reported (Mircea *et al.*, 1977). However, there is no report available in the literature on deep level traps present in MBE-grown $In_{0.53}Ga_{0.47}As$. We have recently made an investigation of deep electron traps in $In_{0.53}Ga_{0.47}As$ layers grown by MBE

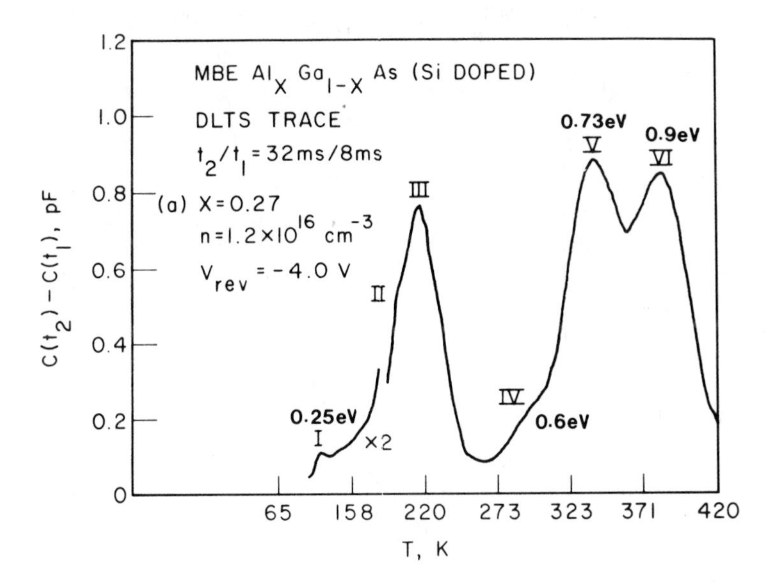

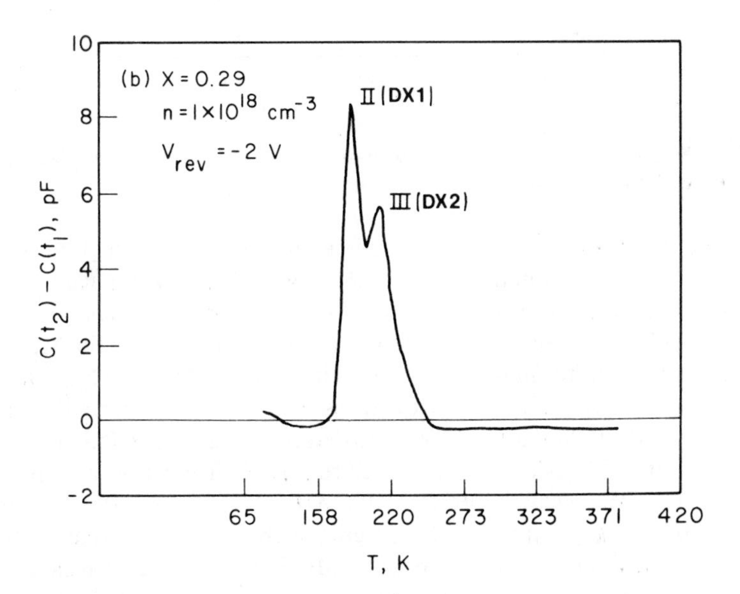

FIG. 39. Typical DLTS obtained from two Si-doped MBE-grown $Al_xGa_{1-x}As$ samples with different values of x and the free carrier concentration, n: (a) with $x = 0.27$ and $n = 1.2 \times 10^{16}\ cm^{-3}$, showing all the traps present in Si-doped $Al_xGa_{1-x}As$, and (b) with $x = 0.29$ and $n = 1 \times 10^{18}\ cm^{-3}$, showing DX1 and DX2 centers separately.

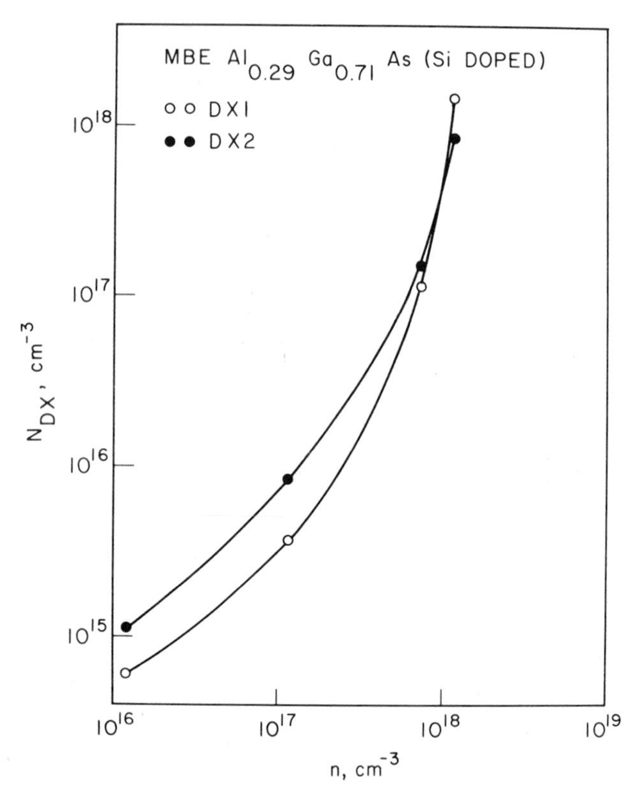

FIG. 40. Variation of the concentrations of DX1 and DX2 centers in Si-doped MBE $Al_{0.29}Ga_{0.71}As$ as a function of free carrier concentration, n.

on InP substrates with and without a buffer layer of high-purity LPE $In_{0.53}Ga_{0.47}As$ (Nashimoto *et al.*, 1986). Figure 45 shows typical DLTS data obtained from a regrown MBE–LPE InGaAs hybrid p–n junction. Data at lower applied reverse bias essentially show traps detected in the MBE layer, whereas the data for higher reverse bias include traps in both MBE and LPE layers. In general, three electron traps A, B, C with thermal activation energies 0.16, 0.20 and 0.37 eV were detected. Careful examination of the depth profiles (Fig. 46) of these traps reveals their location with respect to the MBE–LPE regrowth interface. The MBE layer is from 0 to 0.7 μm, and the LPE layer is given by the depths greater than 0.7 μm. It is apparent that traps B and C are common to both MBE- and LPE-grown layers. This conclusion cannot be made from our data regarding trap A. The characteristics of trap A are similar to a 0.16 eV level detected in LPE InGaAs (Forrest and Kim, 1982). This trap seems to be related to In or impurities contained therein, which is evident from our studies on In-doped GaAs described earlier.

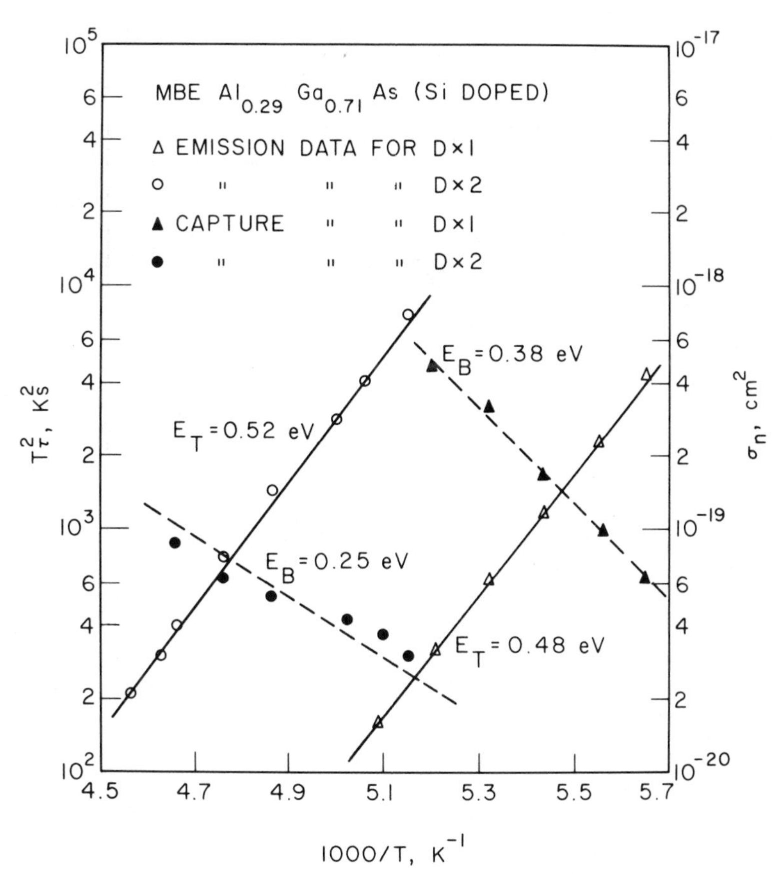

FIG. 41. Temperature-dependent emission and capture data for the two levels DX1 and DX2 identified in MBE-grown Si-doped $Al_xGa_{1-x}As$.

21. $In_{0.52}Al_{0.48}As$

MBE-grown $In_{0.52}Ga_{0.48}As$, lattice-matched to InP (Cheng *et al.*, 1981; Davies *et al.*, 1984), has potential applications in realizing a number of electronic and optoelectronic devices (Ohno *et al.*, 1980; O'Connor *et al.*, 1982; Chen *et al.*, 1982; Tsang, 1981c; Temkin *et al.*, 1983). From Hall measurement data, it is apparent that the material with low Si doping contains a deep donor that might originate from Si (Massies *et al.*, 1983). Based on theoretical models, it has recently been suggested (Tachikawa *et al.*, 1985) that strained layers of $In_{1-x}Al_xAs$ ($0.55 < x < 0.85$) might contain a D–X-like center. The first experimental study of deep level traps in $In_{1-x}Al_xAs$ was made by us recently (Hong *et al.*, 1986). Measurements were made to investigate electron and hole traps present in lattice-matched

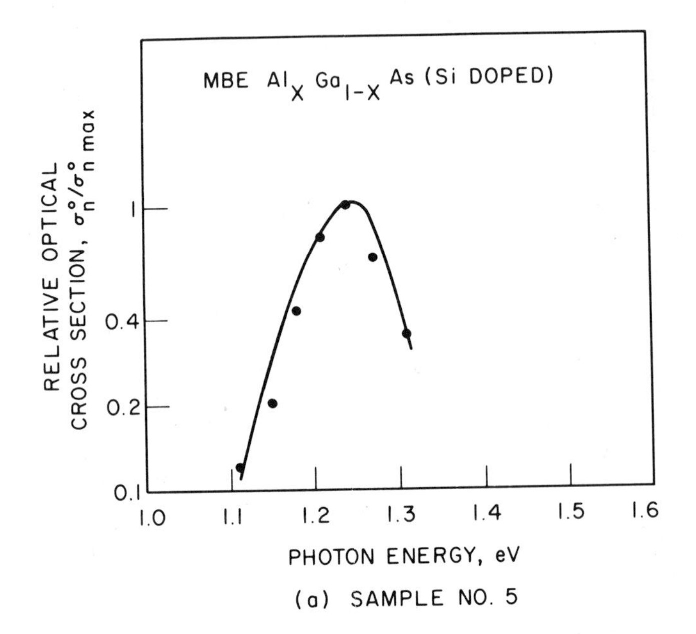

(a) SAMPLE NO. 5

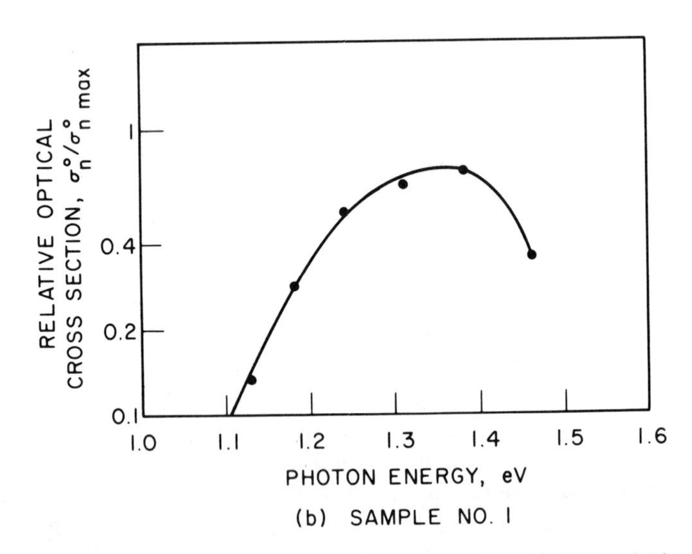

(b) SAMPLE NO. 1

FIG. 42. (a) Relative photoionization cross-sections for DX1 in a Si-doped $Al_{0.29}Ga_{0.71}As$ sample with free carrier concentration, $n = 1 \times 10^{18}\ cm^{-3}$, and (b) relative photoionization cross-sections for DX2 in Si-doped $Al_{0.27}Ga_{0.73}As$ with $n = 1 \times 10^{16}\ cm^{-3}$.

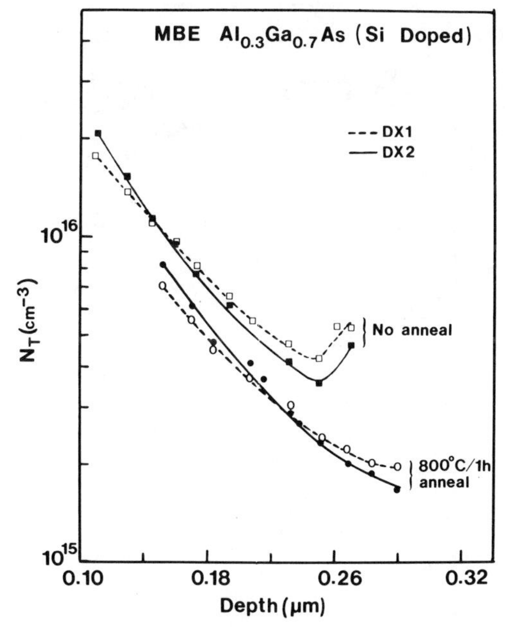

FIG. 43. Depth concentration profiles of DX1 and DX2 below the surface in as-grown and annealed (800°/hr with proximity cap) MBE Si-doped $Al_{0.3}Ga_{0.7}As$.

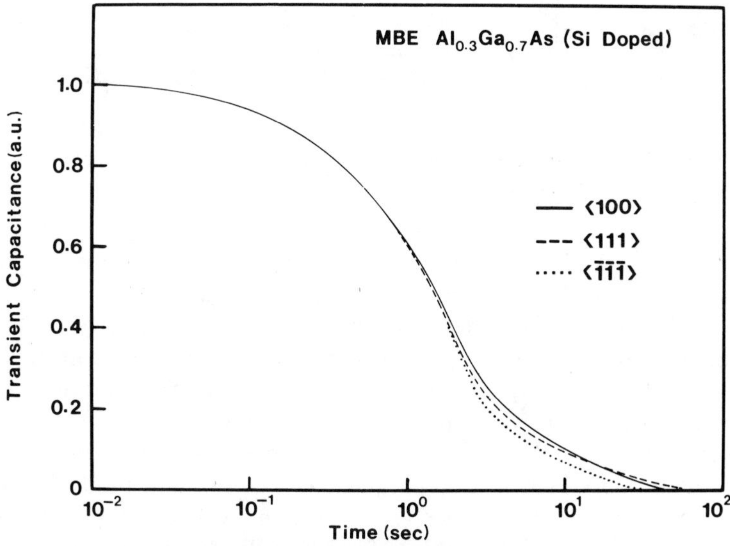

FIG. 44. Transient capacitance signals recorded at 175 K for different orientations of MBE-grown $Al_{0.3}Ga_{0.7}As$ doped with Si. The data was taken at a reverse bias of −6.0 V after a filling pulse of +0.3 V.

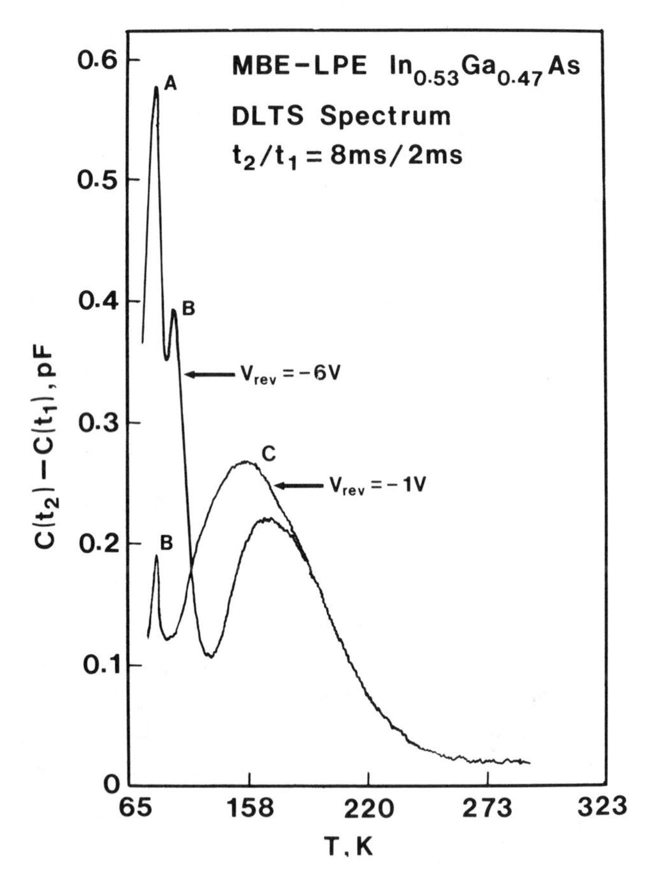

FIG. 45. DLTS data showing electron traps in LPE- and MBE-grown regions of hybrid layers. The interface region is penetrated by increased reverse bias applied to the diode.

InAlAs/InP, doped *n*- and *p*-types with Si and Be, respectively. Figure 47 shows a typical DLTS output for a Si-doped InAlAs sample with $n = 5 \times 10^{16}\,\text{cm}^{-3}$. Four electron traps, labeled EA1 through EA4, are present. In addition, four hole traps, HA1–HA4, are detected in Be-doped samples, as shown in Fig. 48. Hole trap HA3 was observed more clearly in lightly-doped samples. Trap characteristics were determined from the well-resolved peaks, using the temperature dependence of emission rates, depicted in Fig. 49. Table V lists the trap parameters. The concentrations mentioned in this table represent an average value for each layer. Depth profiles of the concentrations were also determined from bias-dependent measurements, to ascertain whether or not some of these levels originate from the substrate or the substrate-epilayer interface. The important points to note are that the

TABLE V

CHARACTERISTICS OF TRAPS IN $In_{0.52}Al_{0.48}As$

Trap type	Trap label	Activation energy E_T (eV)	Capture cross section[a] σ_∞ (cm^2)	Trap concentration[c] N_T (cm^{-3})
Electron trap	EA1	0.25[b]	—	1.3×10^{15}
	EA2	0.56	9×10^{-11}	1.6×10^{15}
	EA3	0.60	1×10^{-12}	2×10^{15}
	EA4	0.71	4×10^{-12}	4×10^{15}
Hole trap	HA1	0.27	3×10^{-13}	6×10^{13}
	HA2	0.45[b]	—	7.6×10^{13}
	HA3	0.65[b]	—	2×10^{13}
	HA4	0.95	4×10^{-14}	4×10^{14}

[a] Determined from the emission rate prefactor.
[b] Approximately determined from the DLTS peak position.
[c] For a sample with $N_D - N_A = 5 \times 10^{16}$ cm^{-3}.

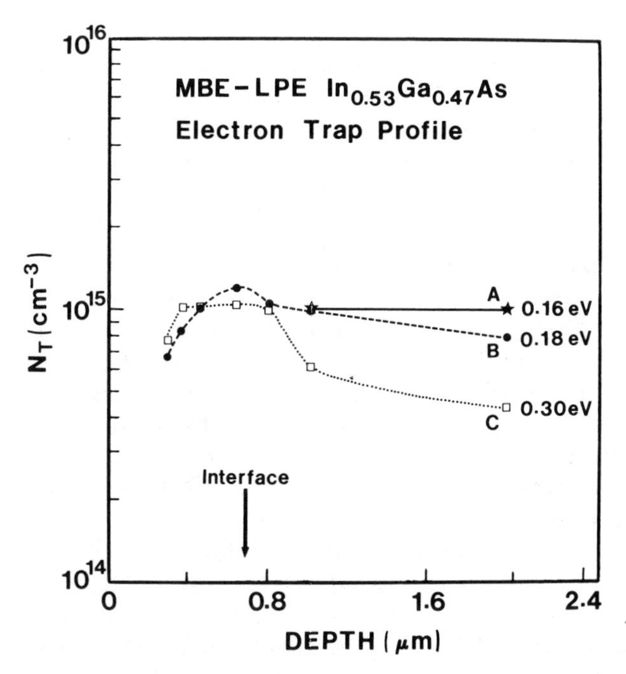

FIG. 46. Concentration profiles of traps observed in MBE- (less than 0.7 μm) and LPE- (greater than 0.7 μm) grown regions of hybrid layers.

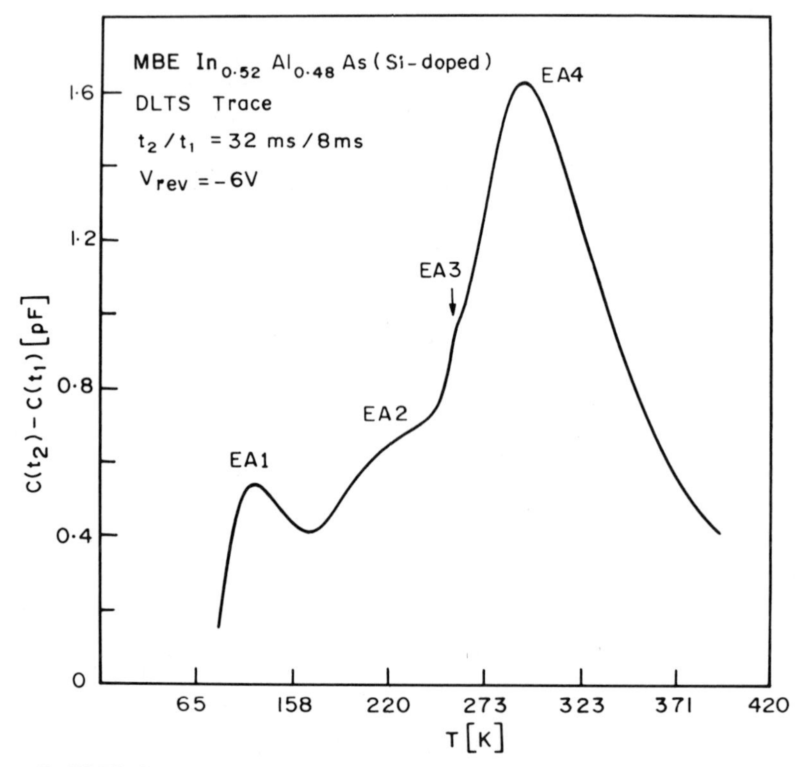

FIG. 47. DLTS data showing electron traps (positive peaks) present in MBE Si-doped $In_{0.52}Al_{0.48}As$.

electron trap concentrations are much larger than those for the hole traps, and the capture cross-sections of the electron traps are very large—almost like attractive centers. We cannot say much regarding the origin of the traps, except for the following observations: traps EA1 and EA2 are detected in both Si- and Be-doped samples; the concentration of traps EA3 and HA3 are independent of the doping levels; and traps EA4 and HA4 probably originate at the substrate-epilayer interface. More detailed work is required to ascertain the origin and microscopic structure of these deep levels and then to take possible steps to eliminate them.

From the data we have just presented, it is apparent that, at least in the lattice-matched alloys, a trap like the D–X center in $Al_xGa_{1-x}As$ is absent. This is borne out by Hall measurement data, where little or no persistent photoconductivity (PPC) is observed at low temperatures. One may then ask the question of whether the centers can be present in compositions close to the lattice-matched one. It is relevant at this point to review some of the well-established properties of the D–X center in $Al_xGa_{1-x}As$. First, the center is believed to be linked to the indirect L and X conduction minima during capture and emission processes. The D–X centers also have very small capture cross-sections for electron capture, due to a large potential barrier to capture. This potential barrier gives rise to large persistent photoconductivity in the material. On observing the conduction band structure of the InAlAs ternary system (Tachikawa *et al.*, 1985), it is seen that, for Al content >55%, the L and X minima are lower in energy than the Γ minimum. It is, therefore, of interest to investigate these compositions for possible existence of the D–X center.

We have performed DLTS measurements on $In_{1-x}Al_xAs$ samples with $0.48 < x \leq 0.57$. These were either grown directly on the InP substrates or with a suitable graded buffer layer, starting with the lattice-matched composition. With the lattice-matched composition, most of the strain is accommodated in the buffer layer. Arrhenius plots of two *new* electron traps,

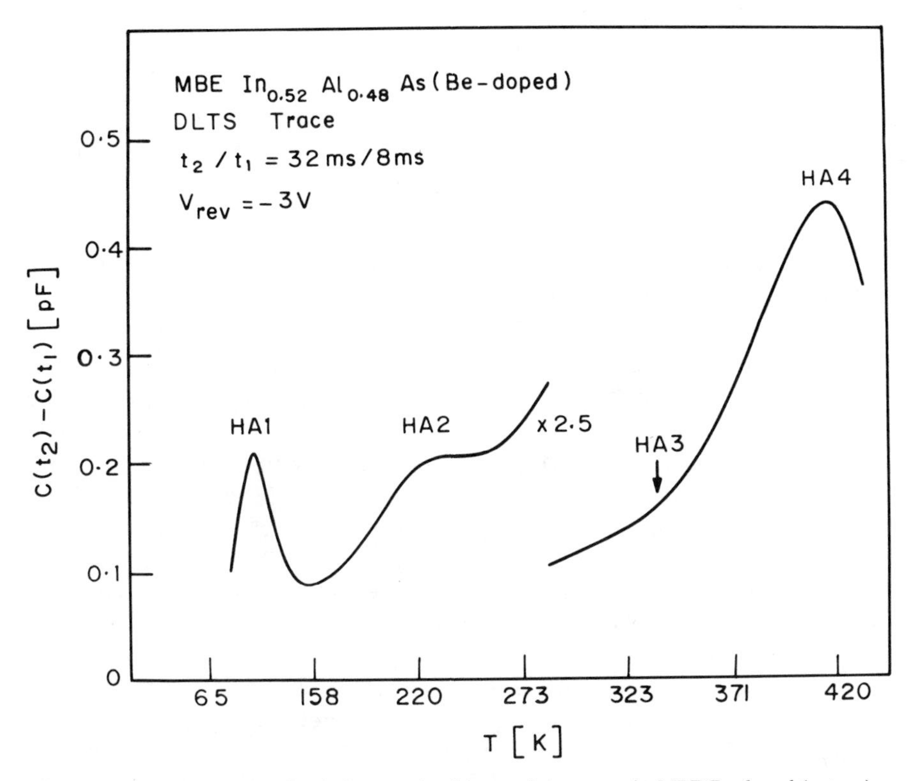

FIG. 48. DLTS data showing hole traps (positive peaks) present in MBE Be-doped (*p*-type) $In_{0.52}Al_{0.48}As$. The experiments were done on n^+–p diodes with majority-carrier pulse filling.

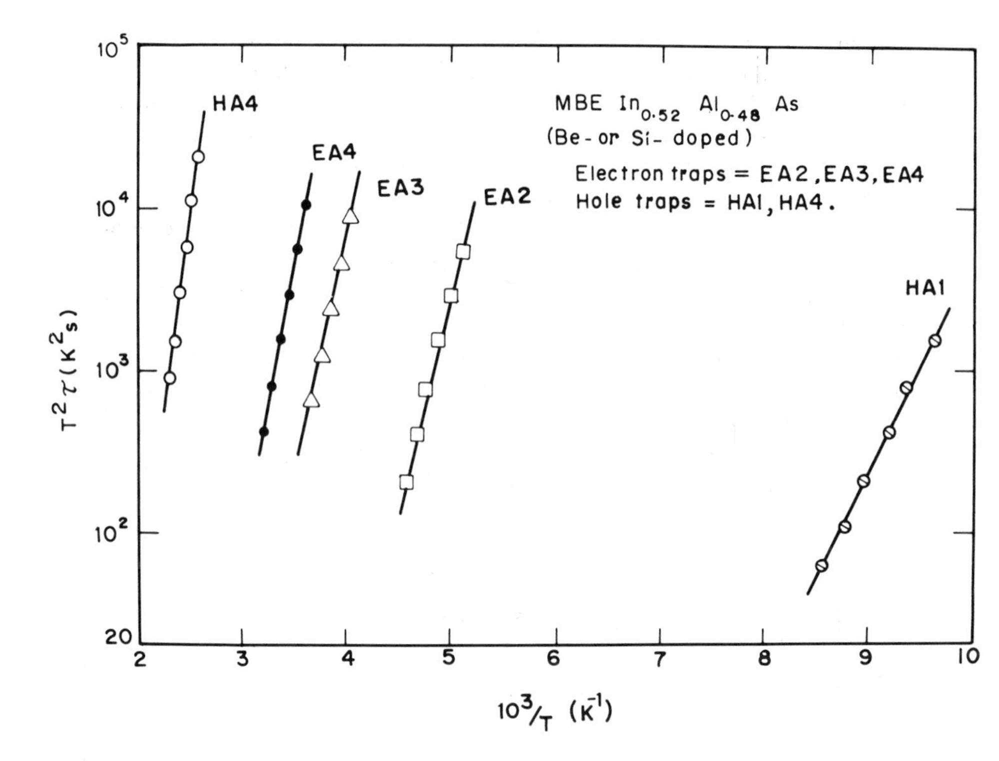

FIG. 49. Temperature dependence of emission time constants for deep level traps identified in Si- and Be-doped MBE $In_{0.52}Al_{0.48}As$.

ES1 and ES2, observed in the directly grown mismatched alloy, are shown in Fig. 50. Note that the extrapolated capture cross-sections σ_∞ are $\sim 10^{-17}$–10^{-18} cm^2. It was also observed that the density of ES1 increased in the range 10^{15} to 10^{16} cm^{-3} with Si-doping level. The ES2 level did not show this trend. On the other hand, the peak labeled ES2 was almost absent ($N_T < 10^{15}$ cm^{-3}) in the layers grown with a buffer layer. This indicates that the trap ES2 is related to strain in the ternary layer. The persistent photoconductivity (PPC) behavior in the samples was studied by cycling the temperature of a Schottky diode and measuring the capacitance under different bias and temperature conditions. A large PPC effect was observed in samples in which only ES1, or in which both ES1 and ES2, were present. By measuring the amount of incremental photocapacitance, it is found that the PPC is due to trap ES1 in samples in which this trap is the only one present, and due to ES1 and ES2 in samples in which both ES1 and ES2 are present. In view of a small σ_∞, strong dependence on Si-doping

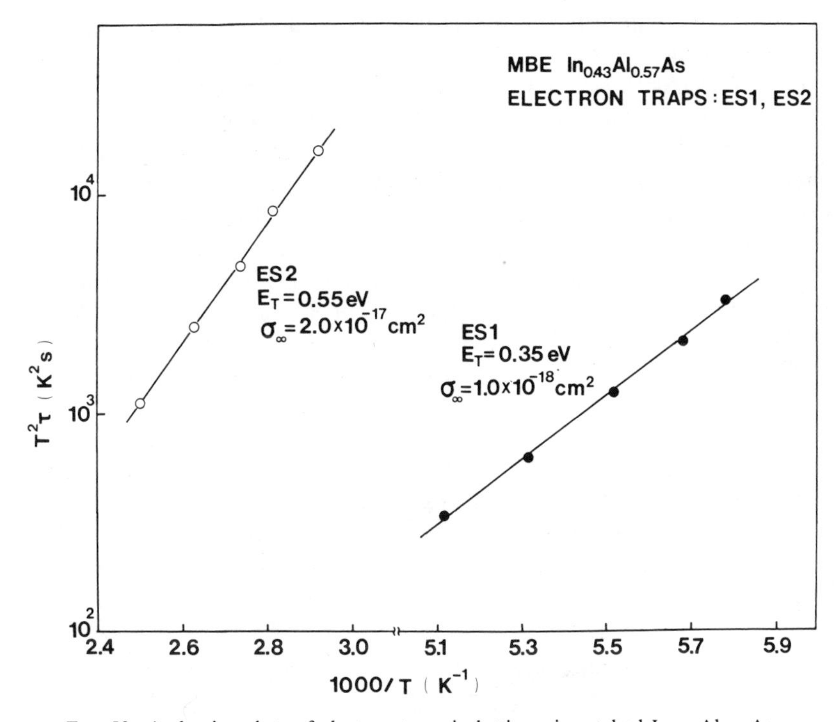

FIG. 50. Arrhenius plots of electron traps in lattice-mismatched $In_{0.43}Al_{0.57}As$.

and the observed PPC, it can be said, with some caution, that the ES1 trap has properties similar to the donor-related D–X center in $Al_xGa_{1-x}As$. It may be mentioned that Nojima and coworkers (1986) have made DLTS measurements on InGaAlP alloys in which the L minima are lowest in energy, and they have identified a trap which exhibits properties similar to the D–X center.

VI. Trapping and Associated Effects in Heterostructures

22. The GaAs–AlGaAs Interface

High-quality GaAs–$Al_xGa_{1-x}As$ heterostructures with abrupt interfaces, grown by MBE, have made it possible to realize many new and promising semiconductor devices. An example is the GaAs–AlGaAs quantum well (QW) structure, which is very important for device applications, as well as from the point of view of fundamental research. The quality of the interface plays a key role in the operation of single and multiple quantum well (MQW) devices. Photoluminescence (PL) techniques have been used to investigate the interface quality in QW structures (Bastard *et al.*, 1984). It has been observed (Miller *et al.*, 1982) that the first few monolayers of GaAs grown over an AlGaAs layer in a QW structure contain an acceptor-like impurity. This impurity, which is probably carbon, is less soluble in AlGaAs than in GaAs, and rides with the AlGaAs-vacuum interface during growth. An interface roughness arises due to the growth-inhibiting nature of the AlGaAs-vacuum interface. Interface roughness at GaAs/AlGaAs heterointerfaces can also arise from the low migration rate of Al during MBE growth and can result in PL line broadening (Singh and Bajaj, 1985a). Using improved growth techniques, however, interface roughness can be reduced to the order of a monolayer, resulting in extremely small (0.3 meV) PL linewidths (Juang *et al.*, 1985). It has been suggested (Petroff *et al.*, 1984) that impurities originate either from the substrate or from the AlGaAs layer and are probably the cause for the different transport properties observed in GaAs–AlGaAs normal and inverted heterointerfaces. McAfee *et al.* (1981, 1982) used DLTS and C–V profiling techniques to investigate traps at the interface of (n)GaAs-(n)$Al_{0.25}Ga_{0.75}As$ double heterostructures. The DLTS signals for various pulse amplitudes are shown in Fig. 51. Four electron

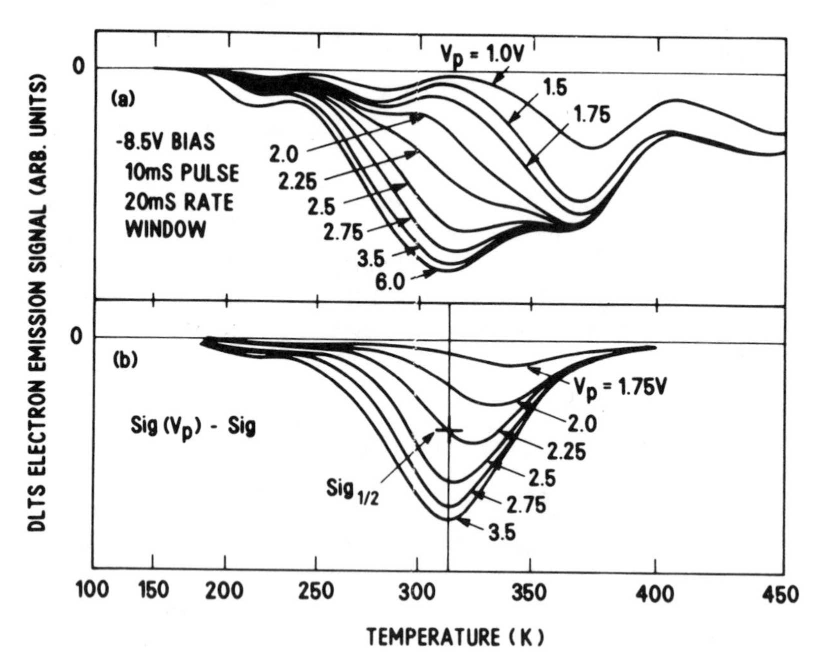

FIG. 51. DLTS spectra for various majority-carrier pulse amplitudes V_P. (a) raw data, and (b) after subtraction of the V_P = 1.5 V data from the other raw data. [From McAfee *et al.* (1982).]

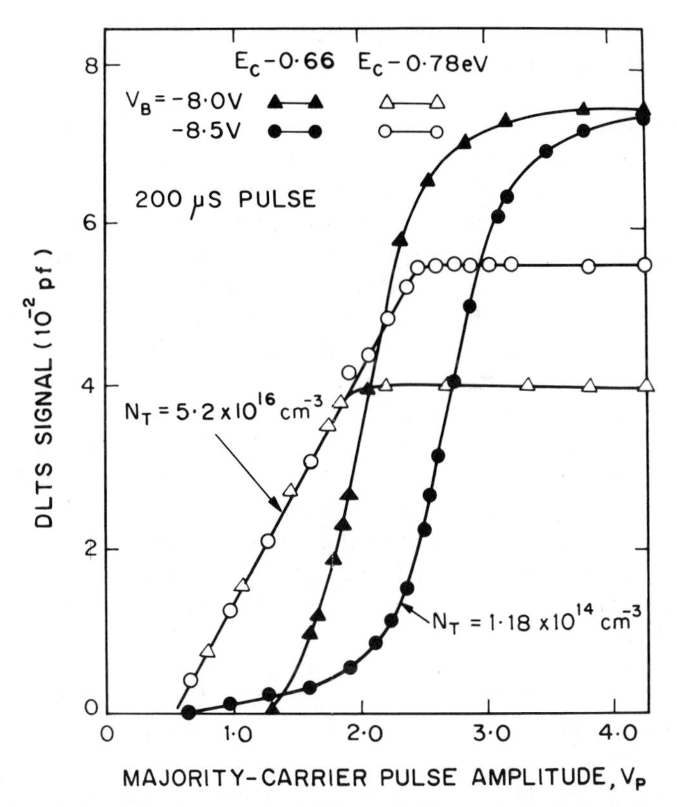

FIG. 52. DLTS signal amplitude at 315 K (E_c − 0.66 eV trap) and at 370 K (E_c − 0.78 eV trap) versus injection pulse amplitude V_P for two different values of the steady-state bias V_B. The open symbols are characteristic of spatially uniform bulk-like behavior, while the solid symbols indicate states with a sharp peak (≲140 Å in the spatial profile). [From McAfee *et al.* (1982).]

traps are seen, of which the broad peak at ~315 K is identified to be an interface trap with an activation energy of 0.66 eV. The interfacial nature of the trap is clearer from Fig. 52, where the variation of concentration of this trap with filling pulse amplitude has been compared to that of the 0.78 eV trap, normally found in single-layer GaAs. From C–V measurements, the 0.66 eV trap has been found to be localized within 140 Å of the GaAs layer next to the interface. The origin of this trap has been related to MBE growth conditions, and it is suspected that it plays a very important role in the operation of DH lasers. Efforts have been made (Meynadier *et al.*, 1985) to improve the quality of the GaAs–AlGaAs interface by introducing thin AlGaAs prelayers with low Al content, which help to trap impurities.

23. $Al_xGa_{1-x}As$–GaAs Modulation Doped Heterostructures

The investigation of trapping and related phenomena in $Al_xGa_{1-x}As$–GaAs modulation-doped (MD) structures has been a matter of considerable interest. $Al_xGa_{1-x}As$–GaAs modulation-doped field effect transistors (MODFET) (Minura *et al.*, 1980) show a number of undesirable effects, such as persistent photoconductivity (Störmer *et al.*, 1979), threshold voltage shift with temperature (Valois *et al.*, 1983), drain I–V collapse (Fischer *et al.*, 1984) and persistent channel depletion by hot electrons under high electric fields (Kinoshita *et al.*, 1985). All these effects, occurring at low temperatures, have been directly or indirectly associated with the presence of the D–X center in the doped $Al_xGa_{1-x}As$ region. It has also been reported that this trap can affect device performance, even at room temperatures (Nathan *et al.*, 1985). Several studies have been made to determine the exact nature of trapping centers in MD structures and their influence on device performance. The highly non-exponential nature of the capacitance transients (Valois *et al.*, 1983), resulting from emission from traps with large densities and effects due to the heterointerface containing the two-dimensional electron gas (2DEG) (Dhar *et al.*, 1986a), makes the usual DLTS analyses of such structures invalid. Measurement of the change of threshold voltage with temperature in long-channel MODFETs (Valois *et al.*, 1983) show the presence of a single trap level with an activation energy of 0.45 eV. A similar result was obtained by measuring drain current transients in identical devices (Valois and Robinson, 1983). The D–X center activation energies determined by these authors are, however, larger than those measured by other groups (Chi *et al.*, 1984; Loreck *et al.*, 1984). It is clear that multi-level D–X centers, as identified in the single ternary layers, affect the performance of MD structures (Valois and Robinson, 1985; Subramanian *et al.*, 1985b; Dhar *et al.*, 1986a; Ohno *et al.*, 1985; Mooney *et al.*, 1985b; Takikawa and Ozeki, 1985). The large scatter in trap activation energy data of MD structures is shown in Table VI (Valois and Robinson, 1985). This scatter is partly due to the problems associated with the interpretation of DLTS data from MD structures.

There are three principal anomalous features that we have observed when DLTS measurements are made on modulation doped structures with the Schottky gate on the AlGaAs layer. The first is shown in Fig. 53(a), which is DLTS data of a typical MD structure. The quiescent reverse bias applied to the diode si -2 V. The data reflect traps in the $Al_xGa_{1-x}As$ doping layer, and a noticeable feature is the absence of the other traps observed in single ternary layers. The second anomalous feature is the observed *decrease* in the emission rate of carriers from D–X centers with increasing depletion layer electric field. This is contrary to that expected or observed for most dominant

TABLE VI

MEASURED VALUES OF THE ACTIVATION ENERGY OF THE DOMINANT TRAP IN Si-DOPED AlGaAs/GaAs MODFETs. [From Valois *et al.* (1985).]

	MODFET DLTS		Standard DLTS	
Wafer	E_1 (eV)	E_2 (eV)	E_1 (eV)	E_2 (eV)
Phi (six wafers)	0.59	0.46–0.51	0.44	0.46
Minn 175	0.55–0.59	0.44–0.47	—	—
Minn 174	—	—	0.40	0.50

traps in single layers (Martin *et al.*, 1981). Similar shifts also occur with increasing filling pulse width. Finally, the real-time capacitance transients, due to electron trap emission under certain bias conditions, are accompanied by a negative-going region superimposed over the usual positive-going capacitance signal. Some of the above observations were also made by Martin *et al.* (1983) on MOCVD grown AlGaAs–GaAs quantum well structures, which they attempted to explain by assuming emissions from the GaAs well. We have formulated a model (Dhar *et al.*, 1986a) that gives a reasonable interpretation of the above anomalous effects and allows the determination of the true trap characteristics. The main features of this model are as follows:

(1) The measured capacitance is actually the sum of two capacitances in series, one due to the Schottky barrier on AlGaAs, and the other due to the charged layers at both sides of the 2DEG heterointerface. This is illustrated in Fig. 54.

(2) In the usual DLTS experiment, if the quiescent reverse bias, after the filling pulse, is less than the flat-band voltage at the heterointerface, some of the electrons emitted from the trap will gradually neutralize part of the accumulated positive charge at the interface. This results in a time-dependent decrease is the interface capacitance.

Including this effect in the normal theory of capacitance transients for a single trap level, it is shown that the true expression for capacitance transient for MD structures will have the form (Dhar *et al.*, 1986a)

$$\frac{\Delta C(t)}{C_0} = \left[a\left(1 + \frac{b}{2}\right) - 1\right] + \left[(1 - a) - b\left(a - \frac{1}{2}\right)\right] \exp(-e_T t) - \frac{b}{2}(1 - a) \exp(-2e_T t), \quad (14)$$

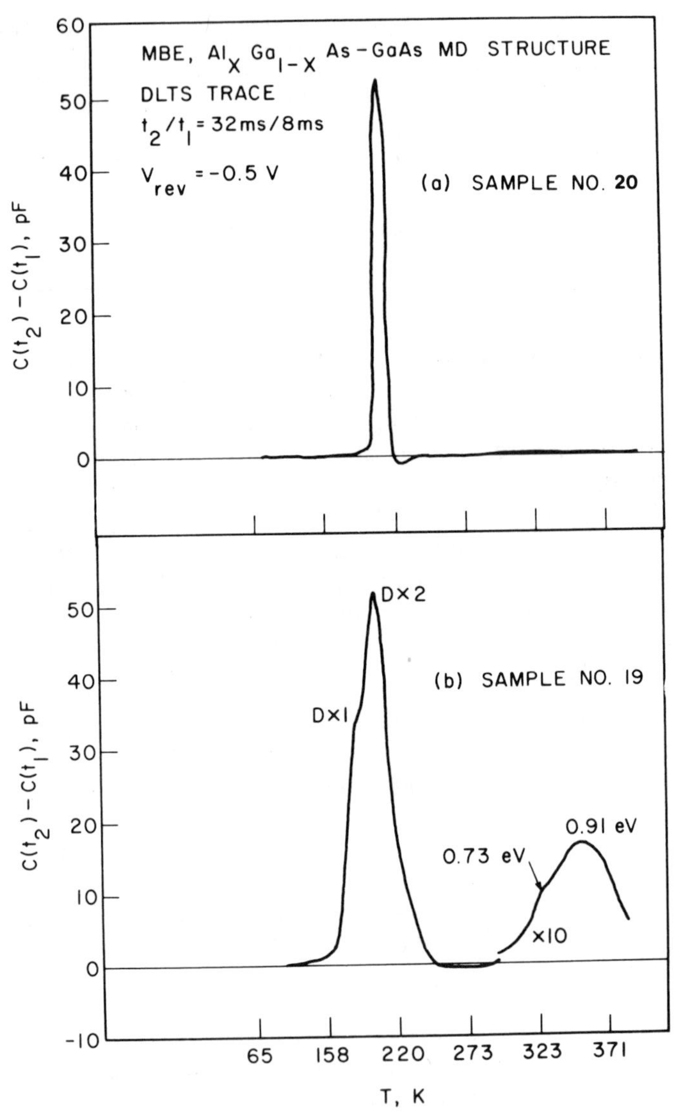

FIG. 53. (a) Typical DLTS data obtained on a MBE-grown $Al_{0.3}Ga_{0.7}As$–GaAs Md structure with 350 Å Si-doped $Al_{0.3}Ga_{0.7}As$ ($n \sim 1 \times 10^{18}\ cm^{-3}$) layer and a 207 Å spacer layer. The 2-DEG concentration at the heterointerface is $2.8 \times 10^{11}\ cm^{-2}$. The single prominent peak is due to the D–X center. (b) DLTS data for a MD structure with 600 Å Si-doped $Al_{0.3}Ga_{0.7}As$ ($n \sim 1 \times 10^{18}\ cm^{-3}$) layer and a 90 Å spacer layer. The 2-DEG concentration is $5.5 \times 10^{11}\ cm^{-2}$. The data clearly show DX1 and DX2, in addition to other prominent peaks observed in single layer $Al_xGa_{1-x}As$.

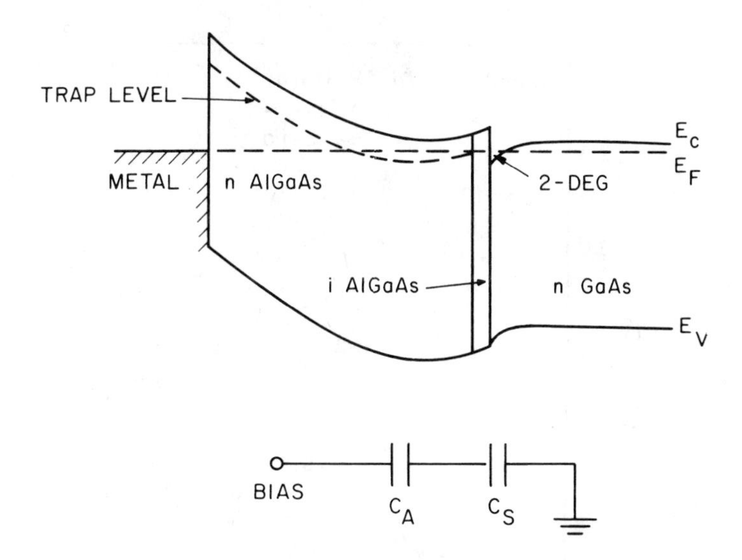

FIG. 54. Band diagram of an $Al_xGa_{1-x}As$–GaAs MD structure, along with its capacitance equivalent circuit. The undoped AlGaAs spacer layer is shown as *i*-AlGaAs.

where C_0 is the measured capacitance at $t = 0$, a is the ratio of the interface capacitance at $t = \infty$ to that at $t = 0$, $b = N_T/N_D$ is the ratio of trap concentration to the shallow donor concentration, and e_T is the thermal emission rate of the trap.

Equation (14) indicates that the observed capacitance of a Schottky barrier on a modulation doped heterostructure is the sum of two components whose relative magnitudes and signs depend on N_T/N_D and the ratio a. It is also apparent from Eq. (14) that, due to the combined effect of a faster transient superimposed over the true transient, the DLTS peak should appear at a lower temperature than normal. Since, with increased reverse bias, $(1 - a)$ gradually approaches zero, the contribution from the faster transient term should correspondingly decrease, and the peak should move to higher temperatures. This is shown in Fig. 55, where DLTS data for different values of a are simulated using Eq. (14). The values of e_T at different temperatures were taken from measured data for DX2 in $Al_xGa_{1-x}As$. As a result, the Arrhenius plot for the trap will also shift to higher temperatures with a corresponding increase in the measured activation energy E_T. This shift is illustrated in Fig. 56 for an $Al_{0.32}Ga_{0.68}As$–GaAs MD structure. At a reverse bias of -0.75 V, the value of E_T is 0.51 eV, and the Arrhenius plot is almost coincident with that for trap DX2 measured in single-layer $Al_xGa_{1-x}As$. The data indicate that the dominant trap in the MD structure is DX2, whose true characteristics can only be obtained by a proper choice of the reverse bias.

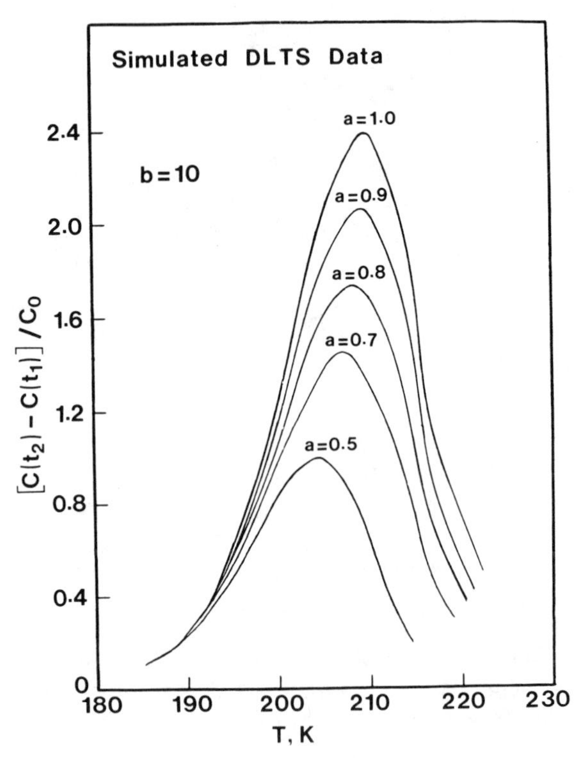

FIG. 55. Theoretically simulated DLTS data based on Eq. (14), showing the shift of the DLTS peak in temperature scale with different values of *a*.

Under the condition that $a \simeq 1$, Eq. (14) reduces to

$$\frac{\Delta C(t)}{C_0} \simeq \frac{N_T}{2N_D}[1 - \exp(-e_T t)], \tag{15}$$

which is the usual form of the capacitance transient for a single trap level. One should expect to see the other traps present in $Al_xGa_{1-x}As$ in the DLTS scan. This is shown in Fig. 53(b) for a structure with a thin AlGaAs spacer layer, so that the 2DEG concentration is high, and the resultant change in the interfacial capacitance is small. The 0.73 eV and 0.91 eV electron traps, which are usually present in MBE-grown AlGaAs, are clearly identified. In addition, both the DX1 and DX2 centers are observed to be present. Photoionization data for a MD structure shown in Fig. 57 also indicate that both the traps DX1 and DX2 are present. These data are in fair agreement with those obtained by Ohno *et al.* (1985) on MD structures, using theoretical curve fitting of the DLTS peaks.

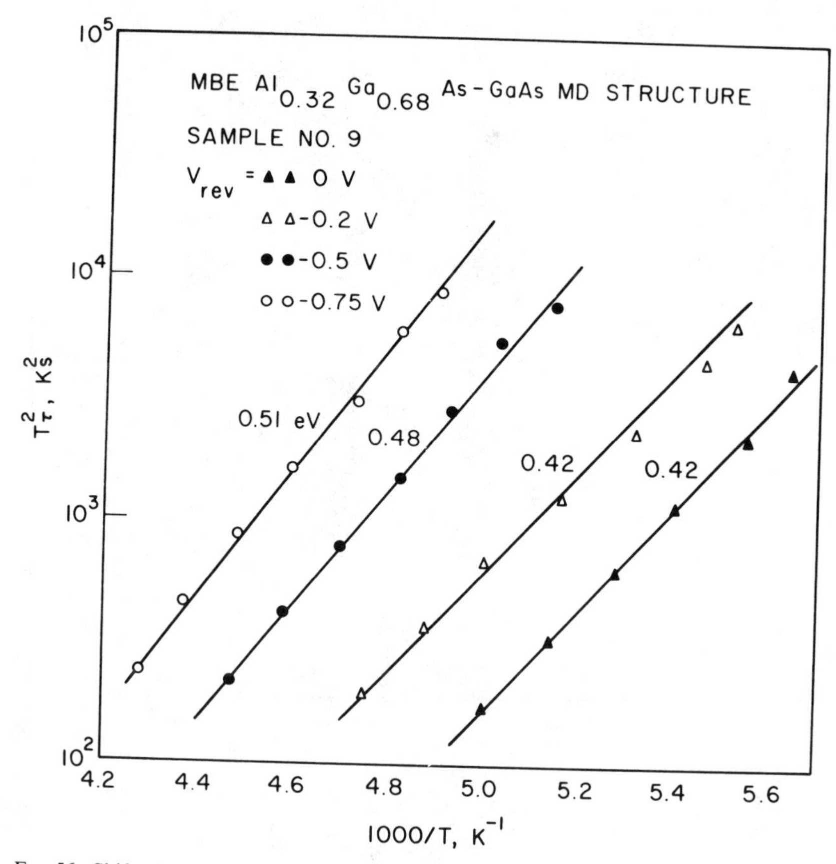

FIG. 56. Shift of the Arrhenius plot of DX center in an $Al_xGa_{1-x}As$–GaAs MD structure with applied reverse bias in capacitance DLTS experiments.

24. Strained Layer Superlattices

Due to the absence of lattice-matching requirements, strained layer superlattices (SLS) (Osbourn, 1982; Osbourn *et al.*, 1982) offer a large tunability in bandgap and other material properties important for device applications (Bhattacharya *et al.*, 1986). The SLS systems of interest are InGaAs–GaAs, GaAsP–GaAs(GaP), $InAs_xSb_{1-x}$–$InAs_ySb_{1-y}$ and GaAsP–InGaAs. The growth and properties of the first two systems have been more intensively studied. Deep levels in such structures may be created due to growth conditions, rearrangement of atoms in the growing strained layer, impurities, and interfacial defects. Deep level traps in MOCVD-grown GaAsP–GaP SLS have been investigated by Barnes *et al.* (1984). The first deep level studies on MBE-grown SLS have been made by us

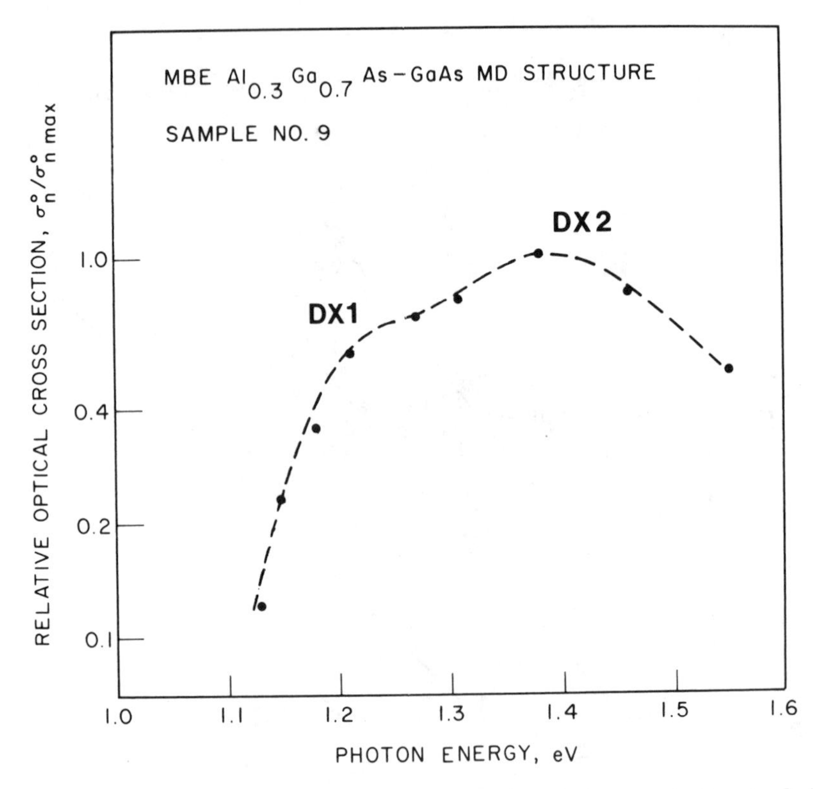

FIG. 57. Relative photoionization cross-sections for DX1 and DX2 in an $Al_{0.3}Ga_{0.7}As$–GaAs MD structure.

(Bhattacharya *et al.*, 1986; Dhar *et al.*, 1986b). We have made a systematic study of deep levels in as-grown and Si-implanted $In_{0.2}Ga_{0.8}As$–GaAs SLS structures suitable for optical guiding (Das *et al.*, 1986). Figures 58(a) and (b) show the electron and hole traps detected in a typical 94 Å $In_{0.2}Ga_{0.8}As$–25 Å GaAs SLS structure. Arrhenius plots for the same are depicted in Fig. 59, and their characteristics are listed in Table VII. It is important to note that all the three electron traps observed in this undoped SLS can be identified with the "M" levels detected in MBE-grown GaAs. Possible origins of these levels have been discussed in Section IV. Traps similar to the 0.20 eV level have also been observed by us (Nashimoto *et al.*, 1986) in lattice-matched InGaAs grown on InP. It may, in general, be concluded that the electron traps observed here exist in the GaAs barrier layer. The 0.42 eV hole trap is possibly related to copper outdiffusing from the substrate, and, similarly, the shoulder near 260 K in Fig. 58 may be due to Cr (Stall *et al.*, 1980). Both traps exist in the GaAs barrier layers.

TABLE VII

CHARACTERISTICS OF ELECTRON TRAPS OBSERVED IN UNDOPED 94 Å $In_{0.2}Ga_{0.8}As$/25 Å GaAs STRAINED LAYER SUPERLATTICE

Trap type	Trap label	Activation energy (ΔE_T) (eV)	Capture cross-section[a] σ_∞ (cm^2)	Trap concentration N_T (cm^{-3})	Possible identity
Electron traps	A	0.20	1.25×10^{-13}	1.8×10^{14}	Trap M1 observed in MBE GaAs
	B	0.37	4.3×10^{-13}	2.6×10^{14}	Trap M3 observed in MBE GaAs
	C	0.75	1.0×10^{-11}	1.0×10^{14}	M6 in MBE GaAs
Hole traps	D	0.24	2.3×10^{-15}	1.0×10^{14}	New
	E	0.42	6.8×10^{-15}	8.0×10^{13}	Related to copper

[a] Determined from the emission rate prefactor.

New levels, an electron trap and a hole trap with activation energies of 0.81 and 0.46 eV, respectively, appear to dominate the samples implanted with Si and pulse annealed under a halogen lamp. The characteristics of the traps are depicted in the Arrhenius plots of Fig. 60. The electron trap is different from the EL2 center produced in MBE GaAs after a half-hour anneal (Day *et al.*, 1981). This trap is detected in the samples annealed at 700°C for 5 sec, but is completely removed after an annealing cycle of 750°C for 5 sec. The depth profile of this trap, shown in Fig. 61, almost replicates the dopant profile. It is, therefore, justifiable to suggest that the 0.81 eV trap originates at implant-induced damages. Annealing at lower temperatures produces a non-discrete band of hole traps with large concentrations that are removed on 750°C/5 sec anneal, leaving only the 0.46 eV trap. It is possible that this trap is also related to implant processes, but not in a simple manner, as for the other traps.

VII. Conclusion

It is of interest to compare the properties of deep level traps commonly identified in MBE-grown materials, as described in the previous sections, with similar semiconducting compounds grown by liquid phase epitaxy (LPE), vapor phase epitaxy (VPE) and metal-organic chemical vapor deposition (MOCVD). We will not go into a lengthy discussion of traps in these materials, but will merely highlight some dominant effects. A quick overall comparative picture can be obtained from the publication of Mircea and coworkers (Martin *et al.*, 1977; Mitonneau *et al.*, 1977).

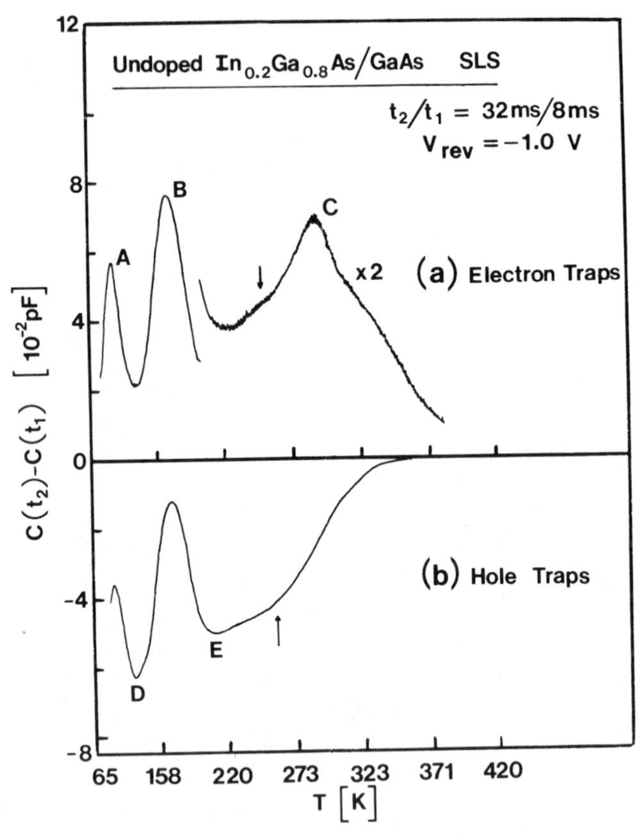

FIG. 58. (a) DLTS, and (b) optical DLTS data obtained from a 95 Å $In_{0.2}Ga_{0.8}As$/25 Å GaAs strained layer superlattice structure showing the electron and hole traps present. The arrows denote shoulders indicating the presence of electron and hole traps of low density.

The well-known EL2 center present in bulk, VPE and MOCVD GaAs is believed to be a complex $As_{Ga}-V_{As}$ antisite defect on a GaAs site (Lagowski *et al.*, 1982). This microstructure has emerged after years of intensive research by several groups of workers. In bulk liquid-encapsulated Czochralski (LEC) GaAs, the defects are believed to be formed during post-growth cooling of the crystal, as a result of Ga vacancy migration (Lagowski *et al.*, 1982). The defects are significantly absent in LPE and MBE materials. In the former, growth occurs in near-equilibrium, Ga-rich conditions, at much lower temperatures, whereas in MBE growth, epitaxy is controlled by the processes of physisorption and chemisorption. Day *et al.* (1981) have reported that EL2 is formed in MBE GaAs encapsulated with SiO_2 after short term (half-hour) annealing. They have attributed this behavior to the preferential diffusion

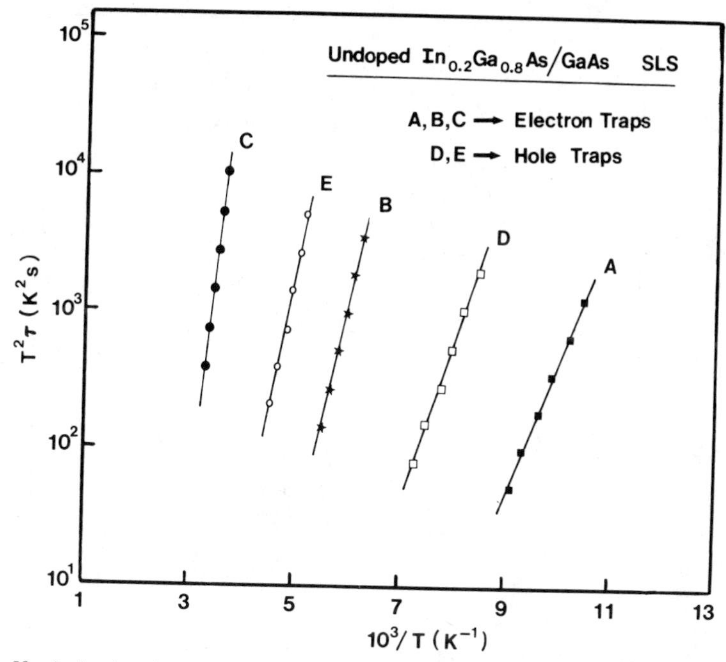

FIG. 59. Arrhenius plots for electron and hole traps observed in 95 Å $In_{0.2}Ga_{0.8}As$/25 Å GaAs strained layer superlattices.

of Ga atoms at the surface into the SiO_2 encapsulation. The exact characteristics of the deep level designated as EL2 by these authors have not been ascertained, since there are other traps in GaAs that behave similar to the dominant trap.

On the contrary, the complex defect labeled as the D–X center is present in $Al_xGa_{1-x}As$ grown by all the epitaxial techniques. The detailed properties, which are intimately related to the dopant specie, are also very similar. What has emerged more recently is that the D–X center may consist of multiple levels closely spaced in energy (Zhou *et al.*, 1982; Ohno *et al.*, 1985; Dhar *et al.*, 1986a). Whether this splitting is an effect of the band structure or of multiple charge states of the same level has yet to be established. The *acceptor* nature of the electron traps (or deep donors) in Si-doped crystals suggest that latter. It was found by Bhattacharya *et al.* (1984c) that multiple level D–X centers are present in MOCVD $Al_xGa_{1-x}As$ grown with traces of oxygen present in the reactor. The density of some of these levels increases with increasing Al content in the crystal. This result suggests that Al–O bonds might be linked with the formation of some types of D–X centers.

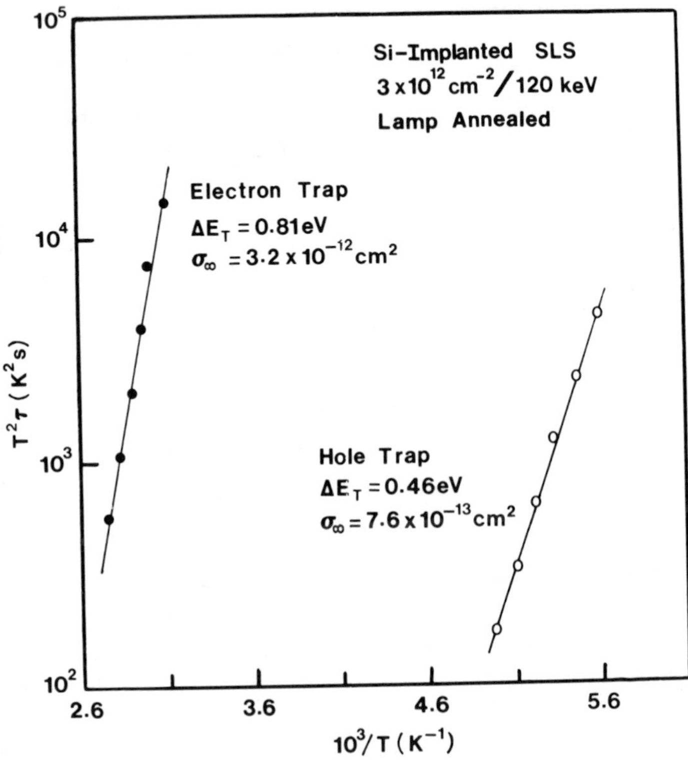

FIG. 60. Arrhenius plots of electron and hole traps observed in a Si-implanted and lamp-annealed $In_{0.2}Ga_{0.8}As/GaAs$ strained layer superlattice.

Deep level densities are usually very low in LPE GaAs. For example, this material has no electron trap, and only one hold trap has been identified (Hasegawa and Majerfeld, 1975). This hole trap has been related to an As-vacancy. With optimum growth conditions, trap densities can be low in MBE GaAs, compared to VPE and MOCVD materials. Most of the identified centers are probably related to point defects. Our recent work demonstrates that trap densities are considerably reduced by the addition of small amounts of In, which suggests the involvement of Ga vacancies. In fact, In-doped MBE GaAs can be an important material for electronic and opto-electronic devices. In the $Al_xGa_{1-x}As$ system also, deep level densities are generally lower in MBE material than in similar material grown by MOCVD (Bhattacharya *et al.*, 1984a and c).

Not much has been reported on the identification of deep levels in MBE-grown heterostructures, particularly at the heterointerfaces. Interface defects can behave as non-radiative centers and deep-level traps and thereby

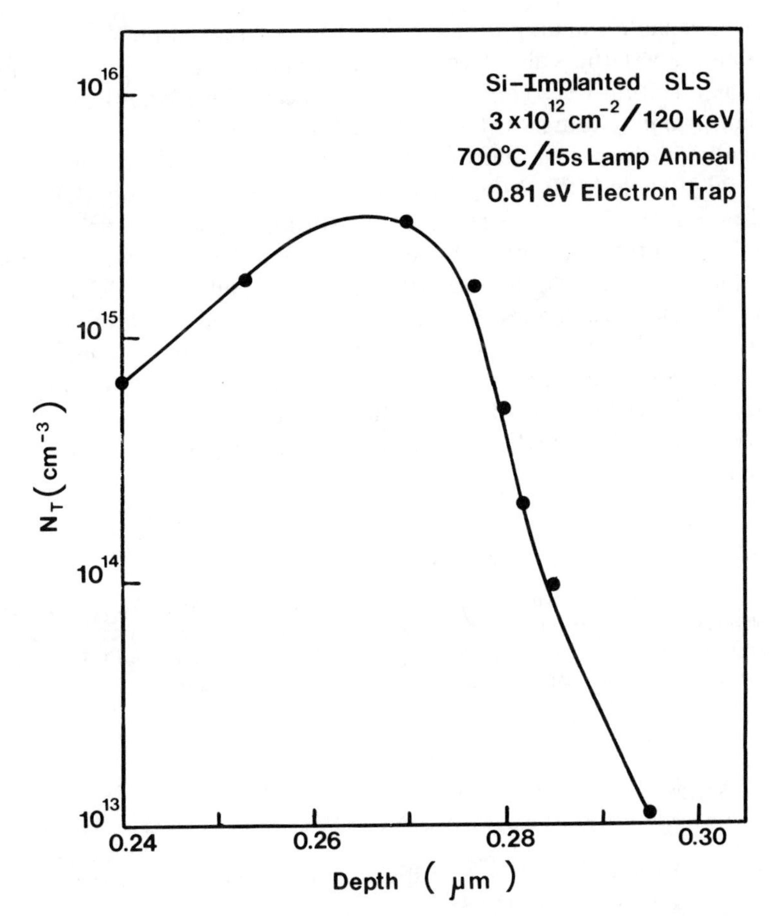

FIG. 61. Depth profile of the 0.81 eV electron trap observed in implanted and annealed $In_{0.2}Ga_{0.8}As/GaAs$ strained layer superlattice.

degrade material properties and device performance. Interface roughness during MBE growth is a consequence of different III–V bond strengths and cation migration rates in the two layers forming the hetero-interface. This gives rise to increased carrier scattering and reduced luminescence efficiency. As described in Section III, we have recently studied this problem in detail by growing InGaAs–InAlAs (lattice-matched to InP) single quantum wells by MBE and effecting interruptions at the heterointerface to reduce interface roughness (Juang *et al.*, 1986). By using the linewidth of the excitonic luminescence in quantum well as the index, it was seen that interruption improved the interface, but decreased the luminescence efficiency. Deep level effects in these regions need to be investigated in detail, as such studies are crucial for understanding device performance. Traps in multiple-quantum wells and superlattices also need to be studied in more detail. It is recognized that the quantum wells confine carriers and can, therefore, themselves act as giant trapping centers (Martin *et al.*, 1983).

Finally, it is apparent that variations of the MBE growth technique, such as metal–organic MBE (MOMBE) and chemical beam epitaxy (CBE), that use gas-source will emerge as technologically important and efficient growth techniques in the near future. Such techniques will allow faster turn-round times, large-area growth and hetero-epitaxy on Si substrates. The materials grown by these techniques thus far have not reached the desired levels of purity, and no identification of deep levels has been made. Undoubtedly, this will emerge as an important area of materials research in the near future. In solid and gas source MBE, the ultimate crystal quality is limited by source purity and growth conditions, and their importance can never be overemphasized.

Acknowledgments

Some of the work described here has been supported by the National Science Foundation, the National Aeronautical and Space Agency, the Department of Energy and the Defence Advanced Resarch Project Agency. The authors would like to thank Debashis Bhattacharya and Rita Szokowski for their careful typing and editing of the manuscript.

References

Adachi, S. (1985). *J. App. Phys.* **58,** R1.

Akimoto, K., Kamada, M., Taira, K., Arai, M., and Watanabe, N. (1986). *J. Appl. Phys.* **59,** 2833.

Amano, C., Shibukawa, A., Ando, K., and Yamaguchi, M. (1984). *Electron Lett.* **20,** 175.

Arthur, J. R. (1974). *Surf. Sci.* **43,** 449.

Arthur, J. R. (1968). *J. Appl. Phys.* **39,** 4032.

Barnes, C. E., Biefeld, R. M., Zipperian, T. E., and Osbourn, G. C. (1986). *Appl. Phys. Lett.* **45,** 408.

Bastard, G., Delalande, C., Maynadier, M. H., Frijlink, P. M., and Voos, M. (1984). *Phys. Rev.* **B 29,** 7042.

Bhattacharya, P. K. (1978). Ph. D. dissertation, The University of Sheffield, United Kingdom.

Bhattacharya, P. K., Buhlmann, H. J., and Ilegems, M. (1982). *J. Appl. Phys.* **53,** 6391.

Bhattacharya, P. K., Majerfeld, A., and Saxena, A. K. (1979). *Conf. Ser.-Inst. Phys.* No. 45, 199.

Bhattacharya, P. K., Das, U., Juang, F. Y., Nashimoto, Y., and Dhar, S. (1986). *Solid-State Electronics.* **29,** 261.

Bhattacharya, P. K., Subramanium, S., and Ludowise, M. J. (1984a). *J. Appl. Phys.* **55,** 3664.

Bhattacharya, P. K., Das, U., and Ludowise, M. J. (1984b). *Phys. Rev.* **B 29,** 6623.

Bhattacharya, P. K., Matsumoto, T., and Subramanian, S. (1984c). *J. Crystal Growth* **68,** 301.

Biefeld, R. M., Osbourn, G. C., Gourley, P. L., and Fritz, I. J. (1983). *J. Electron. Mater.* **12,** 963.

Bleicher, M., and Lange, E. (1973). *Solid-State Electron.* **16,** 375.

Blood, P., and Harris, J. J. (1984). *J. Appl. Phys.* **56,** 993.

Borsuk, J. A., and Swanson, R. M. (1980). *IEEE Trans. Electron Devices* **ED-27,** 2217.

Bourgoin, J., and Lanoo, M. (1983). *In* "Point Defects in Semiconductors II: Experimental Aspects". Springer-Verlag, New York.

Brown, A. S., Palmateer, S. C., Wicks, G. W., Eastman, L. F., and Calawa, A. R. (1985). *J. Electron. Mater.* **14,** 367.

Calawa, A. R. (1981). *Appl. Phys. Lett.* **38,** 701.

Capasso, F. (1985). *In* "Semiconductors and Semimetals" (R. K. Willardson and A. C. Beer, eds.), Vol. **22,** Part D, pp. 1–171. Academic Press, New York.

Capasso, F., Alvi, K., Cho, A. Y., Foy, P. W., and Bethea, C. G. (1983). *Appl. Phys. Lett.* **43,** 1040.

Capasso, F., Tsang, W. T., Hutchinson, A. L., and Williams, G. F. (1982). *Appl. Phys. Lett.* **40,** 38.

Capasso, F., Tsang, W. T., Hutchinson, A. L., and Foy, P. W. (1981). *Conf. Ser. Inst. Phys.* No. 63, 473.

Chang, L. L., Esaki, L., Howard, W. E., Ludeke, R., and Schul, G. (1973). *J. Vac. Sci. Technol.* **10,** 655.

Chang, L. L., and Ludeke, R. (1975). *In* "Epitaxial Growth" (J. W. Matthews, ed.), p. 57. Academic Press, New York.

Chao, P. C., Palmateer, S. C., Smith, P. M., Mishra, U. K., Duh, K. H., and Hwang, J. C. M. (1985). *IEEE Electron Device Lett.* **EDL-6,** 531.

Chen, C. Y., Cho, A. Y., Garbinski, P. A., and Cheng, K. Y. (1982). *IEEE Electron Device Lett.* **3,** 15.

Chen, R. T., Sheng, N. H., and Miller, D. L. (1985). *J. Vac. Sci. Technol.* **B 3,** 652.

Cheng, K. Y., Cho, A. Y., Wagner, W. R., and Bonner, W. A. (1981a). *J. Appl. Phys.* **52,** 1015.

Cheng, K. Y., Cho, A. Y., and Wagner, W. R. (1981b). *Appl. Phys. Lett.* **39,** 607.

Cheng, K. Y., Cho, A. Y., and Bonner, W. A. (1981c). *J. Appl. Phys.* **52,** 4672.

Cheng, K. Y., Cho, A. Y., Drummond, T. J., and Morkoç, H. (1982). *Appl. Phys. Lett.* **40,** 147.

Chi, J. Y., Holmstrom, R. P., and Salerno, J. P. (1984). *IEEE Electron Device Lett.* **EDL-5,** 381.

Cho, A. Y., and Arthur, J. R. (1975). *Prog. Solid State Chem.* **10,** 157.

Cho, A. Y., and Casey, H. C., Jr. (1974). *Appl. Phys. Lett.* **25,** 288.

Cho, A. Y., Dixon, R. W., Casey, H. C., Jr., and Hartman, R. L. (1976). *Appl. Phys. Lett.* **28,** 501.

Cho, A. Y., and Hayashi, I. (1971). *Solid State Electron.* **14,** 125.

Davey, J. E., and Pankey, T. J. (1968). *J. Appl. Phys.* **39,** 1941.

Davies, G. J., Kerr, T., Tuppen, C. G., Wakefield, B., and Andrews, D. A. (1984). *J. Vac. Sci. Technol.* **B 2,** 219.

Day, D. S., Oberstar, J. D., Drummond, T. J., Morkoç, H., Cho, A. Y., and Streetman, B. J. (1981). *J. Electron. Mater.* **10,** 445.

Day, D. S., Tsai, M. Y., Streetman, B. G., and Lang, D. V. (1979). *J. Appl. Phys.* **50,** 5093.

Das, U., Bhattacharya, P. K., and Dhar, S. (1986). *Appl. Phys. Lett.* (accepted for publication).

De Jule, R. Y., Haase, M. A., and Stillman, G. E. (1985). *J. Appl. Phys.* **57,** 5287.

Dhar, S., Hong, W. P., Bhattacharya, P. K., Nashimoto, Y., and Juang, F. Y. (1986a). *IEEE Trans. Electron Devices.* **ED-33,** 698.

Dhar, S., Das, U., and Bhattacharya, P. K. (1986b). *J. Appl. Phys.* (accepted for publication).

Erickson, L. P., Mattford, T. J., Palmberg, P. W., Fischer, R., and Morkoç, H. (1983). *Electron. Lett.* **19,** 632.

Eron, M. (1985). *J. Appl. Phys.* **58,** 1064.

Esaki, L., and Tsu, R. (1970). *IBM J. Res. Develop.* **14**, 61.
Fabre, M. E. (1970). *C.R. Acad. Sc. Paris* **270**, 848.
Fischer, R., Drummond, T. J., Klem, J., Kopp, W., Henderson, T. S., Perrachione, D., and Morkoç, H. (1984). *IEEE Trans. on Electron Devices* **ED-31**, 1028.
Forrest, S. R., and Kim, O. K. (1982). *J. Appl. Phys.* **53**, 5738.
Foxon, C. T., and Joyce, B. A. (1977). *Surf. Sci.* **64**, 293.
Furukawa, Y. (1967). *Jpn. J. Appl. Phys.* **6**, 675.
Furukawa, Y., Kajiyama, K., Seki, Y., and Sugano, K. (1967). *Jpn. J. Appl. Phys.* **6**, 413.
Ghaisas, S. V., and Madhukar, A. (1985). *J. Vac. Sci. Technol.* **B 3**, 540.
Gombia, M., Ghezzi, C., and Mosca, R. (1985). *J. Appl. Phys.* **58**, 1285.
Gossard, A. C. (1982). *Thin Solid Films* **104**, 279.
Grimmeiss, H. J. (1974). *Conf. Ser.-Inst. Phys.* No. 22, 18.
Grimmeiss, H. J. (1977). *Annu. Rev. Mater. Sci.* **7**, 341.
Grimmeiss, H. J., and Ledebo, L.-Å. (1975). *J. Phys. C.* **8**, 2615.
Grimmeiss, H. J., and Olofson, G. O. (1969). *J. Appl. Phys.* **40**, 2526.
Gupta, A., Sovero, E. A., Pierson, R. L., Stein, R. D., Chen, R. T., Miller, D. L., and Higgins, J. A. (1985). *IEEE Electron Device Lett.* **EDL-6**, 81.
Hall, R. N. (1952). *Phys. Rev.* **87**, 387.
Hasegawa, F., and Majerfeld, A. (1975). *Electron. Lett.* **11**, 286.
Hayakawa, T., Suyama, T., Kando, M., Takahashi, K., Yamamoto, S., Yano, S., and Hijikata, T. (1985a). *Appl. Phys. Lett.* **47**, 952.
Hayakawa, T., Suyama, T., Kando, M., Takahashi, K., Yamamoto, S., Yano, S., and Hijikata, T. (1985b). *J. Appl. Phys.* **58**, 4452.
Heiblum, M., Mendez, E. E., and Osterling, L. (1983). *J. Appl. Phys.* **58**, 6982.
Henning, J. C. M., Ansems, J. P. M., and Nijs, A. G. M. De (1984). *J. Phys. C: Solid State Physics* **17**, L915.
Hikosaka, K., Mimura, T., and Hiyamizu, S. (1981). *Conf. Ser.-Inst. Phys.* No. 63, 233.
Holloway, H., and Walpole, J. N. (1979). *Prog. Cryst. Growth Charact.* **2**, 49.
Hong, W. P., Dhar, S., Bhattacharya, P. K., and Chin, A. J. *Electron Materials* (submitted for publication).
Hwang, J. C. M., Temkin, H., Brenan, T. M., and Frahm, R. E. (1983). *Appl. Phys. Lett.* **42**, 66.
Ikeda, K., and Takaoka, H. (1982). *Jpn. J. Appl. Phys.* **21**, 462.
Ilegems, M. (1977). *J. Appl. Phys.* **48**, 1278.
Ilegems, M. (1985). *In* "The Technology and Physics of Molecular Beam Epitaxy" (E. H. C. Parker, ed.), pp. 83–142. Plenum Press, New York.
Ilegems, M., and Dingle, R. (1975). *Conf. Ser.-Inst. Phys.* No. 24, 1.
Ishibashi, T., Tarucha, S., and Okamoto, H. (1982). *Jpn. J. Appl. Phys.* **21**, Part 2, L 476.
Ishikawa, T., Saito, J., Sasa, S., and Hiyamizu, S. (1982). *Jpn. J. Appl. Phys.* **21**, L 675.
Ishikawa, T., Kwon, Y. K., and Kuwano, H. (1985). *Appl. Phys. Lett.* **47**, 1097.
Jacob, G. (1982). *In* "Semi-insulating III–V Materials" (S. Makram-Ebeid and B. Tuck, eds.), pp. 2–18. Shiva Publishing Limited, Nantwich, England.
Jaros, M. (1978). *Solid State Commun.* **25**, 1071.
Joyce, B. A., and Foxon, C. T. (1975). *J. Cryst. Growth* **31**, 122.
Joyce, B. A., and Foxon, C. T. (1977). *Conf. Ser.-Inst. Phys.* No. 32, 17.
Joyce, B. A., Foxon, C. T., and Neave, J. H. (1978). *Nippon Kessho Seicho Gakkaishi* **5**, 185.
Juang, F.-Y., Das, U., Nashimoto, Y., and Bhattacharya, P. K. (1985). *Appl. Phys. Lett.* **47**, 972.
Juang, F.-Y., and Bhattacharya, P. K. (1986a). Unpublished results.
Juang, F.-Y., Bhattacharya, P. K., and Singh, J. (1986b). *Appl. Phys. Lett.* **48**, 290.

Kawamura, Y., Noguchi, Y., Asahi, H., and Hagai, H. (1982). *Electron. Lett.* **18,** 91.
Kinoshita, H., Nishi, S., Akiyama, M., Ishida, T., and Kaminishi, K. (1985). *Jpn. J. Appl. Phys.* **24,** 377.
Kirchner, P. D., Woodall, J. M., Freeouf, J. L., and Petit, G. D. (1981a). *Appl. Phys. Lett.* **38,** 427.
Kirchner, P. D., Schaff, W. J., Maracas, G. N., Eastman, L. F., Chappell, T. I., and Ransom, C. M. (1981b). *J. Appl. Phys.* **52,** 6462.
Kop'ev, P. S., Bar, B. Ya., Ivanov, S. V., Ledenstov, N. N., Mel'tser, B. Ya., and Ustinov, V. M. (1984). *Sov. Phys. Semicond.* **18,** 167.
Kotina, I. M., Mazurik, N. E., Novikov, S. R., and Khusainov, A. Kh. (1969). *Sov. Phys. Semicond.* **3,** 319.
Kukimoto, H., Henry, C. H., and Merritt, F. R. (1973). *Phys. Rev.* **B 7,** 2486.
Kumagai, O., Kawai, H., Mori, Y., and Kaneko, K. (1984). *Appl. Phys. Lett.* **45,** 1322.
Lagowski, J., Gatos, H. C., Parsey, J. M., Wada, K., Kaminska, M., and Walukiewicz, W. (1982). *Appl. Phys. Lett.* **40,** 342.
Laidig, W. D., Peng, C. K., and Lin, Y. F. (1984). *J. Vac. Sci. Technol.* **B 2,** 181.
Lang, D. V. (1974). *J. Appl. Phys.* **45,** 3023.
Lang, D. V. (1979). *Topics in Appl. Phys.* **37,** 93.
Lang, D. V. (1980). *J. Phys. Soc. Jpn.* **49,** Suppl. A, 215.
Lang, D. V., Cho, A. Y., Gossard, A. C., Ilegems, M., and Wiegmann, W. (1976). *J. Appl. Phys.* **47,** 2558.
Lang, D. V., and Henry, C. H. (1975). *Phys. Rev. Lett.* **35,** 1525.
Lang, D. V., and Logan, R. A. (1977). *Phys. Rev. Lett.* **39,** 635.
Lang, D. V., and Logan, R. A. (1979). *Conf. Ser.-Inst. Phys.* No. 43, 433.
Lang, D. V., Logan, R. A., and Kimerling, L. C. (1977). *Phys. Rev.* **5,** 4374.
Lang, D. V., Logan, R. A., and Jaros, M. (1979). *Phys. Rev.* **B 19,** 1015.
Lee, K., Shur, M. S.,, Drummond, T. J., and Morkoç, H. (1983). *IEEE Trans. Electron Devices* **ED-30,** 207.
Lorek, L., Dambkes, H., Ploog, K., and Weimann, G. (1984). *IEEE Electron Device Lett.* **EDL-5,** 9.
Losee, D. L. (1972). *Appl. Phys. Lett.* **21,** 54.
Martin, P. A., Meehan, K., Gavrilovic, P., Hess, K., Holonayak, N., Jr., and Coleman, J. J. (1983). *J. Appl. Phys.* **54,** 4689.
Martin, G. M., Mitonneau, A., and Mircea, A. (1977). *Electron. Lett.* **13,** 191.
Martin, P. A., Streetman, B. G., and Hess, K. (1981). *J. Appl. Phys.* **52,** 7409.
Massies, J., Rochette, J., Delescluse, P., Etienne, P., Chevrier, J., and Linh, N. T. (1982). *Electron. Lett.* **18,** 758.
Massies, J., Rochette, J. F., Etienne, P., Delesclude, P., Huber, A. M., and Chevrier, J. (1983). *J. Crystal Growth.* **64,** 101.
McAfee, S. R., Lang, D. V., and Tsang, W. T. (1982). *Appl. Phys. Lett.* **40,** 520.
McAfee, S. R., Tsang, W. T., and Lang, D. V. (1981). *J. Appl. Phys.* **52,** 6165.
Meynadier, M. H., Brum, J. A., Delalande, C., and Voos, M. (1985). *J. Appl. Phys.* **58,** 4307.
Miller, G. L., Lang, D. V., and Kimmerling, L. C. (1977). *Annu. Rev. Mater. Sci.* **7,** 377.
Miller, B. I., and McAfee, J. H. (1978). *J. Electrochem. Soc.* **125,** 1310.
Miller, R. C., Tsang, W. T., and Munteanu (1982). *Appl. Phys. Lett.* **41,** 374.
Milnes, A. G. (1973). *In* "Deep Impurities in Semiconductors." Wiley Interscience, New York.
Mimura, T., Joshin, K., Hiyamizu, S., Hikosaka, K., and Abe, M. (1981). *Jpn. J. Appl. Phys.* **20,** L 598.
Mimura, T., Hiyamizu, S., Fujii, T., and Nanbu, K. (1980). *Jpn. J. Appl. Phys.* **19,** L 225.
Mircea, A., and Mitonneau, A. (1979). *J. de Physique Lett.* **40,** L 31.

Mircea, A., Mitonneau, A., and Hallais, J. (1977). *Physical Rev.* **B 16,** 3665.
Mitchel, W. C., and Yu, P. W. (1985). *J. Appl. Phys.* **57,** 623.
Mitonneau, A., Martin, G. M., and Mircea, A. (1977). *Electron Lett.* **13,** 666.
Mizuta, M., Tachikawa, M., Kukimoto, H., and Minomura, S. (1985). *Jpn. J. Appl. Phys.* **24,** L 143.
Mizutani, T., and Hirose, K. (1985). *Jpn. J. Appl. Phys.* **24,** L 119.
Mooney, P. M. (1983). *J. Appl. Phys.* **54,** 208.
Mooney, P. M., Fischer, R., and Morkoç, H. (1985a). *J. Appl. Phys.* **57,** 1928.
Mooney, P. M., Solomon, P. M., and Theis, T. N. (1985b). *Conf. Ser.-Inst. Phys.* No. 74, 617.
Morkoç, H., and Cho, A. Y. (1979). *J. Appl. Phys.* **50,** 6413.
Morkoç, H., Cho, A. Y., and Radice, C., Jr. (1980). *J. Appl. Phys.* **51,** 4882.
Morkoç, H., Peng, C. K., Henderson, T., Kopp, W., Fischer, R., Erickson, L. P., Longerbone, M. D., and Youngman, R. C. (1985). *IEEE Electron Device Lett.* **EDL-6,** 381.
Narayanamurti, V., Logan, R. A., and Chin, M. A. (1979). *Phys. Rev. Lett.* **43,** 1536.
Naritsuka, S., Yamanaka, K., Mihara, M., and Ishii, M. (1984). *Jpn. J. Appl. Phys.* **23,** L 112.
Naritsuka, S., Yamanaka, K., Mannoh, M., Mihara, M., and Ishii, M. (1985). *Jpn. J. Appl. Phys.* **24,** 1324.
Narozny, P., and Beneking, H. (1985). *Electron. Lett.* **21,** 1051.
Nashimoto, Y., Dhar, S., Hong, W. P., Chin, A., Berger, P., and Bhattacharya, P. K. (1986). *J. Vac. Sci. Technol.* **B 4,** 540.
Nathan, M. I., Mooney, P. M., Solomon, P. M., and Wright, S. L. (1985). *Appl. Phys. Lett.* **47,** 628.
Neave, J. H., Blood, P., and Joyce, B. A. (1980). *Appl. Phys. Lett.* **36,** 311.
Neumark, G. F., and Kosai, K. (1983). *In* "Semiconductors and Semimetals" (R. K. Willardson and A. C. Beer, eds.), Vol. **19,** pp. 1–74. Academic Press, New York.
Nojima, S., Tanaka, H., and Asahi, H. (1986). *J. Appl. Phys.* **59,** 3489.
O'Connor, P., Pearsall, T. P., Cheng, K. Y., Cho, A. Y., Hwang, J. C. M., and Alan, K. (1982). *IEEE Electron Device Lett.* **EDL-3,** 64.
Ohno, H., Akatsu, Y., Hashizume, T., and Hasegawa, H. (1985). *J. Vac. Sci. Technol.* **B 3,** 943.
Ohno, H., Barnard, J., Wood, C. E. C., and Eastman, L. F. (1980). *IEEE Electron Device Lett.* **EDL-1,** 154.
Ohno, H., Wood, C. E. C., Rathbun, L., Morgan, D. V., Wicks, G. W., and Eastman, L. F. (1981). *J. Appl. Phys.* **52,** 4033.
Okushi, H., and Tokumaru, Y. (1981a). *Jpn. J. Appl. Phys.* **20,** Suppl. 20-1, 261.
Okushi, H., and Tokumaru, Y. (1981b). *Jpn. J. Appl. Phys.* **20,** L 45.
Osbourn, G. C. (1982). *J. Appl. Phys.* **53,** 1586.
Osbourn, G. C., Biefield, R. M., and Gourley, P. L. (1982). *Appl. Phys. Lett.* **41,** 172.
Panish, M. B. (1980). *J. Electrochem. Soc.* **127,** 2729.
Pearsall, T. P., Hendel, R., O'Connor, P., Alvi, K., and Cho, A. Y. (1983). *IEEE Electron Device Lett.* **EDL-4,** 5.
Petroff, P. M., Miller, R. C., Gossard, A. C., and Wiegmann, W. (1984). *Appl. Phys. Lett.* **44,** 217.
Ploog, K. (1980). *Cryst.: Growth, Prop., Appl.* **3,** 73.
Quillec, M., Goldstein, M., Le Roux, G., Burgeat, J., and Primot, J. (1984). *J. Appl. Phys.* **55,** 290.
Reynolds, D. C., Bajaj, K. K., Litton, C. W., Singh, J., Yu, P. W., Henderson, T., Pearah, P., and Morkoç, H. (1985). *J. Appl. Phys.* **58,** 1643.
Robinson, J. Y., and Ilegems, M. (1978). *Rev. Sci. Instrum.* **49,** 205.
Sah, C. T. (1976). *Solid State Electron.* **19,** 975.
Sah, C. T., Forbes, L., Rosier, L. L., and Tasch, A. F., Jr. (1970). *Solid State Electron.* **13,** 759.

Sah, C. T., Ning, T. H., Rosier, L. L., and Forbes, L. (1971). *Solid State Commun.* **9,** 917.
Sah, C. T., Rosier, L. I., and Forbes, L. (1969). *Appl. Phys. Lett.* **15,** 316.
Sah, C. T., and Walker, J. W. (1973). *Appl. Phys. Lett.* **22,** 384.
Sakai, S., Soga, T., Takeyasu, M., and Umeno, M. (1985). *Jpn. J. Appl. Phys. Lett.* **54,** L 666.
Sakamoto, M., Okada, T., and Mori, Y. (1985). *J. Appl. Phys.* **58,** 337.
Saxena, A. K. (1980a). *Appl. Phys. Lett.* **58,** 79.
Saxena, A. K., (1980b). *J. Phys. C: Solid State Phys.* **13,** 4323.
Saxena, A. K. (1981). *Phys. Stat. Sol.* **B 105,** 777.
Schubert, E. F., and Ploog, K. (1984). *Appl. Phys.* **33,** 63.
Seo, K. S.; Bhattacharya, P. K., and Nashimoto, Y. (1985). *IEEE Electron Dev. Lett.* **EDL-6,** 64.
Shockley, W., and Read, W. T., Jr. (1952). *Phys. Rev.* **87,** 835.
Singh, J., and Bajaj, K. K. (1985a). *Appl. Phys. Lett.* **46,** 577.
Singh, J., and Bajaj, K. K. (1985b). *J. Appl. Phys.* **57,** 5433.
Singh, J., and Bajaj, K. K. (1985c). *Appl. Phys. Lett.* **47,** 594.
Singh, J., and Bajaj, K. K. (1985d). *In* "Proceedings of the Large Scale Computational Device Modeling Workshop" (K. Hess, ed.), pp. 91–106. University of Illinois Press, Naperville, Illinois.
Skromme, B. J., Bose, S. S., Lee, B., Low, T. S., Lepkowski, T. R., DeJule, R. Y., Stillman, G. E., and Hwang, J. C. M. (1985). *J. Appl. Phys.* **58,** 4685.
Solomon, P. M., and Morkoç, H. (1984). *IEEE Trans. on Electron Devices.* **ED-31,** 1015.
Stall, R. A., Wood, C. E. C., Kirchner, P. D., and Eastman, L. F. (1980). *Electron. Lett.* **16,** 171.
Störmer, H. L., Dingle, R., Gossard, A. C., Wiegmann, W., and Sturge, M. D. (1979). *Solid State Commun.* **29,** 705.
Stringfellow, G. B., Stall, R. A., and Koschel, W. (1981). *Appl. Phys. Lett.* **38,** 156.
Subramanian, S., Schuller, U., and Arthur, J. R., Jr. (1985a). *J. Vac. Sci. Technol.* **B 3,** 650.
Subramanian, S., Schuller, U., and Arthur, J. R., Jr. (1985b). *J. Appl. Phys.* **58,** 845.
Swaminathan, V., and Tsang, W. T. (1981). *Appl. Phys. Lett.* **38,** 347.
Tachikawa, M., Mizuta, M., Kukimoto, H., and Minomura, S. (1985). *Jpn. J. Appl. Phys.* **34,** L 821.
Takikawa, M., and Ozeki, M. (1985). *Jpn. J. Appl. Phys.* **24,** 303.
Tegude, F. J., and Heime, K. (1984). *Conf. Ser.-Inst.* No. 74, 305.
Temkin, H., Alvi, K., Wagner, W. R., Pearsall, T. P., and Cho, A. Y. (1983). *Appl. Phys. Lett.* **42,** 845.
Thomas, H. (1985). *J. Appl. Phys.* **57,** 4619.
Thorne, R. E., Drummond, T. J., Lyons, W. G., Fischer, R., and Morkoç, H. (1982). *Appl. Phys. Lett.* **41,** 189.
Tsang, W. T. (1981a). *Appl. Phys. Lett.* **39,** 134.
Tsang, W. T. (1981b). *Appl. Phys. Lett.* **38,** 204.
Tsang, W. T. (1981c). *J. Appl. Phys.* **52,** 3861.
Tsang, W. T. (1982). *Appl. Phys. Lett.* **40,** 217.
Tsang, W. T. (1984). *Appl. Phys. Lett.* **45,** 217.
Tsang, W. T. (1985). *In* "Semiconductors and Semimetals" (R. K. Willardson, and A. C. Beer, ed.), Vol. **22,** Part A, pp. 95–207. Academic Press, New York.
Tsang, W. T., and Ditzenberger, J. A. (1981). *Appl. Phys. Lett.* **39,** 193.
Tsang, W. T., Dixon, M., and Dean, B. A. (1983). *IEEE J. Quantum Electron.* **QE-19,** 59.
Valois, A. J., and Robinson, G. Y. (1983). *IEEE Electron Device Lett.* **EDL-4,** 360.
Valois, A. J., and Robinson, G. Y. (1985). *J. Vac. Sci. Technol.* **B 3,** 649.
Valois, A. J., Robinson, G. Y., Lee, K., and Shur, M. S. (1983). *J. Vac. Sci. Technol.* **B 1,** 190.
Van Hove, J. M., Cohen, P. I., and Lent, C. S. (1983). *J. Vac. Sci. Technol.* **A 1,** 546.

Wakefield, B., Halliwell, M. A. G., Kerr, T., Andrews, D. A., Davies, G. J., and Wood, D. R. (1984). *Appl. Phys. Lett.* **44,** 341.

Wood, D. R. (1984). *Appl. Phys. Lett.* **44,** 341.

Walpole, J. N., Calawa, A. R., Hamman, T. C., and Groves, S. H. (1976). *Appl. Phys. Lett.* **28,** 552.

Watanabe, M. O., and Maeda, H. (1984). *Jpn. J. Appl. Phys.* **23,** L734.

Watanabe, M. O., Morizuka, K., Mashita, M., Ashizawa, Y., and Zohta, Y. (1984). *Jpn. J. Appl. Phys.* **23,** L 103.

Welch, D. F., Wicks, G. W., Eastman, L. F., Parayanthal, P., and Pollack, F. H. (1985). *Appl. Phys. Lett.* **46,** 169.

Wicks, G., Wang, W. I., Wood, C. E. C., Eastman, L. F., and Rathbun, L. (1981). *J. Appl. Phys.* **52,** 5792.

Wicks, G., Wood, C. E. C., Ohno, H., and Eastman, L. F. (1982). *J. Electron. Mater.* **11,** 435.

Windhorn, T. H., and Metze, G. M. (1985). *Appl. Phys. Lett.* **47,** 1031.

Wood, C. E. C. (1982). *In* "GaInAsP Alloy Semiconductors" (T. P. Pearsall, ed.). pp. 87–106. Wiley Interscience, New York.

Zhou, B. L., Ploog, K., Gmelin, E., Zheng, X. Q., and Schulz, M. (1982). *Appl. Phys. A.* **28,** 223.

Zylbersztejn, A. (1978). *Appl. Phys. Lett.* **33,** 200.

CHAPTER 4

Semiconductor Properties of Superionic Materials

Yu. Ya. Gurevich

INSTITUTE OF ELECTROCHEMISTRY
U.S.S.R. ACADEMY OF SCIENCES
MOSCOW, U.S.S.R.

and

A. K. Ivanov-Shits

INSTITUTE OF CRYSTALLOGRAPHY
U.S.S.R. ACADEMY OF SCIENCES
MOSCOW, U.S.S.R.

ISBN 0-12-752126-7

List of Symbols

$\mathfrak{A}$	Richardson constant
a_k	activity of particles of species k
C_{dl}	double-layer capacitance
D_k	diffusion coefficient of particles of species k
D_T	thermal diffusion coefficient
$\tilde{D}$	chemical diffusion coefficient
E_a	activation energy
E_g	band gap
E_B	height of the potential barrier from the side of the metal
E_c	energy of conduction band edge in SIM bulk
E_v	energy of valence band edge in SIM bulk
e	absolute value of electron charge (1.6×10^{-19} C)
F	Fermi level
$F_{e,h}$	Fermi quasi level of electrons (e) and holes (h)
J	density of light flux that entered the SIM
i	electric current density
i^o	exchange current density
i_D	limiting current
i_{ph}	density of photocurrent
i_T	thermoemission current density
j	flux density of particles
K	equilibrium constant
L	thickness of sample
L_D	Debye length
m^O	mass of free electron
$\mathfrak{N}_A$	Avogadro number
N_d	concentration of donor impurities

$N_{c,v}$	effective density of states in the conduction band (c) and in the valence band (v)
n	concentration of particles
n_{int}	concentration of electrons (and holes) in intrinsic semiconductor
$n^*_{e,h}$	concentration of electrons (e) and holes (h) in illuminated semiconductor
P	pressure
Q^*	heat of transport
S^*	transport entropy
S	sample cross-section
T	absolute temperature
T_{tr}	temperature of phase transition
t	time
t_k, t'_k	transport numbers of particles of species k
u	mobility
V_m	mole volume
w_M, w_S	thermodynamic work function for electrons from the metal (M) and SIM (S)
$z \gtrless 0$	charge number
α, β	transfer coefficients
γ, δ	deviations from the ideal stoichiometry
η	overvoltage
θ	thermoelectric power
μ	chemical potential
$\tilde{\mu}$	electrochemical potential
ρ(E)	density of states
σ	conductivity
τ	relaxation time
φ	electric potential
χ	affinity to electron

Superscripts

0	equilibrium

Subscripts

c	conduction
e	electron
el	electronic
h	hole
i	ion
s	surface
v	valence

I. Introduction. General Information on Superionic Materials

Superionic materials comprise a special class of solids that has become a subject of ever increasing interest of a large circle of researchers. Publications devoted to the study and to various aspects of superionic material application are counted in hundreds a year, an essential part of many leading physical and chemical journals. Unusual and often paradoxical properties of these substances are of great interest for fundamental problems of both solid state physics and physical chemistry and for a number of applied problems.

Today, superionic materials are already used in the design of power sources, various information transducers, and, in particular, electrochromic displays, capacitors with very high specific capacitance (ionistors), and other functional electric devices. They are also employed in electrochemical sensors used in a number of technical processes, in the analysis of composition, thermodynamic and kinetic characteristics of various substances, etc. The application range of such materials is rapidly increasing (Mahan and Roth, 1976; Hagenmüller and van Gool, 1977; Ukshe and Bukum, 1977; Geller, 1977; Chebotin and Perfil'ev, 1978; Vashishta *et al.*, 1979; Salamon, 1979; Chandra, 1981; Bates and Farrington, 1981; Takahashi *et al.*, 1981; Kleitze *et al.*, 1983; Perram, 1983; Boyce *et al.*, 1986).

One of the most important features of superionic conductors is their anomalously high ionic conductivity, with respect to other solid materials. Thus, ionic conductivity of typical ionic crystals at temperatures not too close to the melting point, does not, as a rule, exceed a value of $\sigma_i \approx 10^{-8}$–$10^{-9}\,(\Omega\,\text{cm})^{-1}$, whereas ionic conductivity of "good" superionic conductors under the same conditions is $\sigma_i \approx 10^{-1}$–$10^{0}\,(\Omega\,\text{cm})^{-1}$. For comparison, electronic conductivity of metals is $\sigma_e \approx 10^{4}$–$10^{6}\,(\Omega\,\text{cm})^{-1}$.

Ionic conductivity of superionic conductors is comparable, in the order of magnitude, to the values typical for melts and concentrated solutions of strong electrolytes. Therefore, superionic conductors are often called solid electrolytes. Note that solids with ionic conductivity exceeding their electronic conductivity are sometimes regarded as solid electrolytes. In such an approach, many "conventional" ionic crystals are also related to solid electrolytes. We shall hold to the definition according to which solid electrolytes are materials with high ionic conductivity, $\sigma_i \gtrsim 10^{-3}\,(\Omega\,\text{cm})^{-1}$.

Below, the compounds possessing mixed properties will be considered. On the one hand, their conductivity is comparable to that of a liquid melt or a solution, while, on the other, their mechanical strength and elasticity are close to those of a solid. In actual fact, superionic materials are a peculiar hybrid of a liquid electrolyte and a solid dielectric or semiconductor.

The explosion of investigations of superionic materials dates from the late sixties. These investigations were triggered, in particular, by the synthesis of Ag_4RbI_5 (Bradley, Green, 1967; Owens, Argue, 1967) and by its subsequent use as a solid electrolyte in electrochemical power sources designed for work in space (the main advantages of such sources are their compactness, mechanical strength and reliability). An essential role has also been played by the fact that Ag_4RbI_5 possesses high ionic conductivity with respect to Ag^+ cations at temperatures as low as 120 K. At room temperature, conductivity is of about $0.3\,(\Omega\,\text{cm})^{-1}$, and with the further increase in temperature, it increases by approximately one order of magnitude, being as high as 3–$5\,(\Omega\,\text{cm})^{-1}$. Note that even $0.3\,(\Omega\,\text{cm})^{-1}$ is very high ionic conductivity for a solid—it exceeds σ_i for sodium chloride by more than seventeen orders of magnitude at the same temperature.

However, this conductivity is still far from the limit. In the late seventies, $Cu_4RbCl_3I_2$, a compound related to Ag_4RbI_5, was synthesized (Takahashi *et al.*, 1979; Geller, 1979; Vershinin *et al.*, 1981) with $\sigma_i \approx 0.5\ (\Omega\ \text{cm})^{-1}$ (with respect to copper cations) at room temperature. This substance is still a "record-holder" among the superionic conductors.

The subject of superionic conductivity had arisen in physics long before the above-mentioned date (late sixties) and has quite a long story. As far back as 1834, Faraday mentioned the existence of crystalline solids with noticeable ionic conductivity. The next studies of this subject were associated with the names of Warburg (1884) and Nernst (1888, 1900), who also investigated ceramic and vitreous solid phases. An important contribution to the field was made by Tubandt and Lorenz (1913) shortly before the First World War. While studying the electrical properties of silver halcogenides, they established that the solid modification of silver iodide, nowadays called the α-phase (α-AgI), possesses rather high ionic conductivity, unlike silver chloride and silver iodide. The state of high conductivity is formed at 420 K and is accompanied by a jumpwise increase in ionic conductivity of more than three orders of magnitude. Then, conductivity continues increasing rather slowly up to the value $2.6\ (\Omega\ \text{cm})^{-1}$ at the melting point (828 K) of the crystal. During melting, ionic conductivity drops again, its value becoming somewhat lower (by approximately 7%) then in the crystalline state. As has been shown by the subsequent studies, the above described behavior is typical of a series of superionic materials.

The same authors made the first attempts to measure electonic conductivity, σ_{el}. They have established that for the superionic conductor α-AgI the condition $\sigma_{el}/\sigma_i \ll 1$ is valid. A similar condition is also fulfilled for many other superionic conductors, although the σ_{el}/σ_i ratio varies over quite a large range and can even exceed unity for some superionic materials, depending on their nature and external conditions.

Some problems associated with electronic conductivity in superionic conductors (solid electrolytes) have been considered by Wagner (1966), Heyne (1973, 1977), Fabry and Kleitz (1976), Weppner and Huggins (1978), Gurevich and Ivanov-Shits (1980), Couturier *et al.* (1983) and Ivanov-Shits (1985).

Experimental data on electronic conductivity in superionic materials (SIMs) is given in Table A in the Appendix.

Without analyzing, at this time, the properties of specific materials, we should like to emphasize one important fact. In contrast to liquid electrolytes, in which electrons contribute a negligible amount to the current flow in the solution, in solid electrolytes the current necessarily has a component associated with electronic carriers, although its absolute value may be much smaller than that of the ionic component (here and in what follows, the term "electronic carriers" implies both electrons proper and holes, i.e., positively charged electronic vacancies). Note that electronic currents in systems with SIMs may play a noticeable, and sometimes even decisive, role, even though in some cases such currents are small.

There is a number of phenomena important in practice whose understanding and description require the analysis of electronic properties of superionic materials. The major factors are:

(i) The origination of electromotive force in batteries and self-discharge in galvanic elements and ionistors (Fig. 1a).

(ii) Heterogeneous chemical and electrochemical reactions at interfaces, with the participation of delocalized electronic carriers in superionic materials (Fig. 1b).

(iii) Nonstationary processes in the range of relatively high frequencies. It is in this range that the characteristic relaxation times in electronic subsystems are, as a rule, much smaller than in ionic ones (Fig. 1c), which is of great importance.

(iv) Excitation of electrons and holes by electromagnetic radiation, which increases the concentration of delocalized electronic carriers, and changes their role in the processes occurring in superionic materials (Fig. 1d).

(v) The formation of radiation absorption and scattering spectra and luminescence spectra for superionic materials (Fig. 1e).

Thus, even if the dc conductivity due to electrons is low, the electronic properties of superionic materials should be thoroughly studied. Also, as has already been noted, there are compounds in which the electronic component of conductivity is comparable with the ionic component, or even exceeds it. Recently, such materials—the so-called ionic-electronic conductors—have been used as elements of various electronic devices. This leads to the subject of deliberate increases of electronic conductivity in materials possessing ionic conductivity. Therefore, the above list of processes caused by the presence of electronic carriers in superionic conductors should be expanded as follows.

(vi) Phenomena occurring in materials with "mixed" ionic-electronic conductivity (Fig. 1f).

Thus, we witness the active formation of a new branch of solid-state physics associated with the study of electronic processes in superionic conductors. High ionic conductivity reflects the electrolytic aspect of the hybrid nature of such materials, whereas their electronic properties are closely related to the semiconductor aspect of their nature. The study of

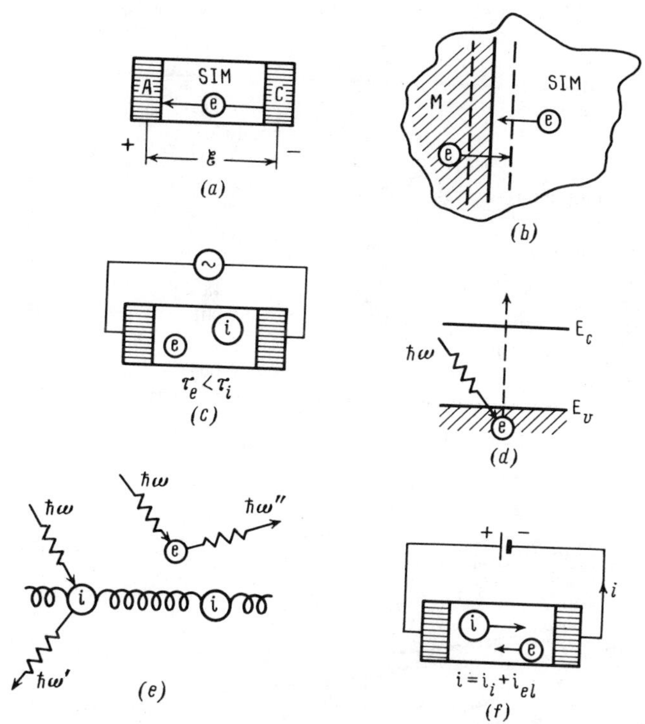

FIG. 1. Various processes in superionic materials with the participation of electronic carriers.

electronic-ionic processes in superionic materials also reflects the intimate relationship between science and technology; "pure scientific" results soon find various practical applications.

The number of synthesized and studied compounds with high ionic conductivity amounts to several dozen. Current carriers in these compounds are mono-, di- and tri-valent cations, such as silver, copper, sodium, potassium, lithium, calcium, lead, barium, neodymium, europium, some other metal cations (Funke, 1976; Seevers *et al.*, 1983; Dunn, 1986) and halide and oxygen anions (Chebotin and Perfil'ev, 1978; Reau and Portier, 1978). There are also superionic compounds with proton conductivity (Chandra, 1984; Goodenough *et al.*, 1985). Some examples of superionic materials, that have typical high conductivity at room temperature, are listed in Table I.

In turn, temperatures at which high ionic conductivity ($\sigma_i \gtrsim 10^{-3}\ \Omega^{-1}\ \text{cm}^{-1}$) is reached vary over the wide range for different compounds. The full temperature range of superionic conductivity covered by the known superionic materials exceeds 1500°C (Fig. 2).

TABLE I

IONIC CONDUCTIVITY OF SOME SIMs

Compound	Ionic charge carrier	Ionic conductivity, σ_i, at 298 K, (Ω^{-1} cm^{-1})	Activation energy, E_a, (eV)
$Na_2O \times mAl_2O_3$	Na^+	0.033	0.14
Ag_4RbI_5	Ag^+	0.30	0.10
$Cu_4RbCl_3I_2$	Cu^+	0.48	0.15
Li_3N	Li^+	0.001	0.29
$H_3PO_4(WO_3) \times 29H_2O$	H^+	0.17	0.14
Pb-β''-Al_2O_3	Pb^{2+}	0.004	—

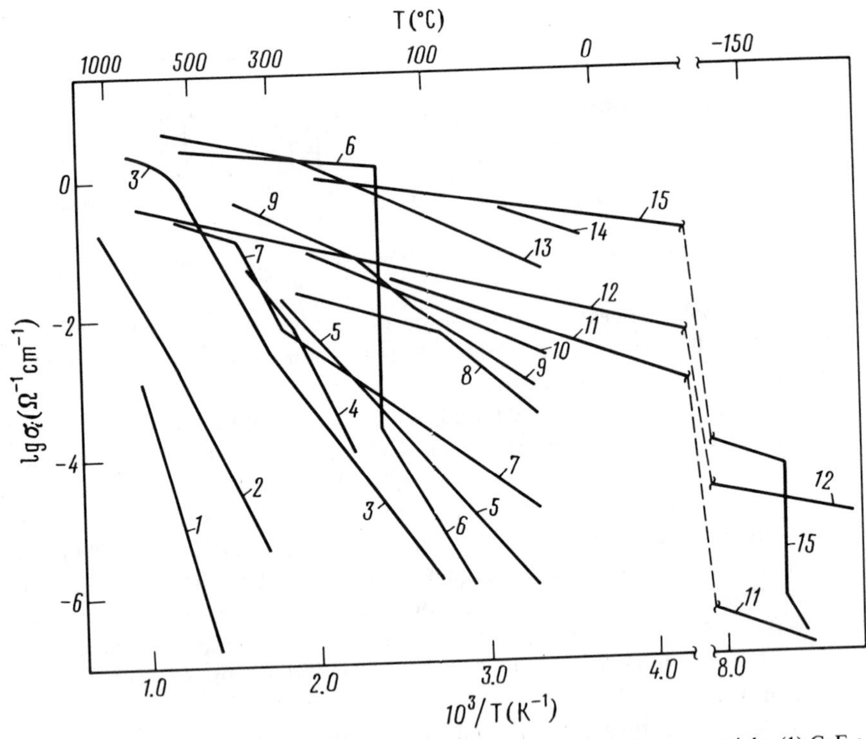

FIG. 2. Ionic conductivity, σ_i, versus temperatures of some superionic materials. (1) CaF_2; (2) ZrO_2 + 16 mole % Y_2O_3; (3) PbF_2; (4) $Li_3Sc_2(PO_4)_3$; (5) $Li_{14}Zn(GeO_4)_4$ (LISICON); (6) AgI; (7) CuI; (8) $PbSnF_4$; (9) $Na_3Zr_2PSi_2O_{12}$ (NASICON); (10) Li_3N; (11) Cu_2RbCl_3; (12) Na-β-Al_2O_3; (13) Na-β''-Al_2O_3; (14) $H_3Mo_{17}PO_{40} \times 29H_2O$; (15) Ag_4RbI_5.

The physical mechanism providing superionic conductivity is associated with high mobility of one ionic species in the rigid framework of another species. Such high ion mobility is attained if the following conditions are met:

Firstly, the number of vacant positions in the rigid framework should essentially exceed that of ions able to occupy the vacant positions. Only the fulfillment of this condition can prevent "competition" between ions to occupy these positions.

Secondly, the positions should be such that the activation energy of the transition between these positions should not be too high, or, more exactly, that the dimensionless E_a/kT ratio (k is the Boltzmann constant) should not exceed a value of two–three at a temperature below the melting point or that of the compound decomposition.

Thirdly, there should exist a net of ion trajectories in a rigid framework that penetrates the whole structure; otherwise, no direct ionic current can flow. Note here that a certain arrangement of such trajectories may result in ionic conductivity of reduced dimensionality—one- or two-dimensional conductivity.

If the above-formulated conditions are met, ionic mobility in a solid is of the same order of magnitude as that observed in water at room temperature ($\sim 10^{-3}$ cm^2/V s). Anomalously high ion mobility (or fast ionic transport) is observed in polymer structures, ion-exchange resins and membranes, and, along with electronic conductivity, in some oxide films and even in metals (LiAl-alloy). If, along with the three above indicated conditions, the fourth condition is also fulfilled, i.e., if the number of fast ions in a material is sufficiently large, then superionic conductivity arises, and such a material becomes a superionic conductor.

Proceeding from the specific character of the structure and character of ionic conductivity, superionic materials may conventionally be divided into several types. The first type is built by crystals with the so-called intrinsic structural disorder. These are, in particular, classical α-AgI, a series of compounds with the α-AgI motive described by the general formula Ag_4MI_5 (where M = Rb, K, NH_4, $Cs_{0.5}K_{0.5}$), and also fluorides of some di- and trivalent metals (CaF_2, YF_3, LuF_3) in which high ionic conductivity is due to F^- anions.

Figure 3 schematically shows the well-studied structure of an α-AgI crystal. The rigid framework of the structure is built by I^- anions forming a cubic body-centered sublattice. The unit cell includes two I^- ions, namely, the I^- ion in the cube center and the I^- ion that is a sum of eighths of ions occupying eight cube vertices (the lattice as a whole is built by the unit cell translations in three mutually orthogonal directions). Silver cations, which are of comparatively small size, may occupy positions in lattice voids. These voids have approximately equal volumes but different shapes and

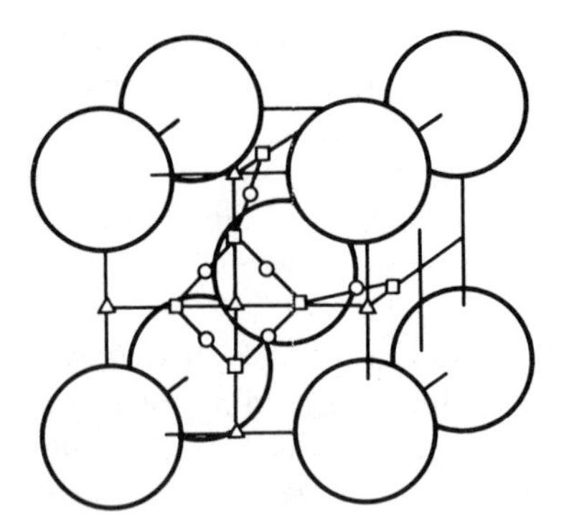

FIG. 3. Schematic representation of the α-AgI structure: (△) b-sites, (□) d-sites, (○) h-sites.

coordination numbers. In the unit cell, there are six equivalent, sixfold-coordinated octahedral positions (b-sites), twelve fourfold-coordinated tetrahedral positions (d-sites), and 24 threefold-coordinated positions (h-sites). Deformed tetrahedrals are of the maximum volume, whereas distorted octahedrals have the minimum volume.

Thus, altogether there are 42 sites in the unit cell that may be occupied by Ag^+ ions. Since there are two cations and two anions in the unit cell, each silver cation may formally occupy 21 sites (Strock, 1936; Burley, 1967).

At the same time, the detailed structure investigations carried out recently indicate (Schulz, 1982) that, in actual fact, silver ions do not occupy all 42 possible positions, but are statistically distributed over twelve tetrahedral d-sites.

The strong anharmonicity of thermal vibrations observed for Ag^+ ions indicates their possible motion, mainly within the chain of tetrahedra with the shared planes.

As a result, in the existence range of the α-phase, silver cations continuously travel over the allowed positions in the rigid anion sublattice. Thus, α-AgI is a typical superionic conductor with intrinsic structural disorder with respect to Ag^+ ions. This structure may be pictorially represented as one consisting of a rigid anion sublattice "immersed" into a "cation liquid".

A crystal structure of the α-AgI type is also possessed by a number of other superionic materials—in particular, by silver chalcogenides and ternary silver salts (Ag_3SI, Ag_3SBr). In these compounds, Ag^+ cations are also distributed over tetrahedral positions, but the unit cell contains not two (as in α-AgI), but three (as in Ag_3SI) or four (as in Ag_2S), cations.

Superionic material CaF_2 with high conductivity with respect to fluorine anions (F^-) is of the fluorite type (Fig. 4), in which Ca^{2+} ions occupy vertices

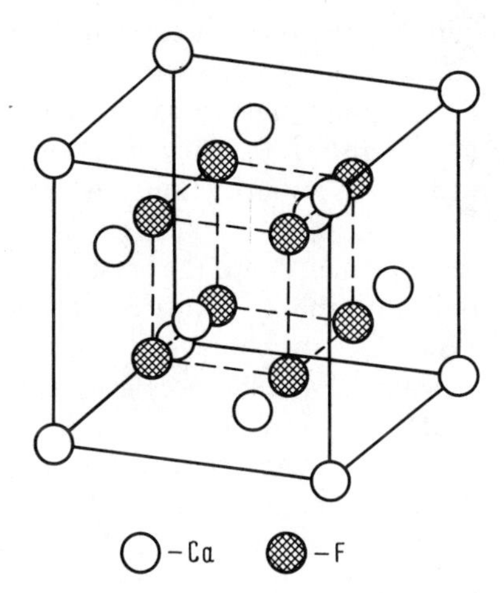

FIG. 4. Schematic representation of the CaF_2 structure. Dotted line indicates a cube formed by fluorine ions.

and the centers of the cube faces. In the ordered low-temperature phase, eight fluorine anions of the unit cell, occupying all the tetrahedral sites—the octahedral ones being free—form a primitive cubic sublattice.

In the high-temperature superionic phase, these anions are distributed over the positions of two types—eight tetrahedral and four octahedral. In distinction from the α-AgI structure, in fluorite, tetrahedra do not share their faces; they share faces with octahedra, forming a chain of alternating tetrahedra and octahedra. It is along this chain that fluorine anions are moving.

A distinctive feature of many superionic materials with intrinsic structural disorder is the existence of a certain critical temperature for each substance at which a jumpwise change in ionic conductivity is observed. At the same time, anomalies in the temperature behavior of some thermodynamic and kinetic characteristics are also observed. In the final analysis, these anomalies are due to specific (in other words, jumpwise) disordering—partial or complete—of a sublattice formed by the ions of one kind. At the same time, the other sublattice experiences, as a rule, a certain rearrangement (polymorphous transition), preserving its rigidity and, thus, providing the mechanical strength of the crystal as a whole. This jumpwise structural disordering is a first-order phase transition. The thermodynamic theory of phase transitions in superionic conductors was stated by, e.g., Gurevich and Kharkats (1977, 1978, 1986a).

Thus, the above considered superionic materials may have two qualitatively different states. At temperatures below the critical, they behave as usual ionic crystals (dielectric or semiconductor phase); above these temperatures, they acquire a superionic state (electrolytic phase). In the latter state, these materials are superionic conductors.

The second type of superionic materials is formed by compounds with impurity-induced structural disorders. Their high ionic conductivity is due to the high concentration of impurity ions, which promote structural disordering.

Let CaO monoxide be introduced into the crystal lattice of ZrO_2 dioxide. Then, Ca^{2+} ions are built into the Zr^{4+} lattice, whereas O^{2-} ions build up the oxygen sublattice. But since the charges of zirconium and calcium ions are different, oxygen vacancies should necessarily appear in such a mixed crystal. If the number of such vacancies is sufficiently large, the mobility of oxygen ions (O^{2-}) markedly increases, and they start moving from one vacancy to another. Typical superionic materials (solid electrolytes) with impurity-induced disorder are compounds described by the general formulae $MO_2 - M'O$ and $MO_2 - M''_2O_3$, where M is a tetravalent metal (Zr, Hf, Ce), M' is a divalent metal (Ca, Sr, Ba), and M'' is a trivalent metal (Sc, Y).

Unlike crystals possessing intrinsic structural disorders, compounds with impurity-induced disorders do not show a drastic temperature jump of ionic conductivity, although the ionic conductivity markedly increases with the temperature rise.

Also, since the activation energy E_a of ions is high, temperatures T, corresponding to the fast ion transport, are, as a rule, rather high ($T \gtrsim 1000$ K). Therefore, the majority of superionic materials with impurity-induced disorders are related to the so-called high-temperature solid electrolytes.

And, lastly, some noncrystalline substances may also be regarded as superionic materials. These are, in particular, glassy solids possessing relatively high ionic conductivity due to modifiers (or electroactive impurities) introduced into their composition and supercooled amorphous phases of some compounds that, in the crystalline state, are superionic conductors (Ivanov-Shits and Tsvetnova, 1982; Liu *et al.*, 1985; Minami, 1985).

Special attention should be payed to the so-called oxide bronzes of transition metals—or, for brevity, simply "bronzes," since they possess, as a rule, high electronic conductivity.

Transition metals (vanadium, tungsten, molybdenum and others) may form chemical compounds with oxygen. The main structural element here is an octahedron formed by six oxygen atoms with a transition-metal atom M in the center (Fig. 5). Regular octahedra connected by vertices may, in principle, form regular crystal structures corresponding to an ideal stoichiometric composition MO_3, with six oxygen "half-atoms" per each M atom.

The above situation is ideal. In fact, the metal-oxygen distances and angles between chemical bonds in the above oxides may vary over wide ranges. Therefore, generally speaking, the initial octahedra are not regular

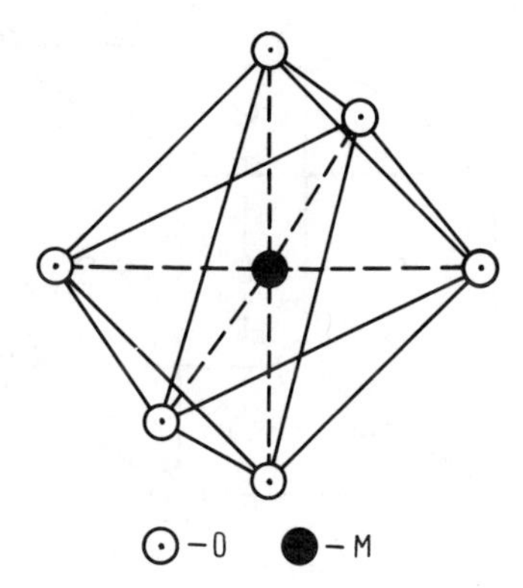

FIG. 5. Octahedral structure formed by oxygen (O) and transition metal (M) atoms.

geometrical figures. Moreover, transition metals may form such compounds with oxygen, in which case they are oxidized to different degrees. Thus, the valence of tungsten may vary from six or five, and, therefore, the oxide formula should be written as WO_{3-y}, where $y > 0$ characterizes the fraction of W atoms with the reduced valence. Accordingly, the initial octahedra in complicated structures may be connected by vertices or share edges, the latter case corresponding to the reduced degree of oxidation, since each oxygen atom is now bound not to two, but to a larger number of tungsten atoms.

Thus, octahedron deformation and different ways of connecting octahedrons result in the formation of peculiar three-dimensional frameworks in transition-metal oxides. These oxides may have relatively wide channels, appropriate for fast ionic transport.

Oxide bronzes are the compounds of variable compositions containing, along with transition metal and oxygen, electropositive (as a rule, alkali or alkali-earth) metal. Thus, sodium-tungsten bronze is described by the formula Na_xWO_{3-y}, where $x, y > 0$. If $x = 0$, the above formula describes tungsten oxide, WO_{3-y}; if $x = 1$ and $y = 0$, the compound corresponds to the ideal stoichiometric composition $NaWO_3$, with the perovskite-type structure schematically shown in Fig. 6. It should be emphasized that, in this structure, oxygen atoms form regular octahedra, with the centers being occupied by tungsten atoms. In other words, the mutual arrangement of O and W atoms is as in the ideal tungsten oxide (cf. Figs. 5 and 6).

Thus, in the perovskite-type structure, sodium atoms situated between the WO_3-octahedra make the octahedra rigid. If there is a slight deficiency of sodium atoms, some sites are empty and the crystal lattice becomes slightly deformed. At the same time, at low sodium concentration, the whole bronze lattice becomes distorted—Na^+ cations, now located in channels formed by

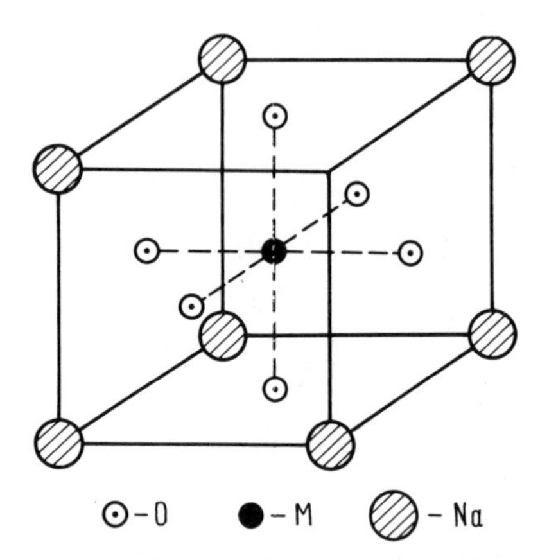

FIG. 6. Schematic representation of the perovskite structure.

deformed octahedra, possessing high mobility. Therefore, sodium-tungsten bronze is a good ionic conductor, with σ_i being a nonmonotonic function of x. With an increase in x from the zero value, σ_i also increases, since the number of carriers, Na^+ ions, monotonically increases. On the other hand, as x approaches unity, the structure becomes more ordered, and the Na^+-ions loose their mobility.

With the variation in x, electronic conductivity and some other properties of bronzes vary. In particular, at high sodium content ($x \simeq 1$), these compounds are of a golden-yellow color and have metallic lust, hence their name "bronzes."

On the whole, oxide bronzes of transition metals comprise quite a vast class of compounds, many of which possess, along with electronic conductivity, conductivity with respect to sodium, potassium, or lithium ions. Note also that the term "bronze" in the physics of superionic materials often applies to any compound formed by the introduction of additional ions into the structures of transition-metal oxides. Thus, H_xWO_3, which may be called "hydrogen bronze," has a high proton conductivity.

Thus, the class of superionic materials may be conventionally divided into three main types: compounds with intrinsic structural disorders, compounds with impurity-induced structural disorders, and compounds with noncrystalline structures. Along with bronzes, high electronic conductivity (the property that is typical for materials with mixed ionic-electronic conductivity) is also possessed by a number of superionic materials of the two former types.

Thus, among superionic conductors with structural disorder, silver and copper chalcogenides (Ag_2S, Ag_2Se, Ag_2Te, Cu_2S) are typical ionic-electronic conductors. Special searches for ionic-electronic conductors at room temperature resulted in the synthesis of $AgCrSe_2$, $CuGaS_2$, and other compounds of this type. Additional examples can be found in Table A of the Appendix.

The stoichiometry of superionic materials with impurity-induced disorder, in particular, that of high-temperature solid electrolytes, may vary over a wide range of values. Under such conditions, the electronic component of conductivity also changes, in some cases even being prevalent over the ionic component. These problems will be discussed in more detail in the following sections.

II. Delocalized Electrons in Superionic Materials

Let us consider origination of electronic carriers in superionic materials and analyze their equilibrium concentration as a function of external conditions.

Electronic properties of crystalline materials, including those with ionic conductivity, e.g., ionic crystals, are described in terms of the well-known concept of the band theory of solids. Such an approach presents no problems, since the relative number of ions occupying lattice interstitials for "ordinary" crystals is very small. On the contrary, in superionic conductors, under the condition that one of the sublattices is "melted," it is impossible to exclude *a priori* the effect of ionic disordering on the electronic properties of the material. The disturbance of the structure periodicity (the absence of long-range order) may make the electronic energy spectra of SIMs similar to the spectra of disordered semiconductors.

Another feature of equilibrium electronic properties of SIMs is associated with the fact that the variation in the external conditions (e.g., ambient atmosphere and temperature) may change the bulk concentration of the chemical components forming the compound, which, in turn, results in the change of the electronic-carrier concentration.

Finally, electronic characteristics, especially of thin samples and films, may be strongly affected by phenomena occurring in the vicinity of the interfaces.

1. Structural Disordering and Electronic Spectra

The study of electronic phenomena in disordered systems (liquid and amorphous semiconductors, strongly doped semiconductors and dielectrics, and liquid metals) is one of the most important directions in the physics of condensed media. A series of monographs and reviews are devoted to the modern theory of the electronic properties of disordered materials and corresponding experiments (see, e.g., Mott and Davis, 1979; Shklovskii and Efros, 1979; Bonch-Bruevich *et al.*, 1981; Bonch-Bruevich, 1983).

The energy spectra of electrons and holes in crystalline semiconductors have been studied quite well. There are regions of continuous spectra—the conduction (c) and valence (v) bands—separated by the so-called forbidden band, in which only some discrete levels are allowed, usually associated with impurities or defects. It is important that the continuous-spectrum states are delocalized—in other words, electrons and holes occupying such levels may participate in the charge transfer at any arbitrary low temperature. On the contrary, the states of the discrete spectrum in the forbidden band are localized; i.e., electron carriers occupying these states may participate in the charge transfer only under conditions of thermal activation (hopping conductivity).

An increase in the impurity concentration, structural disorder, or some other factors may result in the disturbance of the crystal lattice periodicity, with the electronic energy spectrum becoming more complicated. But the spectrum still has ranges corresponding to delocalized and localized states. The above conclusions, drawn from a large set of experimental data on the electrical and optical properties of noncrystalline materials, also follow from the theory.

At present, it may be taken as a fact that the spectra of three-dimensional disordered semiconductors have continuous ranges (the situation for two- and one-dimensional systems is more complicated). Such ranges of continuous spectrum are similar to the c- and v-bands in crystalline materials (in fact, they are often referred to in this way). At the same time, in distinction from the ideal crystals, one cannot consider here the dispersion laws in the $E(p)$ bands; i.e., one cannot consider the energy of electron carriers as a function of quasi-momentum p: if there is no periodicity in the bulk, the components of quasi-momentum p cannot be introduced as "good" quantum numbers.

Between c- and v-bands, there exists an energy region with electron levels corresponding to the localized states of electrons and holes. On the energy scale, these levels may be located at any arbitrary close distances from one another, and, therefore, their energy distribution is characterized by the continuous density of states, $\rho(E)$. Despite the continuity of the $\rho(E)$ function, electrons are localized, since the levels with equal energy E are at sufficiently large distances from one another.

It should be emphasized that such localization levels may form even in the absence of impurities. Their appearance is associated with the spatial potential-energy fluctuations in the disordered structure, and, therefore, they are sometimes called fluctuation levels. By virtue of this, the density of electronic states in disordered materials does not vanish at the boundaries of the c- and v-bands. The disturbance of the ideal structure leads to the diffusion of the edges of the allowed E_c and E_v bands and the formation of the "tails" of the state density in the forbidden band.

This is illustrated by Fig. 7. In crystalline semiconductors (Fig. 7a), the density of states $\rho(E)$ in the vicinity of the band boundaries becomes zero, in accordance with the square root law, $\rho_{c,v}(E) \sim \sqrt{|E_{c,v} - E|}$. Unlike this, in disordered systems, the $\rho(E)$ function has nonzero values for E_c and E_v (Fig. 7b). Depending on the nature of a noncrystalline substance, the "tails" related to the fluctuation states may either break off for some energies (as depicted in Fig. 7b) or overlap with one another (Fig. 7c).

Moreover, within the same energy range, peaks may form due to the presence of introduced defects, e.g., of some dopants. In the regular structure, such peaks should be described by a δ-function-type energy distribution. An application of a random field would result in the broadening of such a peak. The number of such peaks may be different, and the above-described distributions may overlap (Fig. 7d).

Then the term "forbidden band" becomes somewhat inadequate in the above-described situation. Instead of the "band gap," we should introduce the concept of the "mobility slit." The analogues of the bottom of the conductivity band and the top of the valence band are now the

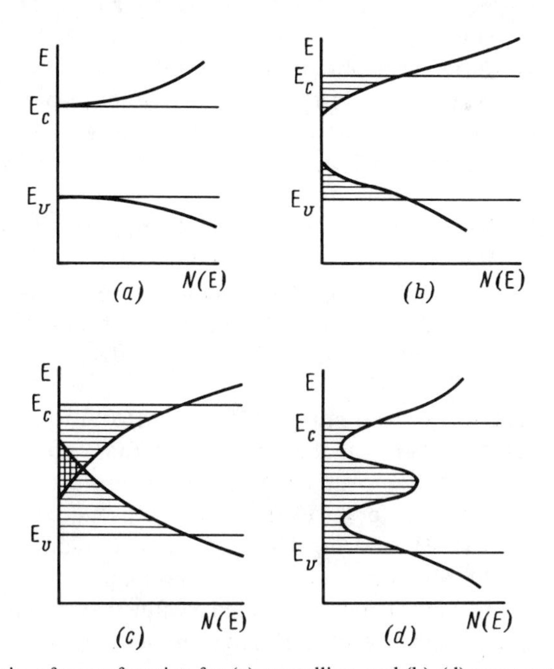

FIG. 7. The density of states function for (a) crystalline, and (b)–(d) non-crystalline materials.

mobility thresholds, in accordance with the real physical meaning of these quantities. The above-indicated threshold energies are usually denoted by the symbols E_c and E_v, used earlier to denote the band edges in ordered systems.

In connection with the above statements, the results of numerical calculations of electronic energy structures of SIMs should be considered. The achievements here are rather modest, which is partly explained by the intricacy of the problem. It is because of this intricacy that, in all the models, structural disorder has been treated as a disturbance in a strictly ordered crystal.

The first attempts to describe the band structures refer to the "classical" solid electrolyte, the α-phase of silver iodide (Smith, 1976). In the structural model used, all silver cations occupy b sites alone (Fig. 3). A similar approach was later applied to silver chalcogenides (Hasegawa, 1985). The quantitative analysis is based on the adiabatic approximation. It is assumed that cations, although mobile, move at a very low speed in comparison with electrons. Therefore, when calculating the energy spectra, one may assume cations are "frozen"in some sites associated with the structure of the rigid anion sublattice. Then, the electronic structure of such hypothetic compounds with a band structure may be calculated, within the accepted model, by conventional methods—e.g., using the tight-binding approach (Smith, 1976). It turns out that the band gap of silver chalcogenides depends mainly on 5s- and 4d-states of silver-atom electrons. It is these states that determine the band gap. The model also indicates possible deviations of the dispersion curve $E(p)$ from the parabola in the conduction band. Obviously, this approach does not take into account, to a sufficient degree, the effect of the disordered ionic structure.

In this connection, work by Bauer and Huberman (1976) should be mentioned. They made an attempt to take into account the effect on the energy spectrum of electrons of mobile silver ions in Ag_4RbI_5. According to Bauer and Huberman (1976), disordering of silver atoms results in the diffusion of the electron-state spectrum at the edge of the valence and conduction bands.

Electronic structure of the superionic conductor Li_3N was calculated by different methods, being made of the fact that, unlike other SIMs, Li_3N is almost completely ordered at low temperatures (Kerker, 1981; Blaka *et al.*, 1984). Thus, Kerker (1981) used the pseudopotential method, and Blaka *et al.* (1984) the linearized augmented plane-wave (LAPW) method. The results obtained indicate predominant ionic bonding in the substance. The band gap calculation yielded a value that is somewhat smaller than that obtained from the optical measurements ($\simeq 1.5$ and ≈ 2.2 eV, respectively).

In a number of works (Starostin and Ganin, 1973; Starostin and Shepilov, 1975; Nemoshkalenko *et al.*, 1976; Albert *et al.*, 1977), the electron band structure of fluorite phases of binary alkali-earth fluorides, MF_2, was calculated by different methods, including quantum-mechanical ones (Evarestov *et al.*, 1984). It has been shown that the position of the E_v level is genetically associated with the energy level corresponding to the 2*p*-states of initial fluorine ions, whereas the location of the conduction band bottom E_c corresponds to the 4s states of metal atoms.

Thus, taking into account the somewhat different physical meanings of the quantities used, the main qualitative concepts of the standard band model may be effectively used also for the interpretation of some electronic processes in disordered media.

In particular, it is possible to extend the explanation of the differences observed between the dielectrics, semiconductors, and metals used in the conventional solid-state theory to materials without long-range order. The observed difference is also related, to a large extent, to the description of the electron transport and optical absorption (where, in particular, a sharp boundary analogous to the fundamental absorption edge is observed). On the whole, the above approach is justified if the range of relatively low temperatures, where the specific features of irregular structures are seen most clearly, is not taken into consideration.

Now, bearing in mind the above discussion, the main mechanisms providing the formation of delocalized electrons and holes in superionic materials may be formulated. These are:

(1) The transition of valence-band electrons into the conduction band under the action of temperature or irradiation (Fig. 8a).

(2) Ionization of defects and/or impurities in the material bulk, accompanied by the formation of electronic carriers in the bands (Fig. 8b).

(3) Injection of electronic carriers from the electrode (metal or semiconductor) through the electrode-superionic material interface (Fig. 8c).

The above mechanisms will be considered in more detail later on.

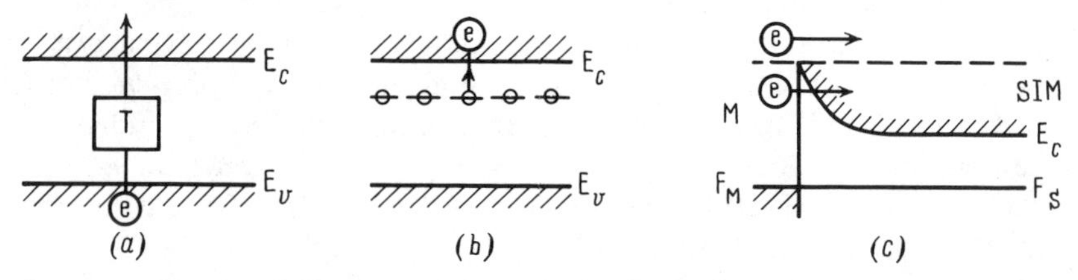

FIG. 8. Basic mechanisms of the formation of delocalized electronic carriers in SIMs: (a) band-band excitation; (b) impurity ionization; (c) injection.

2. Materials with Intrinsic Conductivity

The mechanism of electron carrier generation, including interband transitions (mechanism 1), in its nature is analogous to the mechanism well known for usual semiconductor materials. Physically, it is associated with the fact that at $T \neq 0$, a nonzero probability exists for valence band electrons to be injected into the conduction band. This is accompanied by the formation of a delocalized electron in the c-band and a hole in the v-band. Thus, two-types of carriers are formed that provide the so-called intrinsic electronic conductivity.

The equilibrium concentrations of electrons in the c-band, n_e^o, and holes in the v-band, n_h^o, for nondegenerate materials are related as follows (see, e.g., Seeger, 1973; Smith, 1978):

$$(n_e^o n_h^o) = (N_c N_v) \exp\left(-\frac{E_g}{kT}\right) \equiv n_{int}^2, \tag{1}$$

where $N_{c,v}$ is the effective density of the electron states for energies close to E_c or E_v, corresponding to the energy thresholds of electrons and holes, respectively, $E_g = E_c - E_v$.

For the simplest model, $N_{c,v} = 2(m_{c,v}kT/2\pi\hbar^2)^{3/2}$, where $m_{c,v}$ is the effective mass of electron carriers in the c- or v-bands, respectively. In general, it is rather difficult to determine $N_{c,v}$, but the above equations may be used for rough estimates.

If no donor-acceptor impurities are present (intrinsic semiconductors), $n_e^o = n_h^o = n_{int}$. The latter equality, given Eq. (1), uniquely determiones the concentrations of electron carriers of both species. Note that Eq. (1), relating n_e^o and n_h^o, is analogous to the law of mass action, the quantity n_{int}^2 playing the part of the chemical equilibrium constant. From the above discussion, it becomes clear that Eq. (1) is also valid for nondegenerate materials in the presence of impurities, although the concentrations n_e^o and n_h^o may essentially vary. In order to determine these concentrations, the impurity characteristics should be known.

Table 2 gives the experimental values of E_g for a series of superionics, as well as the values of n_{int} at 300 K calculated by Eq. (1). As is seen, concentration n_{int} varies over a wide range. For materials with a relatively narrow gap, the intrinsic conductivity may be quite high when $n_{int} = n_e^o = n_h^o$. In this case, it is possible to attain a satisfactory agreement with the independent data obtained in electronic-conductivity measurements. Such an agreement can hardly be expected for large E_g values, since, in such cases, the role of intrinsic conductivity is unimportant.

The electronic conductivity of a semiconductor, σ_{el}, for the simplest model is a sum of the conductivities of electrons, σ_e, and that of holes, σ_h.

TABLE II

BAND GAP, E_g, AND CONCENTRATION, n_{int}, FOR SOME SIMs (AT 300 K)

SIM	E_g (eV)	n_{int} (cm^{-3})
Ag_2S	0.9	7×10^{11}
Ag_3SI	1.8	1×10^4
Li_3N	2.2	8×10^0
α-AgI	2.5	2.5×10^{-2}
β-AgI	3.0	2×10^{-6}
Ag_4RbI_5	3.2	3×10^{-8}
PbF_2	5.8	—
Na-β-Al_2O_3	6.0	—

If their mobilities are denoted as u_e and u_h, σ_{el} may be written as

$$\sigma_{el} = e(n_e u_e + n_h u_h), \tag{2}$$

where e is the absolute value of the electron charge. For materials with intrinsic conductivity, $\sigma_{el} = en_{int}(u_e + u_h)$.

Proceeding from the latter equation, it is possible to find σ_h for the superionic conductor Li_3N: $\sigma_h = 10^{-12}$ $(\Omega\ cm)^{-1}$ (the hole mobility was taken to be much higher than the electron mobility, namely, $u_h = 10^{-1}$ $cm^2/V\ s$, as it takes place for materials close to Li_3N). The above value of σ_h corresponds to the data listed in Table A in the Appendix.

If we assume that electron mobility prevails in Ag_2S, which is equal to $u_e = 10^3$ $cm^2/V\ s$, then the calculated value is $\sigma_e = 10^{-3}$ $(\Omega\ cm)^{-1}$ at 400 K and lies within the range of experimental σ_e values measured at the same temperature (see Appendix).

The generation of electronic carriers by interband transitions in SIMs may be more complicated than in usual semiconductors, owing to disordering in the crystal lattice, which may be illustrated by atomic disordering in compounds with covalent bonding (Chebotin, 1982).

Some A atoms leave the crystal bulk and come to the surface, resulting in the formation of neutral vacancies V_A in the bulk. It may readily be shown that the equilibrium concentration n_V of such vacancies is given by the expression

$$n_V = N_A \chi_V \exp\left(-\frac{w_{AV}}{kT}\right). \tag{3}$$

Here, N_A is the number of initial A-atoms per unit volume of the crystal ($N_A \gg n_V$); $w_{AV} > 0$ is the energy necessary for the removal of a neutral atom from the bulk to the surface; the factor χ_V takes account of the entropy changes in such a process. In fact, $\chi_V \exp(-w_{AV}/kT) = \exp(-f_{AV}/kT)$, where f_{AV} is the free energy of the process under consideration.

Vacancies formed in the material bulk may possess the donor-acceptor properties that influence the electronic carrier concentration. For definiteness, it is assumed that vacancies have the donor properties, i.e., transit to the ionized state by the quasichemical reaction

$$V_A \leftrightarrows V_A^+ + e^-. \tag{4}$$

In accordance with the law of mass action, it follows from reaction (4) that $n_v^+ + n_e^o = K_V n_V$, where n_V^+ is the equilibrium concentration of ionized vacancies. Quantity K_V is the equilibrium constant of reaction (4) and is given by equation $K_V = N_c \chi_e \exp(-w_{Ve}/kT)$, where w_{Ve} is the ionization energy of a neutral vacancy, V_A, and χ_e is the entropy factor. The condition of electroneutrality in the bulk of a superionic material in this case has the form

$$n_e^o = n_V^+ + n_h^o. \tag{5}$$

Multiplying all terms of Eq. (5) by n_e^o, and taking into consideration Eqs. (1) and (3), we obtain, for the electron concentration in the conduction band,

$$n_e^o = \sqrt{n_{int}^2 + n_{Ae}^2}. \tag{6}$$

Here, $n_{Ae}^2 = N_A N_c \chi_V \chi_e \exp[-(w_{AV} + w_{Ae})/kT]$. Depending on the relation between the N_c and N_A, and between $w_{AV} + w_{Ve}$ and E_g, two limiting cases should be considered. If $n_{int} \gg n_v$, we return to the situation described by Eq. (1); if $n_{int} \ll n_{Ae}$, then $n_e^o \approx n_{Ae}$.

It should be noted that if $n_e^o \approx n_{Ae}$, then inequality $n_h^o \ll n_e^o$ is valid; i.e., the material is an n-type semiconductor. But, unlike conventional n-type semiconductors, in this case concentration $n_e^o = n_{Ae}$ exponentially depends on the reciprocal temperature T^{-1}. Physically, this difference is explained by the fact that the temperature rise increases the number of defects playing the part of donors.

The transition between the two, above considered limiting cases occurs when $n_{int} = n_{Ae}$, i.e., at temperature $kT_* = (w_{AV} + w_{Ve} - E_g)/\ln(N_A \chi_V \chi_e / N_v)$. Note here an interesting possibility—if, in the temperature range corresponding to intrinsic conductivity ($T < T_*$), electron mobility is relatively low ($u_e \ll u_h$), then, at $T \gtrsim T_*$, an increase in n_e^o results in the change of the type of majority electronic carriers.

If neutral vacancies V possess acceptor properties, then, as earlier, an increase in the temperature and the number of vacancies may make the hole conductivity predominant.

The materials in which the above-mentioned mechanism of the electronic carrier generation plays the key role are called intrinsic-defect semiconductors.

3. NONSTOICHIOMETRY

Now let us consider another mechanism (mechanism 2), according to which electrons (holes) may appear in the conduction (valence) band, due to the presence of impurities playing the role of electron donors (acceptors). [The formation of electronic carriers through ionization of intrinsic defects may also be regarded, to a certain extent, as the electron (hole) generation by mechanism 2.]

It is important to note that donor- and acceptor-type impurities in superionic conductors may appear when the composition of the material deviates from the stoichiometry. As an idealized example, consider here the binary compound of the composition $M_{n+\delta}X_{l+\gamma}$, where M and X are the electropositive (metal) and electronegative (non-metal) elements, respectively. The ideal stoichiometric composition corresponds to $\delta = \gamma = 0$. Now let quantities δ and γ be nonzero, but still lying within the homogeneous range of the substance under consideration. Then, if $\delta > 0$ or $\gamma < 0$ (excess of the component M), electron conductivity may be expected because of the ionization of electropositive atoms (especially if bonding in the compound under consideration is ionic). If $\delta < 0$ or $\gamma > 0$ (excess of the component X), hole conductivity should appear. Finally, if the parameters δ and γ are of the same sign, both donors and acceptors are present. For certain nonzero values of these parameters, the equation $n_e^o = n_h^o$ is valid; i.e., the impurities are mutually compensated.

It has been shown that, in accordance with the above stated concepts, an increase in parameter δ results in higher electronic conductivity. Moreover, at sufficiently large values of δ, the conductivity may change from the nonmetallic to metallic type. These problems will be considered in more detail in Sec. 5.

Let excessive atoms (or atomic vacancies) be uniformly distributed over the crystal bulk. Then, in order to determine the equilibrium concentration of electronic carriers, one may use the relationships describing the effect of donors or acceptors on equilibrium concentrations (Seeger, 1973), well known in the physics of semiconductors. Let us assume, for example, that $\delta > 0$ and $\gamma = 0$. Then, the concentration n_e^o of electrons in the conduction band ($n_e^o \gg n_{int}$) is

$$n_e^o = \frac{n_1}{2}\left(\sqrt{1 + \frac{4N_d}{n_1}} - 1\right) = \begin{cases} \sqrt{n_1 N_d}, & n_1 \ll N_d, \quad (7a) \\ N_d, & n_1 \gg N_d. \quad (7b) \end{cases}$$

Here, N_d is the concentration of donor impurities, which, for small δ, is given by the relationship

$$\mathfrak{N}_d = \frac{N_A \delta}{V_m}, \tag{8}$$

where $\mathfrak{N}_A$ is the Avogadro number and V_m is mole volume of the M_nX_l compound. In the simplest case, parameter n_1 in Eq. (7) is equal to $n_1 = N_c \exp(-\Delta E_d/kT)$, where $\Delta E_d > 0$ is the ionization energy of a donor impurity.

At relatively low temperatures (or large ΔE_d), n_1 is small, and, in accordance with Eq. (7a), the electron concentration is $n_e^o \sim \exp(-\Delta E_d/2kT)$. At sufficiently high temperatures (or small ΔE_d), n_1 is quite large: $n_1 \approx N_c$ and $n_1 \gg N_d$. In this case, donors are completely ionized, and, in accordance with Eq. (7b), $n_e^o = N_d$, and n_e^o is independent of T.

Note also the essential difference in the temperature dependences $n_e^o(T)$ described by Eq. (6) (intrinsic-defect semiconductors) and Eq. (7) (impurity-defect semiconductors).

It has been assumed that the electron carrier concentration is relatively low ($\lesssim N_{c,v}$), so that the Boltzmann statistic is valid, whereas the effects due to degeneration may be neglected. At the same time, for sufficiently large deviations from stoichiometry, degeneracy in SIMs may take place with respect to only one carrier species. Under such conditions, the equilibrium concentrations n_e^o and n_h^o are given by the relationships (Kittel, 1969)

$$
\begin{aligned}
n_e^o &= N_c \Phi_{1/2}(\zeta), \\
n_h^o &= N_v \Phi_{1/2}(\xi).
\end{aligned}
\tag{9}
$$

Here, $\zeta \equiv (F - E_c)/kT$, and $\xi \equiv (E_v - F)/kT$, where F is the electrochemical potential of electronic carriers in the material. The dimensionless function $\Phi_{1/2}$ is determined by the so-called Fermi-Dirac integral with subscript 1/2,

$$
\Phi_{1/2} \equiv \frac{2}{\sqrt{\pi}} \int_0^\infty \frac{z^{1/2}\,dz}{1 + \exp(z - t)}. \tag{10}
$$

For large negative and positive t values, the above integral may be calculated analytically, yielding

$$
\Phi_{1/2}(t) = \begin{cases} e^t, & t < 0, \quad |t| \gg 1, \qquad \text{(11a)} \\ \dfrac{4t^{3/2}}{3\sqrt{\pi}}, & t \gg 1. \qquad \text{(11b)} \end{cases}
$$

The function $\Phi_{1/2}$ is also described by the analytical interpolation formulae, and its values are tabulated (McDougall, Stoner, 1938, and others).[1] Formula (11a) corresponds to the limiting case in which there is no degeneracy.

[1] See, for example, the discussion by A. C. Beer (1963) in *Galvanomagnetic Effects in Semiconductors*. Academic Press, New York.

Then, in accordance with Eq. (9), we have

$$\begin{aligned} n_e^o &= N_c \exp\left\{\frac{(F - E_c)}{kT}\right\}, \\ n_h^o &= N_v \exp\left\{\frac{(E_v - F)}{kT}\right\}, \end{aligned} \tag{12}$$

Multiplying Eqs. (12) together, we arrive at Eq. (1).

An important practical case occurs when nonstoichiometry depends on the atomic exchange between the crystal and the surrounding medium (usually gas).

As an example, consider the binary nonstoichiometric compound $MX_{l+\gamma}$, which is an intrinsic semiconductor or dielectric when its composition corresponds to stoichiometry ($\gamma = 0$). It is assumed that the crystal of the composition $MX_{l+\gamma}$ is in thermodynamic equilibrium with the gas phase containing an electronegative component. This component is in the form of diatomic molecules X_2 (molecular oxygen in the case of oxides) with the controlled partial gas pressure P_{X_2}. Stoichiometry is upset either because of excessive absorption of the component X from the gas phase (the region of high pressures of X_2), or because a certain amount of the component X passes from the crystal into the gas phase (the region of low pressure of X_2).

The ideal stoichiometry at a constant temperature is observed under definite partial pressure $P_{X_2}^{\text{stoic}}$ when $n_e^o = n_h^o = n_{\text{int}}$. With the variation in pressure P_{X_2}, the values of γ and, therefore, of the concentrations n_e^o and n_h^o change.

Under such conditions, the change in the composition is described by the following quasi-chemical reactions (Kröger, 1974):

$$X_X \rightleftarrows V_X + \frac{1}{2}X_2, \tag{13a}$$

$$\frac{l}{2}X_2 \rightleftarrows lX_X + V_M. \tag{13b}$$

Reaction (13a) describes either the transition of an X atom from its site in the crystal (structural element X_X or a zero-defect) into the gas phase, with the simultaneous formation in the crystal of a neutral vacancy (structural element V_X), or the reverse process of the incorporation of an atom from the gas phase into vacancy V_X, with the formation of element X_X.

The direct reaction (13b) (→) describes the completion of the lattice at the crystal surface: component M, necessary for the formation of a new formula unit MX_l, arrives to the surface from the crystal bulk, with the simultaneous formation of a neutral vacancy (structural element V_M). The same reaction (13b), proceeding in the reverse direction (←), describes the destruction of the formula unit MX_l.

If the vacancies of component X possess the donor properties and those of the component M acceptor properties, they may be ionized in correspondence with the following reactions [cf. Eq. (4)]:

$$V_X \rightleftarrows V_X^{z_X^+} + z_X e^-, \tag{14a}$$

$$V_M \rightleftarrows V_M^{z_M^-} + z_M h^+, \tag{14b}$$

where h^+ denotes an electron hole, and parameters z_X and z_M characterize the degree of vacancy ionization.

Substitution of (14) into (13) yields the following quasichemical reactions:

$$X_X \rightleftarrows V_X^{z_X^+} + z_X e^- + \frac{1}{2} X_2, \tag{15a}$$

$$\frac{l}{2} X_2 \rightleftarrows V_M^{z_M^-} + z_M h^+ + l X_X. \tag{15b}$$

Assuming that substance X_2 outside the crystal is an ideal gas under pressure P, and using the law of mass action, we obtain from (15)

$$\begin{aligned} n_{V_X} n_e^{z_X} P^{1/2} &= K_X, \\ n_{V_M} n_h^{z_M} &= K_M P^{l/2}, \end{aligned} \tag{16}$$

where n_{V_X} and n_{V_M} are the concentrations of charged structural elements $V_X^{z_X^+}$ and $V_M^{z_M^-}$ in reactions (15a) and (15b), and K_X and K_M are the equilibrium constants of these reactions.

Relationships (16), together with Eq. (1) and the condition of electroneutrality

$$n_e^o + z_M n_{V_M} = n_h^o + z_X n_{V_X}, \tag{17}$$

comprise four equations for the determination of four concentrations of charged components as functions of pressure P and temperature (via constants K_X, K_M and n_{int}).

In the case $z_X = z_M = 1$, the general solution of the set of Eqs. (1), (16) and (17) may be obtained in the explicit form as

$$\begin{aligned} n_e^o &= n_{int}\sqrt{\frac{1 + l\alpha}{1 + \beta}}, & n_h^o &= n_{int}\sqrt{\frac{1 + \beta}{1 + l\alpha}}, \\ n_{V_M} &= \beta n_e^o, & n_{V_X} &= l\alpha n_h^o, \end{aligned} \tag{18}$$

where dimensionless parameters α and β are given by the relationships

$$\alpha = K_X n_{int}^{-2} P^{-1/2},$$
$$\beta = K_M n_{int}^{-2} P^{l/2}, \tag{19}$$

and may vary over a wide range, with the variation in P and T.

As is seen from Eq. (18), in terms of electronic properties, the material under consideration is similar to intrinsic semiconductors or dielectrics ($n_e^o \approx n_h^o \approx n_{int}$) if $\alpha, \beta \ll 1$. It follows from Eq. (19) that both latter inequalities are satisfied if gas pressure is maintained within the range

$$P_- < P < P_+, \tag{20}$$

where $P_- = (K_X)^2 n_{int}^{-4}$, $P_+ = n_{int}^{4/l}(K_M)^{-2/l}$.

For relatively low pressures of the X_2 gas, when $P < P_-$, we have $\alpha \gg 1$ and $\beta \ll 1$. Then, according to Eqs. (18) and (19), n_e^o is

$$n_e^o = P^{-1/4}(lK_X)^{-1/l}, \qquad P < P_-. \tag{21}$$

Under the given conditions, $n_h^o \ll n_e^o$; in other words, the material is an electronic semiconductor. Physically, it is associated with the fact that, in this case, the main reaction is that described by Eq. (15a).

However, in the range of relatively high pressures, when $P > P_+$, reaction (15b) plays the key role, and the material becomes a p-type semiconductor, so that $n_h^o \gg n_e^o$. According to Eqs. (18) and (19),

$$n_h^o = P^{l/4}(K_M)^{1/2}. \tag{22}$$

Thus, if conditions (20) are met, there are three ranges of gas pressures (P_{X_2}) under which electronic, intrinsic, or hole conductivity is observed. At the same time, the characteristic pressures P_- and P_+ depend on the energy, entropy and thermal parameters of the system. Therefore, a situation is possible when $P_+ < P_-$. Under such conditions, the range of intrinsic electronic conductivity "shrinks" into a narrow region in the vicinity of $P = P^{stoic}$. Outside this region the values of n_e^o and n_h^o may be determined from Eq. (18), on the simplifying assumption that $\alpha, \beta \gg 1$. Namely,

$$n_e^o = n_{int}\left(\frac{lK_X}{K_M}\right)^{1/2} P^{-(1+l)/4};$$
$$n_h^o = n_{int}\left(\frac{K_M}{lK_X}\right)^{1/2} P^{(1+l)/4}. \tag{23}$$

In general, the role of donors and acceptors with pressure-dependent concentrations may be performed not only by vacancies, but also by interstitial ions of a nonstoichiometric compound. The degree of donor and acceptor ionization may also vary. Thus, variants, which do arise, may be considered in accordance with the above-mentioned scheme. The approximate solutions for the concentrations of predominant electron carriers are always written in the form (Chebotin, 1982)

$$n = B(T)P^{1/r}\exp\left(-\frac{E_e}{kT}\right), \tag{24}$$

where $n = n_e^o$ if $n_e^o \gg n_h^o$, and $n = n_h^o$ if $n_h^o \gg n_e^o$. Quantity $B(T)$ in Eq. (24) is a power function of temperature, and $E_e > 0$ is the effective activation energy of the formation of predominant electronic carriers, which includes various additive combinations of energies characterizing quasi-chemical reactions. Finally, parameter r ($|r|$ always exceeds unity) depends on the range of pressures under consideration, the type and the ionization degree of structural defects, and the type of electronic carriers.

In particular, for an important case of metal oxides, MO_l ($l = 1, 3/2, 2$), with the formed defects fully ionized, $r = -4$ for $n = n_e^o$, and $r = 4$ for $n = n_h^o$.

Note here that relationships of type (24) differ notably from the corresponding relationships for n_e^o and n_h^o from the theory of usual semiconductors.

Experimentally, the values of r may be, in particular, determined if we consider the functions $n_{e,h}^o(P)$ in logarithmic coordinates.

As a rule, in superionic conductors, where the total concentration of mobile ions is sufficiently high, the variation of this concentration (and, hence, of ionic conductivity) with pressure is not important. At the same time, the concentration of electron carriers and the corresponding conductivity may change by many orders of magnitude, especially for materials with small n_{int}. In this case, there exists such a range of gas pressure, $P_e < P < P_h$, within which ionic conductivity prevails. Outside this range, called the electrolytic pressure range, either electronic ($P < P_e$) or hole ($P > P_h$) conductivity is predominant.

Under pressures $P_{e,h}$, the ionic and the corresponding electronic (of e- or h-type) components of conductivity are equal. In virtue of the exponential dependence of $P_{e,h}$ on T, the width of the electrolytic pressure range (on a logarithmic scale) for metal oxides is (Schmalzried, 1962)

$$\ln P_h - \ln P_e = \text{const} + \frac{4E_*}{kT}. \tag{25}$$

For the simplest model, $E_* = E_g - 2E_a$, where E_a is the activation energy of ionic conductivity.

As an example illustrating the above-considered laws, we should like to discuss the results obtained in the conductivity measurements of a superionic material that is a solid solution, ThO_2 + 0.1 mole % Ce at various temperatures, see Fig. 9 (Fujimoto and Tuller, 1979). Under sufficiently high partial pressure of oxygen, $P_{O_2} \gtrsim 10^{-6}$ atm, the observed dependence of σ on P_{O_2} obeys the law $\sigma \sim P_{O_2}^{1/4}$. The obtained r value ($r = 4$) may be understood on the assumption that such pressures provide the formation of mainly interstitial oxygen anions and electron holes by the reaction

$$\tfrac{1}{2}O_2(\text{gas}) \rightleftarrows O_i'' + 2h^+ \qquad \text{[cf. (15b)].}$$

Since the ionic carrier concentration is rather high and therefore, only weakly depends on P_{O_2}, it follows that $n_h^o \sim P_{O_2}^{1/4}$. Further on, if u_h is independent of P, we arrive at such a dependence for σ_h, which was, in fact, observed in the experiment.

A decrease in P_{O_2} results in the formation of a segment on the $\sigma(P)$ curve, where total conductivity is purely ionic and independent of oxygen pressure. Finally, with a still greater decrease in P_{O_2}, σ starts increasing, in accordance with the power law, $\sigma \sim P_{O_2}^{-1/4}$. The above law may be understood from the facts that now $\sigma \approx \sigma_e$ and numerous oxygen vacancies and electrons are formed in the material

$$O_o \rightleftarrows V_o^{2+} + 2e^- + \tfrac{1}{2}O_2(\text{gas}) \qquad \text{[cf. (15a)]}$$

Then $n_e^o \sim P_{O_2}^{-1/4}$, and, therefore, $\sigma_e \sim P_{O_2}^{-1/4}$.

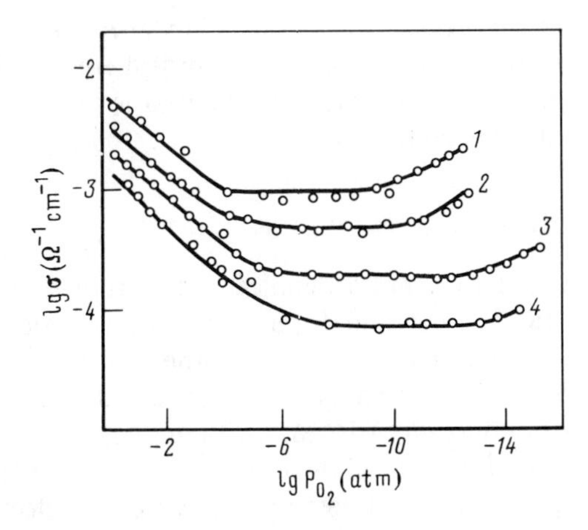

FIG. 9. Conductivity of the ThO_2 + 0.1 mole % Ce_2O_3 superionic material versus partial oxygen pressure: (1) 1300, (2) 1200, (3) 1073, (4) 1000°C. [From Fujimoto and Tuller (1979).]

At the same time, the situation here turns out to be more complicated (Fujimoto and Tuller, 1979), so that conductivity at very low pressures may still remain mainly ionic.

In this connection, we should like to underline that the previous analysis was carried out on the assumption of thermodynamic equilibrium in the system. But under sufficiently low partial pressures of gas, some effects may appear, due to very slow exchange processes between the gas phase and the sample. Therefore, it is necessary to control the conditions for the attainment of the equilibrium state of the objects under consideration.

In conclusion, impurities not related to nonstoichiometry should be considered. The atoms of foreign elements may be introduced into a superionic material for changing its ionic conductivity or imparting some specific physical-chemical properties (Weppner, 1977). Such doping results, in turn, in the change of electronic conductivity.

Of course, not all atomic inclusions necessarily possess donor-acceptor properties. Thus, doping may give rise not to the formation of electronic carriers, but to changes in the valence state of other ions in the lattice (Moizhes, 1983). Then the corresponding electronic level of the dopant may lie far from the band edges, and the dopant will not be ionized, etc.

Finally, the role of electron donors or acceptors may also be played by foreign, often uncontrolled impurities, which also change electronic conductivity. Some such cases were analyzed by Kröger (1974), Patterson (1974) and Stoneham (1985). Small amounts of impurities can be introduce during the process of superionic conductor synthesis. Moreover, the analysis of impurity content is much less accurate than that achieved in the evaluation of most semiconductors. Uncontrolled impurities are also introduced during sample preparation for measurements (Shirokov *et al.*, 1972; Kukoz *et al.*, 1977), e.g., in the process of annealing in atmospheres of different gases. These facts explain, to a great extent, a relatively large scatter in the measured low values of electronic conductivities.

4. Effects Associated with Interfaces

Now let us proceed to a more detailed description of some processes occurring at interfaces. First and foremost is the injection of electronic carriers into a superionic material through the interface (mechanism 3). Consider first a contact between a metal electrode and a superionic material with no electronic conductivity (the M/SIM contact).

The energy diagram for both solids prior to their contact is schematically represented in Fig. 10, which shows the energy of an electron at rest in vacuum, E_0, the band edges in the superionic material, E_c and E_v, and the location of the Fermi levels (electrochemical potentials) of electrons in the

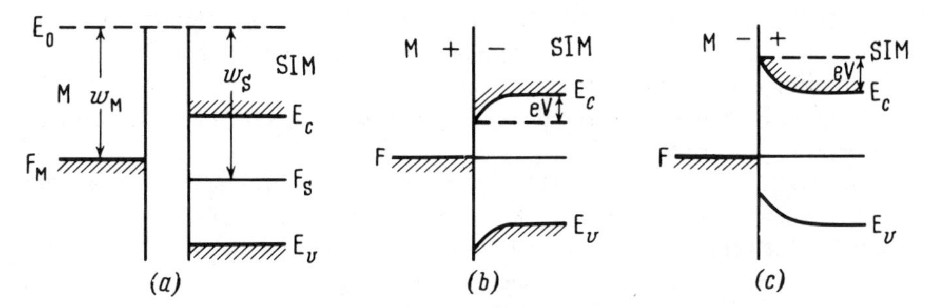

FIG. 10. Energy diagram of a metal-superionic material contact: (a) no contact; (b) a contact in thermodynamic equilibrium, $w_M < w_S$, enriched layer; (c) the same, $w_M > w_S$, for a depleted layer.

metal F_M, and in the superionic material F_s. The affinity of a superionic material to electrons is denoted by $\chi = E_o - E_c$. When the solids are brought into contact, electrons in the region near the contact start redistributing, providing the attainment of a thermodynamic equilibrium. Upon the attainment of such an equilibrium, F_M and F_S levels will coincide.

For definiteness, we assume that the thermodynamic work function for electrons coming into a vacuum from the metal w_M is smaller than that from the superionic material w_S. In this case, for solids which have not been brought into contact, the level F_M is higher than F_S, and electrons "flow" from upper levels to lower ones and, thus, enrich the region near the contact with electrons (the band edges E_c and E_v are bent downwards) (Fig. 10b). If $w_M > w_S$, then, upon the attainment of thermodynamic equilibrium, the band edges are bent upwards (Fig. 10c). In both cases, the potential difference V is located between the bulk and the surface of the superionic material, so that $eV = w_M - w_S$ ($e > 0$). If E_B is the height of the potential barrier from the side of the metal, then using the above definition of χ, $E_B = w_M - \chi$. Such a simple equation is valid if the band edges of the superionic material were straight prior to the contact with the metal (Fig. 10a)—in other words, if there were no charges on the surface.

Let us now consider the role of the surface states that may be formed at the SIM surface, as the formation takes place for the conventional semiconductor. For ideal crystal surfaces, surface states may be associated with the Tamm levels (Davison and Levine, 1970). For real surfaces, the surface levels are also created by impurities (adsorbed and chemically-bound foreign atoms and molecules). These levels are located in the forbidden band and may be either donor or acceptor levels.

In the process of surface-state charging, the bands in the subsurface layer start bending. If the surface (surface states) of the compound being studied is charged negatively, then the bands bend upwards; if the surface is charged positively, the bands bend downwards. Thus, the contact characteristics in this case only weakly depend on the properties of the metal and SIM (on the work function for an "ideal" metal and electron affinity for an "ideal" compound), being determined by the presence of electronic donor and/or acceptor surface levels.

In general, where the bands are initially bent by eV_o, $E_B = w_M - \chi + eV_o$; i.e., for large $|eV_o|$, E_B may weakly depend on the nature of the metal (more complicated models have been analyzed, e.g., by Rhoderick, 1978).

From the equality of electrochemical potentials, $F_M = F_S = F$, it follows that the equilibrium electron concentration in the conduction band at the interface $n_e(0)$ is given by the relationship

$$n_e(0) = N_C \exp\left(-\frac{E_B}{kT}\right). \tag{26}$$

Assuming $E_B \simeq 0.4$ to 0.9 eV for $\sigma_e(0)$ from Eq. (26), we have $\sigma_e \simeq 10^{-6}$ to 10^{-14} $(\Omega\ \mathrm{cm})^{-1}$, which, on the whole, is in accordance with the tabulated data (see Appendix).

Strictly speaking, Eq. (26) gives the thermodynamic equilibrium value, $n_e(0)$, but, in actual fact, the range of its validity is much wider. As is shown by numerical calculations, the value of $n_e(0)$ only slightly changes for currents corresponding to the real fields.

Physically, such a small change is explained by the dynamical nature of the above-considered equilibrium. In other words, continuous electron exchange proceeds at the interface maintaining the constant value of n_e.

The absence of the resulting electronic current ($i_e = 0$) in the state of dynamic equilibrium is described by the equation $\vec{i}_e - \overleftarrow{i}_e = 0$, where $\vec{i}_e$ and $\overleftarrow{i}_e$ are currents flowing through the surface from left to right, and vice versa. The quantity $i_e^o = |\vec{i}_e| = |\overleftarrow{i}_e|$, called the exchange current with respect to electrons, is an important kinetic characteristic of the interface.

Under nonequilibrium condition, $\vec{i}_e \neq \overleftarrow{i}_e$, and the resulting current i_e flowing through the interface is nonzero; $i_e \neq 0$. If current i_e satisfies the inequality $|i_e| \ll i_e^o$, the concentration $n_e(0)$ is approximately equal to its equilibrium value, corresponding to the condition $i_e = 0$.

For rough estimates, we may assume that the exchange current i_e^o is determined by the thermoemission current i_T through the metal-superionic material interface. The expression for i_T in the simplest model has the form (see, e.g., Rhoderick, 1978)

$$i_T = \mathfrak{A} T^2 \exp\left(-\frac{E_B}{kT}\right). \tag{27}$$

Here, $\mathcal{A} = 4\pi m_0 ek^2/(2\pi\hbar)^3$, where m_0 is the mass of a free electron. The numerical value of $\mathcal{A}$ (the so-called Richardson constant) equals 120 A/cm^2 k^2 (real $\mathcal{A}$ values may slightly differ from the above given, since in the derivation of Eq. (27), tunneling, the deviation of the effective mass from m_0, and some other effects have been neglected).

As is shown by the estimates, for $E_B = 0.4$ eV and $T = 300$ K, we have, in accordance with Eq. (27), $i_T = 1$ A/cm^2. Therefore, the condition $|i_e| \ll i_T$ is not a limiting one. At the same time, if the above condition is fulfilled, the given equilibrium concentration of electronic carriers is maintained at the interface, even if a current is flowing through the sample. Under the action of an external field, electronic carriers may migrate into the crystal depth, thus participating in the charge transport and providing electronic conductivity. The sites of carriers migrated into the material depth are immediately occupied by newly-arrived ones.

Thus, the noticeable electronic conductivity, especially in the case of sufficiently thin samples and films, may be provided by the mechanism of electron injection.

III. Phenomenological Description of Transport Phenomena

The microscopic picture of electron transport in superionic materials (i.e., the allowance for polaron effects and band and hopping mechanisms) is far from complete. In comparison with the situation in most crystals, it is essentially complicated by the effects associated with partial disorder of one of the sublattices and the flow of the ionic current. Some aspects of the microscopic approach will be considered in more detail later on. On the whole, it is premature to consider the details of electron transport in SIMs at the present stage, and, therefore, the theoretical treatment is based mainly on the phenomenological approach. Such an approach coincides with that used in the treatment of the charged particle transport in electrolyte solutions, semiconductors and dielectrics, slightly ionized gases, etc. (Friauf, 1972; Newman, 1974; Balescu, 1975).

5. General Form of the Transport Equation

The phenomenological transport equation is derived from the assumption that the densities of partial currents i_k of species k (ions, vacancies, and electronic carriers) are proportional to the gradients of the corresponding electrochemical potentials $\tilde{\mu}_k$; namely,

$$i_k = \kappa_k \frac{\partial \tilde{\mu}_k}{\partial x}. \tag{28}$$

Here, x is the coordinate (for simplicity, we restrict ourselves to the one-dimensional case), and κ_k is a certain phenomenological coefficient.

The electrochemical potential $\tilde{\mu}_k$ is defined by the relation

$$\tilde{\mu}_k = \mu_k^{\circ} + kT \ln n_k + ez_k\varphi, \tag{29}$$

where n_k is the concentration (activity) of species k, $z_k \gtrless 0$ is their charge number, and φ is the electrical potential. The term μ_k° in Eq. (29) is independent of n_k and φ, but is a function of temperature and pressure. The quantity μ_k° depends on the choice of the electric potential origin and the standard concentration. In particular, with the appropriate choice of μ_k°, the electrochemical potential $\tilde{\mu}_k$ for electrons coincides with the Fermi level F in the solid.

Quantity $\tilde{\mu}$ at $\varphi = 0$ is called the chemical potential μ_k. Then

$$\tilde{\mu}_k = \mu_k + ez_k\varphi. \tag{29'}$$

In the limiting case of zero concentration and temperature gradients, Eq. (28) describes Ohm's law $i_k = -\sigma_k\,\partial\varphi/\partial x$, where σ_k is the conductivity of particles of species k. Allowing for the above, it follows from Eqs (28) and (29) that the phenomenological coefficient κ_k is related to σ_k by the equation $\kappa_k = -\sigma_k/ez_k$.

In order to determine the exact form of σ_k, we must choose a certain model. For the simplest model of noninteracting particles, σ_k is written as a product

$$\sigma_k = ez_k^2 u_k n_k, \tag{30}$$

where u_k is the mobility related to the diffusion coefficient D_k by the Einstein relation

$$u_k = \frac{eD_k}{kT}. \tag{31}$$

Then, it follows from Eq. (30) that

$$\sigma_k = \frac{e^2 z_k^2 D_k n_k}{kT}. \tag{32}$$

Substituting Eq. (29) into Eq. (28), we obtain, using Eq. (32), that, at constant temperature, the following equation is valid:

$$i_k = -ez_k D_k \frac{dn_k}{dx} - \sigma_k \frac{d\varphi}{dx}. \tag{33}$$

The first term in the right-hand side of Eq. (33) describes the diffusion component of the current due to the concentration gradient; the second term describes its migrational component, determined by the electric field.

Note that, proceeding from Eq. (33), it is possible to obtain the Einstein relation [with account of Eq. (30)] from the condition that, for $i_k = 0$, concentration n is described by Boltzmann's distribution, $n_k \sim \exp(-ez_k\varphi/kT)$.

The equation for the flux of neutral ($z_k = 0$) particles, $j_k = i_k/ez_k$, also follows from Eqs. (28) and (29) with the use of Eq. (30): $j_k = -D_k\,dn/dx$.

It should be mentioned that, generally speaking, the transport equations may have a more complicated form than that following from Eqs. (30) and (33). In particular, in the description of the ionic transport, the effect of lattice saturation may be important (Gurevich and Kharkats, 1976b); ion (vacancy) concentrations cannot increase infinitely—they are limited by the finite number of sites in the crystal lattice available for ions (vacancies). Denoting the maximum possible concentration of carriers of species n_k as $n_k^{\max}$, we obtain the following formula for the ionic current (Gurevich, 1980; Gurevich and Kharkats, 1980):

$$i_k = -eD_k z_k\left[\frac{dn_k}{dx} + \frac{ez_k n_k}{kT}\left(1 - \frac{n_k}{n_k^{\max}}\right)\frac{d\varphi}{dx}\right]. \tag{34}$$

Note the appearance of factor $(1 - n_k/n_k^{\max})$ in the term, describing the migration component of the current.

If we use, in Eq. (29), activities a_k instead of concentrations n_k (which, in fact, is equivalent to including interaction between particles), then the quantity D_k in the main Eq. (33) should be replaced by $D_k(\partial \ln a_k/\partial \ln n_k)$. Taking into account the correlation between individual ion hoppings in the diffusion process also requires renormalization of D_k. Under these conditions, Einstein's relation, in its usual sense [see Eq. (31)], is not valid any more.

In a similar way, for electronic carriers, Eq. (31) becomes invalid if the electronic carriers are degenerate. Taking into account that, under the equilibrium conditions, electron concentration n_e depends on $\tilde{\mu}_e$ alone, we obtain, from Eq. (29) (where $z_k = -1$), $dn_e/dx = -(dn_e/d\tilde{\mu}_e)(d\tilde{\mu}_e/dx)$. Substituting the latter expression into the right-hand part of Eq. (33), we have for $i_e = 0$

$$eD_e\frac{dn_e}{d\tilde{\mu}_e}\frac{d\varphi}{dx} - n_e u_e\frac{d\varphi}{dx} = 0,$$

whence the extended Einstein relation

$$u_e = eD_e\frac{d\ln n_e}{d\tilde{\mu}_e}. \tag{31$'$}$$

In the absence of degeneration, when $n_e \sim \exp(\mu_e/kT)$, Eq. (31′) transforms into Eq. (31).

The set of equations describing the distribution of l components $n_k(x, T)$, where $1 \leq k \leq l$, generally has the form

$$\mathrm{e} z_k \frac{\partial n_k}{\partial t} = \frac{\partial i_k}{\partial x} + q_k(n_1, \ldots, n_l, \varphi), \qquad k = 1, \ldots, l, \tag{35}$$

$$\frac{\partial^2 \varphi}{\partial x^2} = -\frac{\mathrm{e}}{\varepsilon_0 \varepsilon} \sum_{k=1}^{l} z_k n_k . \tag{36}$$

Each of Eqs. (35) is the balance equation for the concentration of the carriers of species k. The functions q_k in Eq. (35) describe the possible processes of carrier generation and recombination. In addition to electron-hole recombination, typical for semiconductor materials, other types of recombination are possible, e.g., recombination of electronic and ionic carriers, in accordance with the quasi-chemical reactions

$$\begin{gathered} V_M^- + h^+ \rightleftarrows V_M, \\ V_X^+ + e^- \rightleftarrows V_X, \\ M^+ + e^- \rightleftarrows M, \\ X^- + h^+ \rightleftarrows X, \end{gathered} \tag{37}$$

and recombination of ionic carriers and corresponding vacancies.

We assume now that the carriers of species $k = 1, 2$ (for the sake of definiteness, M^+ ions and e^- electrons) may recombine with the formation of the neutral component M_l [the third reaction in Eqs. (37)]. Under the stationary conditions $q_1 = q_2 = q$, we have the simplest case of the excess of the neutral component M,

$$q = K_g - K_r n_1 n_2 , \tag{38}$$

where $K_g > 0$ and $K_r > 0$ are the constants of the generation and recombination rates for species 1 and 2. For small deviations of concentrations n_1 and n_2 from their equilibrium values n_1^0 and n_2^0 (so that $\Delta n_{1,2} \ll n_{1,2}^0$, where $\Delta n_{1,2} \equiv n_{1,2} - n_{1,2}^0$), Eq. (38) may be linearized. Neglecting the term $\Delta n_1 \Delta n_2$ and bearing in mind that $K_g - K_r n_1^0 n_2^0 = 0$, we arrive at

$$q = -K_r(n_1^0 \Delta n_2 + n_2^0 \Delta n_1). \tag{38$'$}$$

If $\Delta n_1 = \Delta n_2 = \Delta n$, then $q = -\Delta n/\tau$, where $\tau \equiv [K_r(n_1^0 + n_2^0)]^{-1}$, τ being the lifetime (recombination time) of carriers. A similar expression for q is used in the theory of semiconductors for the description of electron-hole recombination.

In the above, simple example, q_1 and q_2 are equal. In more general cases, the relation between different q_k functions is more complicated, but they are always related by virtue of the law of conservation of charge.

Equation (36), where ε is the dielectric permissivity of the material, is the Poisson equation, which closes the set of equations for a potential and concentrations.

The drastic change in the potential φ occurs in the narrow subsurface layers forming near the interfaces, the so-called space-charge regions (Smith, 1978; Bonch-Bruevich and Kalashnikov, 1977). Space-charge regions are of the order of (or smaller than) the Debye length L_D, given by the equation

$$L_D \equiv \left(\frac{\varepsilon_0 \varepsilon k T}{e^2 \sum_k z_k^2 n_k} \right)^{1/2}. \tag{39}$$

For concentrations of the order of $10^{18} - 10^{19}\ \mathrm{cm}^{-3}$, L_D is of the order of $10^2 - 10^1$ nm.

To describe the transport processes out of space-charge regions one may use, instead of Eq. (36), the relation

$$\sum_{k=1}^{l} n_k z_k = 0, \tag{40}$$

which is the condition of the local electroneutrality.

It should be emphasized that Eq. (40) is much simpler than the Poisson Eq. (36), which, in many instances, essentially facilitates the analytical solution of the problem.

The small parameter providing the transition from Eq. (36) to Eq. (40) is the dimensionless $(L_D/L)^2$ ratio, where L is the size of the sample or the electrochemical cell.

Thus, for finding values of concentrations n_k and potential φ (altogether, $l + 1$ unknown functions), we have l equations of type (35) and one equation of type (36) or (40).

6. Analysis of Transport Equations

In some special cases, the description of the ionic-electronic transport may be essentially simplified. Dividing each side of Eq. (33) by eD_k, we have, by use of Eq. (32), l equations

$$-z_k \frac{dn_k}{dx} - e z_k^2 \frac{n_k}{kT} \frac{d\varphi}{dx} = \frac{i_k}{eD_k}, \qquad k = 1, \ldots, l. \tag{41}$$

Summing up Eqs. (41) and bearing in mind the use of (40),

$$\sum_{k=1}^{l} z_k \frac{dn_k}{dx} = 0,$$

the following expression for the potential gradient, is obtained:

$$\frac{\mathrm{e}}{kT}\frac{d\varphi}{dx} = -\frac{\sum_{k=1}^{l} i_k/\mathrm{e}D_k}{\sum_{k=1}^{l} z_k^2 n_k}. \tag{42}$$

Now, let us consider the case where all currents i_k but one, i_m, are zero and the concentration of the mth component is small in comparison with the concentrations of other components. Substituting Eq. (42) into Eq. (41) for component m, we arrive at

$$-\mathrm{e}z_m D_m \frac{dn_m}{dx} = i_m\left(1 + \frac{z_m n_m}{\sum_k z_k^2 n_k}\right). \tag{43}$$

The second term in brackets on the right-hand side of Eq. (43) depends on the concentrations of all components n_k, and, generally speaking, it also varies with the coordinate. But, since n_m is small in comparison with the concentrations of other components, this term may be neglected.

Thus, the transport of the component with relatively low concentration proceeds in accordance with the diffusion mechanism. This fact is widely used in the classical electrochemistry of liquid electrolytes; in some cases, large amounts of ions not discharging on the electrode (the so-called supporting electrolyte) are deliberately introduced into the initial solution. It may be readily assumed that immobile ions of the supporting electrolyte screen the electrical field near the electrodes. Therefore, the field providing migration in the solution bulk is negligibly small, and the current is caused by diffusion alone.

Obviously, that simultaneous transport of several components may also be described within the pure diffusional mechanism if their concentrations are low enough.

The total current i due to all the carriers in a multicomponent system is the sum of the partial currents i_k of the components,

$$i = \sum_{k=1}^{l} i_k. \tag{44}$$

The relative fraction of each component—in particular, of the mth component—in the total current is characterized by the so-called transport numbers t_m, defined by the equation

$$t_m = \frac{i_m}{\sum_{k=1}^{l} i_k}. \tag{45}$$

It is methodically important that, along with Eq. (45), the transport numbers are sometimes obtained from the relation

$$t'_m \equiv \frac{\sigma_m}{\sum_{k=1}^{l} \sigma_k} \tag{46}$$

(where σ_m and σ_k are conductivities of the corresponding components), which implies that $t_m = t'_m$. It should be emphasized that, in general, t'_m does not coincide with t_m.

This is clearly seen in the case where the current of one of the components (e.g., the mth component) is blocked at the electrode. Under such conditions $i_m = 0$, and, therefore, in accordance with Eq. (45) and the physics of the process, $t_m = 0$. At the same time, it is seen from Eq. (46) that $t'_m \neq 0$, since σ_m [obtained, e.g., from Eq. (30)] does not vanish.

But even in cases where carriers can freely cross the interface, $t_m \neq t'_m$. Moreover, t'_m is a function of coordinates, $t'_m(x)$, since all k concentrations $n_k(x)$ determining t'_m (through σ_k) are not constant in the sample bulk.

In other words, quantities t'_m vary in the sample from point to point as a function of the concentration n_k of all the components. Therefore, they may be regarded as local transport numbers [in distinction from the true transport numbers given by Eq. (45)].

The local transport numbers t'_m are constant and equal to t_m only in some special occasions. This is the case if, for example, all mobile carriers in the given material were of the same sign (e.g., two cation species, cations and holes, anions and electrons), or if all the carrier species pass sufficiently freely over all interfaces. In a similar way, $t_m = t'_m$ in a binary system where carriers are differently charged and may recombine, the corresponding recombination time τ being sufficiently short so that $\sqrt{D_{1,2}\tau} \ll L$ (Gurevich and Kharkats, 1980). Nevertheless, in general, $t'_m \neq t_m$.

From the above discussion, it follows, in particular, that one should be very careful in attempts to obtain conductivities from experimental data on the transport numbers.

In the simplest case, the transport numbers may be found from the Faraday law, by determining the substance mass transported by different ions and then comparing these values with the total amount of electricity passed through an electrochemical cell (the electronic component of the current does not participate in substance transport). Historically, this first method was suggested by Tubandt (1932) and described in detail by Lidiard (1957).

Another method for determining the transport numbers, from measurements of the electromotive force of the cell, has been suggested by Wagner (1933) and developed by a number of scientists (Kiukkola and Wagner, 1957;

Wagner, 1966; Vecher and Vecher, 1968; Sudarikov, 1973). The method is based on "short circuiting" of a galvanic element, caused by the flow of electronic carriers in the SIM bulk.

To illustrate the method, we consider a concentrational galvanic cell

$$M'|M_nX_l|M''.$$

Here, M_nX_l is the material under consideration, with conductivity provided, for simplicity, by monovalent M^+-cations and electrons. M' and M'' are electrodes (reversible with respect to M^+-cations) with different contents of element M (e.g., alloys with different component ratios or solid solutions of different concentrations).

If no current flows through the cell, the obvious relation holds: $i_e + i_i = 0$. Substituting Eq. (33) for currents i_k into this equation, we find the electric-potential gradient

$$\frac{d\varphi}{dx} = -\frac{1}{e}\left(t_i'\frac{d\mu_i}{dx} - t_e'\frac{d\mu_e}{dx}\right), \tag{47}$$

where

$$t_{i,e}' \equiv \frac{\sigma_{i,e}}{\sigma_i + \sigma_e}.$$

Assuming that at each SIM point a local equilibrium between ions and electrons is attained, i.e., $\mu_i + \mu_e = \mu_M$, we may transform Eq. (47) to the form

$$\frac{d\varphi}{dx} = -\frac{1}{e}\left[(1 - t_e')\frac{d\mu_M}{dx} + \frac{d\mu_e}{dx}\right]. \tag{47'}$$

In order to determine the emf of the cell, $\mathcal{E}$, it is necessary to integrate Eq. (47′) over SIM thickness and take into account the potential jumps at the interfaces. These potential jumps are calculated proceeding from the condition of equality of electrochemical potential for electrons of M', M'' and the SIM.

Then, the emf of the cell is determined by the equation

$$\mathcal{E} = -\frac{1}{e}\int_{\mu_M(o)}^{\mu_M(L)} (1 - t_e')\, d\mu_M, \tag{48}$$

where the integration limits are the values of μ_M in the regions close to the electrodes. For small currents (in the absence of electrode polarization), these values are taken to be equal to the values of μ_M at the electrodes

$$\mathcal{E} = -\frac{1}{e}\int_{\mu_{M'}}^{\mu_{M''}} (1 - t_e')\, d\mu_M, \tag{49}$$

If $t_e' = 0$, Eq. (49) is transformed into the conventional Nernst formula, determining the thermodynamic value $\mathcal{E}_{th}$ of the emf. Comparing the values of $\mathcal{E}$ and $\mathcal{E}_{th}$, it is possible to determine the transport numbers t_e'. If ionic conductivity in the sample is provided by a different ion species k, then the emf of the concentrational cell is determined by the relationship (Heyne, 1977)

$$\mathcal{E} = -\frac{1}{e}\int_{\mu_{M'}}^{\mu_{M''}} \sum_k \frac{t_k'}{z_k}\, d\mu_k, \tag{50}$$

where μ_k is the chemical potential of neutral atoms. The integration limits are μ_k values in the electrode material.

7. Boundary Conditions

Depending on the properties of contacting substances, the type of current carriers, and the value of the potential drop at the interface, the penetrability of the interface between a superionic material and an electrode may vary over a wide range of values.

If the interface is impenetrable to species k (the exchange current is zero), the corresponding electrode is considered to be ideally polarizable or blocking. [The first term, introduced by M. Planck, relates to the metal-electrolyte solution interface and is of common use in electrochemistry. In solid-state physics, the second term is more common]. Obviously, if the interface is impenetrable, the current does not flow through the interface, and the boundary condition for k-species is $i_k = 0$.

Now, let us analyze the reverse situation. Let us assume that the process at the electrode (with the participation of species k, which either pass over the interface or discharge there) proceeds at a high rate (the exchange current is strong). In this case, the equilibrium at the interface is not upset, i.d., $\tilde{\mu}_{ed}^{M} = \tilde{\mu}^{S}$, where $\tilde{\mu}_{ed}^{M}$ and $\tilde{\mu}^{S}$ are the electrochemical potentials of k-species in the metal electrode and in the superionic material at points in the direct vicinity of the interface.

It is assumed that $\tilde{\mu}_{ed}^{M}$ does not change with the varying particle concentration n_k in the superionic material (which is the case for metal electrodes). Then, using Eq. (29),

$$\varphi_{ed}^{o} = \varphi_{st} - \frac{kT}{ez_k} \ln n_{k,s}^{0}. \tag{51}$$

Here $n_{k,s}^{o}$ is the equilibrium concentration of species k in the vicinity of the interface from the side of the superionic material, and φ_{ed}^{o} is the equilibrium electrical potential of the electrode (measured, as a rule, from a certain reference electrode). Quantity φ_{st}, the so-called standard electrical potential, depends on the difference between the standard chemical potentials of the electrode and the superionic material.

If a current is flowing, the concentration n_k at the interface, $n_{k,s}$, does not coincide with the equilibrium value n_k^o, and the electrode potential φ_{ed} differs from φ_{ed}^o. But, if the equilibrium at the electrode is not upset, φ_{ed} and $n_{k,s}$ are related by a relationship analogous to Eq. (51),

$$\varphi_{ed} = \varphi_{st} - \frac{kT}{ez_k} \ln n_{k,s}. \tag{52}$$

The relation between the change of the electrode potential $\Delta\varphi_{ed}$ and the near-electrode concentration of particles participating in the fast reaction at the electrode follows from Eqs. (51) and (52):

$$\Delta\varphi_{ed} = \frac{kT}{ez_k} \ln \frac{n_{k,s}}{n_{k,s}^o}. \tag{53}$$

The sign of $\Delta\varphi_{ed}$ is chosen in accordance with the rules used in electrochemistry of solutions, in such a way that $\Delta\varphi_{ed} < 0$ if metal ions discharge at the electrode (in other words, the electrode is functioning as a cathode) and $\Delta\varphi_{ed} > 0$ if metal ions migrate from the electrode into the bulk of the electrochemical cell (the electrode is functioning as as anode).

If, during current flow, $\Delta\varphi_{ed}$ is small (in particular when $|e\,\Delta\varphi_{ed}/kT| \ll 1$), then Eq. (53) is physically equivalent to the condition that, in the vicinity of the electrode surface, a certain concentration, $n_{k,s} \approx n_{k,s}^o$, is maintained.

Relationships (51)–(53) were first suggested by Nernst for a metal-liquid electrolyte interface. As follows from the derivation, they may also be used to describe quasi-equilibrium processes occurring at interfaces formed with superionic materials (with respect to both ionic and electronic carriers).

Now, let us consider particles passing through an interface under non-equilibrium conditions when the electrochemical potentials of these particles are different in different phases.

In particular, let us consider a monovalent cation that moves from a certain initial position with the coordinate x_1 in condensed phase I to then position with the coordinate x_2 in condensed phase II. The potential-energy profile along the trajectory of ion motion is schematically shown in Fig. 11. Since phases I and II have different properties, the profile may be essentially asymmetric. Let the ratio of the distances from the barrier top (coordinate x_*) to a stable position in phase I (coordinate x_1) to the distance from position x_2 in phase II to the barrier top be described as

$$\frac{\alpha}{\beta} = \frac{x_* - x_1}{x_2 - x_*}. \tag{54}$$

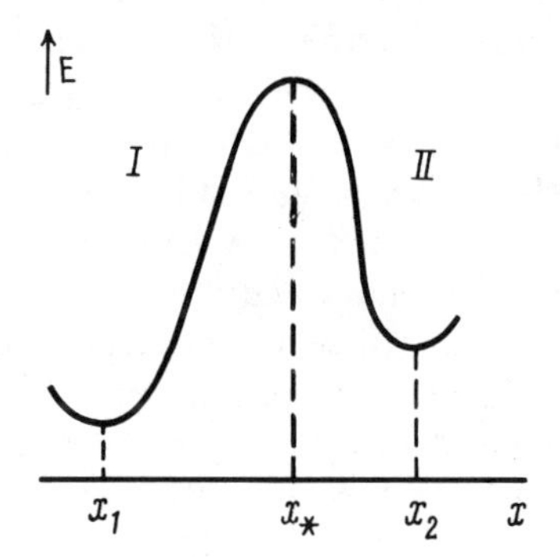

FIG. 11. Potential energy profile versus coordinate of a moving ion.

Dimensionless parameters α and β are related as

$$\alpha + \beta = 1, \tag{55}$$

characterizing the degree of asymmetry of the position of the potential-barrier maximum inside sector $[x_2, x_1]$. If point x_* lies exactly in the middle of this sector, then $\alpha = \beta = \frac{1}{2}$.

If the applied electric-potential difference between points x_1 and x_2 is η, then, in the simplest case, the change in the activation-barrier height, caused by an additional electric field in the transition from x_1 to x_2, is $e\alpha\eta$. For the reverse transition, from x_2 to x_1, it is $e\beta\eta$. Assuming that particle current $\vec{i}$, flowing from left to right, is proportional to particle concentration n_{I} at point x_1 in phase I, and that particle current $\overleftarrow{i}$, flowing from right to left, is proportional to concentration n_{II} at point x_2 in phase II, we arrive at the following equation for the resulting current $i = \vec{i} - \overleftarrow{i}$ through the interface (for simplicity, we omit the subscript indicating the particle sort):

$$i = i_{\mathrm{o}}\left[\left(\frac{n_{\mathrm{I}}}{n_{\mathrm{I}}^{\mathrm{o}}}\right)\exp\left(\frac{e\alpha\eta}{kT}\right) - \left(\frac{n_{\mathrm{II}}}{n_{\mathrm{II}}^{\mathrm{o}}}\right)\exp\left(-\frac{e\beta\eta}{kT}\right)\right]. \tag{56}$$

Here, i^{o} is the exchange current of the particle under consideration, which is determined from the condition that, in the equilibrium state, $i^{\mathrm{o}} = |\vec{i}| = |\overleftarrow{i}|$, $n_{\mathrm{I}}^{\mathrm{o}}$ and $n_{\mathrm{II}}^{\mathrm{o}}$ being the equilibrium values of concentrations n_{I} and n_{II}. By definition, η is the deviation of the potential difference $\Delta\varphi_{1,2}$ between points x_2 and x_1 from the equilibrium value $\Delta\varphi_{1,2}^{\mathrm{o}}$,

$$\eta \equiv \Delta\varphi_{1,2} - \Delta\varphi_{1,2}^{\mathrm{o}}. \tag{57}$$

Quantity η is widely used in electrochemistry and is the called overvoltage of the electrode reaction or the activation overvoltage. [Letter η is also used to denote the total deviation of the potential from its equilibrium value in the electrochemical cell during current flow.] The latter quantity is called the total overvoltage or, simply, overvoltage. It is obvious that the activation overvoltage is a part (sometimes the main part) of the total overvoltage. Parameters α and β in Eq. (56) are called the transfer coefficients.

Gurevich and Kharkats (1986b) gave a consistent description of the ion transport kinetics through an interface within the stochastic approach. The explicit expression for the exchange current i° was derived in terms of the microcharacteristics of the interface. Gurevich and Kharkats also studied the function $i(\eta)$ and showed that, under certain conditions (a gently sloping wide barrier), quantity i should depend on η in a more complicated way than is given by Eq. (56) formally, this result may be interpreted as the dependence of α and β on η, with condition $\alpha(\eta) + \beta(\eta) = 1$ preserving its validity).

Relationship (56) coincides, in essence, with the well-known relationship of chemical kinetics derived within the slow discharge theory (Vetter, 1961; Petrii and Damaskin, 1983). In the latter case, it describes the electron transport when, e.g., the following reaction proceeds at the solution-metal electrode interface:

$$M \rightleftarrows M^{+} + e^{-}. \tag{58}$$

Activation energy entering the equation for i° does not have such a simple physical meaning in the case of electron transfer. It is now essentially related to the energy necessary for the reorganization of the surrounding medium in the fast electron transfer (Ulstrup, 1980). Neither can transfer coefficients α and β now be interpreted in a simple and obvious way. Note also that in metals (medium I, for definteness) we may take $n_{\mathrm{I}} = n_{\mathrm{I}}^{\circ}$.

If the exchange current is strong ($i^{\circ} \gg |i|$), a condition that physically corresponds to the quasi-equilibrium situation, the expression in square brackets in Eq. (56) is almost zero. Whence, with account of the fact that $n_{\mathrm{I}} = n_{\mathrm{I}}^{\circ}$ and with the validity of Eq. (55), we arrive, as should be expected, at the relationship of type (53).

Equation (56) may be somewhat extended, if we introduce into consideration the deviation of the chemical potential μ of particles from its equilibrium value μ° at points x_1 and x_2,

$$\Delta\mu_{\mathrm{I}} = \mu(x_1) - \mu^{\circ}(x_1) \tag{59}$$

and

$$\Delta\mu_{\mathrm{II}} = \mu(x_2) - \mu^{\circ}(x_2). \tag{60}$$

Then, with account of Eqs. (56), (59), (60) and (29), we have, for the resulting current,

$$i = i_{\mathrm{o}}\left\{\exp\left[\frac{(\Delta\mu_{\mathrm{I}} + e\alpha\eta)}{kT}\right] - \exp\left[\frac{(\Delta\mu_{\mathrm{II}} - e\beta\eta)}{kT}\right]\right\}. \tag{61}$$

Using Eq. (61), it is also possible to take into account a more complicated dependence of μ on n_k, e.g., one caused by the effect of lattice saturation (Gurevich and Kharkats, 1980) or by the changes of the entropy factors during the passage of particles through the interface, which may be associated, for example, with the changes in the local particle oscillations in the vicinity of points x_1 and x_2.

If the deviations from equilibrium are small

$$\left|\frac{\Delta\mu_{\mathrm{I,II}}}{kT}\right| \ll 1, \qquad \left|\frac{e\eta}{kT}\right| \ll 1,$$

then, linearizing Eq. (61) and making simple transformations, we arrive at Eq. (28), which is the phenomenological thermodynamic equation for irreversible processes. Thus, Eq. (61) is an extension of linear equation (28), which also establishes the correspondence between the thermodynamics of irreversible processes and the slow discharge theory.

If different particles cross the interface, each species is characterized by its own exchange current i_k^0 and transfer coefficients α_k and β_k.

At the same time, it should be kept in mind that, in real solid-state systems, transport through the interface may be complicated by a series of other processes (Chebotin, 1982), such as deposition or dissolution of the substance at the sites of contacts, structural rearrangements of the lattice near the interface, diffusion of particles (atoms or ions) along the surface, etc.

The experimental values of exchange currents for a series of SIMs are listed in Table 3.

TABLE 3

EXCHANGE CURRENTS OF SOME SIM/ELECTRODE INTERFACES

Interface	Ionic charge carrier	Exchange current, mA/cm^2	Reference
$Cu \mid Cu_4TlCl_3I_2$	Cu^+	0.013 at 297 K	Vershinin *et al.* (1985)
$Cu \mid Cu_4RbCl_3I_2$	Cu^+	0.07 at 298 K	Vershinin *et al.* (1985)
$Ag \mid Ag_4RbI_5$	Ag^+	3–20 at 300 K	Armstrong and Dickinson (1976)
$Ag \mid Ag_6WO_4I_4$	Ag^+	1–3 at 300 K	Armstrong and Dickinson (1976)
$Na \mid Na\text{-}\beta''\text{-}Al_2O_3$	Na^+	500 at 420 K	Armstrong *et al.* (1973)
$Na \mid Na\text{-}\beta\text{-}Al_2O_3$	Na^+	0.03 at 298 K 7 at 410 K	Bukun *et al.* (1974)

IV. Stationary Electron-Hole Currents in Superionic Materials

Under the condition that a superionic material being studied is in the "normal" state, with ionic conductivity being almost zero, the currents of electronic carriers are described by a relationship well known in the theory of semiconductors and dielectrics.

The situation here is as follows: if a material possesses a noticeable electronic conductivity, the usual Ohm's law is valid up to voltages at which the concentration of injected electrons becomes comparable with their initial equilibrium concentration in the material under consideration.

With the further rise of voltage, a drastic (almost vertical) increase of the current is observed, which is explained by the filling of separate groups of trapping levels. At still higher voltages, the transition to the so-called quadratic law (Lampert and Mark, 1970) is observed. The equation relating electronic current i_{el} and the applied constant voltage V has the form $i_{\mathrm{el}} = bV^2/L^3$, where L is the sample length and coefficient b depends on the material properties. Such a current-voltage characteristic coincides with a characteristic well known in the theory of currents limited by the space charge (the Mott–Gurney law).

Electronic currents in SIMs in the superionic state obey laws qualitatively different from those considered above. These laws will be discussed later.

8. Current-Voltage Characteristics

If ionic conductivity of the material being studied is high, electron-hole transport is essentially determined by the behavior of mobile ions. In particular, the existence of mobile charged particles, other than electrons and holes, results in additional screening of the external electric field. At a sufficiently high equilibrium concentration n_{i} of mobile ions, the voltage drops almost to zero in the space-charge region, the size of which is of the order of L_{D} [see Eq. (39)] and is determined by the concentration n_{i} ($L_{\mathrm{D}} \sim n_{\mathrm{i}}^{-1/2}$).

a. Electrochemical Cells for Measuring σ_{el}

For example, we assume mobile carriers in the material being studied are monovalent ions, electrons, and electron holes. In accordance with Eqs. (2) and (33), the density of the total electron-hole current i_{el} is:

$$i_{\mathrm{el}} = \left[-D_{\mathrm{e}}\mathrm{e}\frac{dn_{\mathrm{e}}}{dx} + D_{\mathrm{h}}\mathrm{e}\frac{dn_{\mathrm{h}}}{dx} \right] + \left[(\mathrm{e}u_{\mathrm{e}}n_{\mathrm{e}} - \mathrm{e}u_{\mathrm{h}}n_{\mathrm{h}})\frac{d\varphi}{dx} \right]. \quad (62)$$

We shall consider electrochemical cells of two main types:

$$C_r|SIM|A_b, \tag{A}$$

$$C_b|SIM|A_r. \tag{B}$$

Here, A and C denote anode and cathode (both with electronic conductivity), and subscripts r and b characterize the properties of A and C electrodes—reversible or blocking relative to conducting ions.

Depending on the sort of mobile carriers (cations or anions), either the A- or B-type cells are used. Since, in both types of cells, one of electrodes blocks ions, the resulting current flowing in the system under the stationary conditions is pure electronic.

An example of an A-type electrochemical cell is the system

$$(-)Ag|Ag_4RbI_5|C(+),$$

with solid electrodes (a silver electrode reversible with respect to cations and a cation-blocking graphite electrode).

The system $(-)Pt|PbF_2|F_2(gas)(+)$ is an example of type-B cells, with an anion-reversible gaseous electrode. Gas electrodes contain an electrochemically active gas element in a free state. They are fabricated from a chemically inert electroconductive material (as a rule, platinum or carbon) that plays the part of a material-carrier and is saturated with the gas. Such electrodes are named either after reversible potential-inducing elements (oxygen, fluorine electrodes) or after inert materials (platinum, carbon electrodes).

b. Diffusion Mode of Electronic Current Flow

Now consider current-voltage characteristics (CVCs) of A- and B-type cells. Since the SIMs being studied are characterized by high concentrations of mobile ions, then, in accordance with earlier discussions [see Eq. (43)], we may assume that the total electronic current is due to diffusion. In this case, instead of Eq. (62), we have

$$i_{el} = -eD_e \frac{dn_e}{dx} + eD_h \frac{dn_h}{dx}. \tag{63}$$

Note also here that the function of the supporting electrolyte is performed by mobile ions, with the discharging particles being electrons and holes.

The direct integration of Eq. (63) yields

$$i_{el} L = eD_e[n_e(O) - n_e(L)] + eD_h[n_h(L) - n_h(O)], \tag{64}$$

where $n_{e,h}(O)$ and $n_{e,h}(L)$ are the electron-carrier concentrations at the cathode ($x = O$) and anode ($x = L$). Under the same assumptions, the space-charge regions at the interfaces (x = O and x = L) may be neglected. Assuming also that at both the cathode and the anode the local equilibrium is attained relatively by electrons and holes (in other words, the electrochemical potentials are continuous at the interfaces), it may readily be shown that, in accordance with Eq. (51), the following relationship takes place:

$$-eV = kT \ln\left[\frac{n_e(L)}{n_e(O)}\right] = kT \ln\left[\frac{n_h(O)}{n_h(L)}\right], \tag{65}$$

where V is the external potential difference applied to the sample.

A virtue of the properties inherent in A- and B-type cells is that the concentration of electrons (and also holes) at the reversible electrode is constant, independent of V, and determined by the equilibrium conditions at this interface [Eq. (26)].

Therefore, using Eqs. (65) and (32), we may rewrite Eq. (64) in the form

$$i_{el} = i_{e,O}\left[1 - \exp\left(-\frac{eV}{kT}\right)\right] + i_{h,O}\left[\exp\left(\frac{eV}{kT}\right) - 1\right] \tag{66}$$

for an A-type cell, and in the form

$$i_{el} = i_{e,L}\left[\exp\left(\frac{eV}{kT}\right) - 1\right] + i_{h,L}\left[1 - \exp\left(-\frac{eV}{kT}\right)\right] \tag{67}$$

for a B-type cell.

Here,

$$i_{e,h,O} \equiv \frac{eD_{e,h}n_{e,h}(O)}{L} = \frac{\sigma_{e,h}(O)kT}{eL}, \tag{68}$$

$$i_{e,h,L} \equiv \frac{eD_{e,h}n_{e,h}(L)}{L} = \frac{\sigma_{e,h}(L)kT}{eL}. \tag{69}$$

Relationships (66) and (67) were first derived by Wagner (1956, 1957), who used a somewhat different approach. The concepts were developed by Hebb (1952).

We shall illustrate the above approach by way of an example of an A-type cell. We assume that the SIM under test has only mobile singly-charged cations and electrons, and that the local equilibrium for carriers is attained at each point of the crystal and is preserved during current flow through the cell.

Under the equilibrium conditions, we have

$$\tilde{\mu}_M = \mu_M = \tilde{\mu}_i + \tilde{\mu}_e, \tag{70}$$

where $\tilde{\mu}_M(\mu_M)$ is the electrochemical (chemical) potential of a neutral atom M.

Since the concentration of mobile ions is high, it is almost constant, and, therefore, it may be assumed that

$$\frac{d\tilde{\mu}_i}{dx} = 0. \tag{71}$$

Then, it follows from Eq. (70) that

$$\frac{d\tilde{\mu}_M}{dx} = \frac{d\tilde{\mu}_e}{dx}. \tag{72}$$

Now, making use of the general equation (28), we may write the equation for the electronic current

$$i_e = \frac{\sigma_e}{e}\frac{d\tilde{\mu}_e}{dx} = \frac{\sigma_e}{e}\frac{d\tilde{\mu}_M}{dx}. \tag{73}$$

Integrating (73) along the sample, we arrive at

$$i_e = \frac{1}{L}\int_{\mu_1}^{\mu_2} \frac{\sigma_e}{e}\, d\tilde{\mu}_M, \tag{74}$$

where μ_1 and μ_2 are the chemical potentials of the M atom at the anode and cathode, respectively.

On the other hand, the potential difference V between the electrodes may be written in the form [see also Eq. (65)]

$$V = -\frac{1}{e}\int_{\mu_1}^{\mu_2} d\tilde{\mu}_e = -\frac{1}{e}\int_{\mu_1}^{\mu_2} d\tilde{\mu}_M = \frac{1}{e}(\mu_1 - \mu_2). \tag{75}$$

Using Eq. (30) for σ_k and the equilibrium expressions for electronic carriers

$$n_e = n_e' \exp\left(\frac{\tilde{\mu}_e}{kT}\right) = n_e'' \exp\left(\frac{\tilde{\mu}_M}{kT}\right) \tag{76}$$

(here, n_e'' is the electron concentration at $\tilde{\mu}_M = 0$, i.e., with respect to the chosen reference electrode), we may rewrite Eq. (74), taking into account Eq. (76), in the form

$$i_e = \left(\frac{kT}{L}\right)\sigma_e(\mathrm{O})\left[1 - \exp\left(-\frac{eV}{kT}\right)\right], \tag{77}$$

where $\sigma_e(\mathrm{O}) = eu_e n_e'' \exp(\mu_1/kT)$ is the electronic conductivity at the SIM-reversible electrode interface. It may readily be seen that Eq. (77) is equivalent to Eq. (66), which describes the CVC of an A-cell (in the absence of the hole component of the current).

Note also that the above relationships are analogous to the well-known expression of the theory of liquid electrolytes, which describes CVCs of reversible electrode (see, e.g., Damaskin and Petrii, 1983). Current $i_{e,h,O}$ (or $i_{e,h,L}$) coincides with the limiting current in a capillary of length L, at one end of which the constant concentration $n_{e,h}(O)$ [or $n_{e,h}(L)$] is maintained. Equation (65) is analogous to the equation for concentration polarization, well known in electrochemistry of "conventional" electrolytes.

The method for determining $\sigma_{e,h}$, based on the principle of the ionic current suppression in A- and B-type cells, is often called the Hebb–Wagner polarization technique (Joshi and Wagner, 1972; Kennedy, 1977a).

c. Current-Voltage Characteristics for Diffusion Mode

Let us consider several typical current-voltage characteristics (CVCs) (Mizusaki and Fueki, 1980).

If, for A-type cells, the condition $\sigma_e(O) \ll \sigma_h(O)$ is fulfilled, or if voltage V is high ($|V| \gg kT/e$), then, according to the above-stated concepts, the electron-hole current should increase at a very high rate (exponentially) (Fig. 12, curve 1),

$$i_{el} = i_{h,O} \exp\left(\frac{eV}{kT}\right) = i_{h,L}. \tag{78}$$

Therefore, a "ln i_{el} versus eV/kT" plot in a straight line with a slope of 45°. The intercept on the ordinate axis determines the value of ln $i_{h,O}$ and, hence, $\sigma_h(O)$.

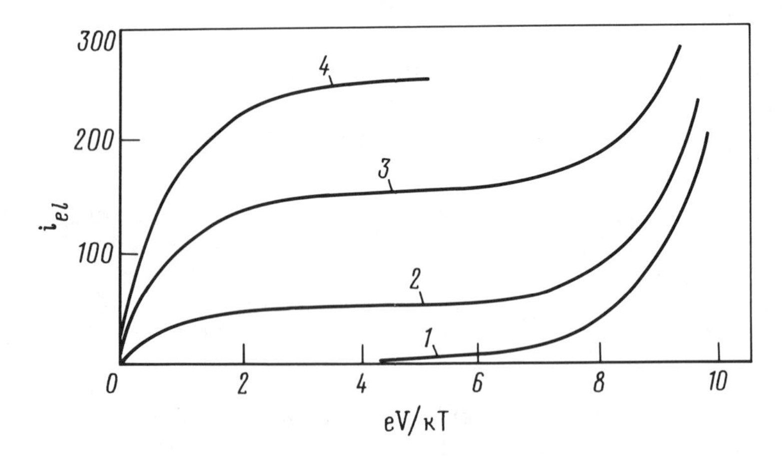

FIG. 12. Current-voltage characteristics of an A-type cell calculated by Eq. (66). Current (arb.un.): $i_{h,O} = 10^{-2}$, (1) $i_{e,O} = 1$; (2) $i_{e,O} = 50$; (3) $i_{e,O} = 150$; (4) $i_{e,O} = 250$.

In the reverse, limiting case, when $\sigma_e(O) \gg \sigma_h(O)$, the curve $i_{el}(V)$ has a portion that corresponds to the saturation current $i_{e,O}$ (Fig. 12, curve 2). Therefore, the CVC should be constructed in the i_{el}, eV/kT coordinates. Current $i_{e,O}$ determined, we may find conductivity $\sigma_e(O)$.

At last, if electron and hole conductivities are approximately equal, $\sigma_e(O) \approx \sigma_h(O)$, the CVC has the shape schematically shown in Fig. 12 (curves 3, 4). In order to determine $i_{e,O}$ and $i_{h,O}$ graphically, the CVC should be constructed in the $i_{el}/[\exp(eV/kT) - 1]$, $\exp(-eV/kT)$ coordinates, as has been suggested by Patterson *et al.* (1967). Then, $\sigma_e(O)$ is given by the slope of the thus-obtained lines (at different values of T), and the intercept on the ordinate axis yields $\sigma_h(O)$.

If the concentrations of electronic carriers are not fixed at any of the interfaces, the current flowing through the cell is still determined by the diffuse component alone, and Eq. (64) is valid. But the chemical potentials of the components are not fixed at the electrodes, and, therefore, σ_{el} depends on the method used for compound synthesis.

In order to obtain the dependence of type (66), some additional assumptions should be made, in accordance with the specific properties of the electrodes and material under consideration (Weppner and Huggins, 1977a; Kennedy, 1977a). Thus, for a symmetric cell with inert electrodes, the application of a potential difference to the cell results in the formation of a specific reversible electrode at one of the interfaces, due to deposition of mobile ions forming an intermediate phase (it is also possible that the chemical potential of ions in the "deposited phase" differs from the corresponding standard value).

We should like to emphasize that the results obtained by the polarization method may be analyzed only if the applied voltage is essentially lower than the decomposition potential and if the amount of isolated component does not exceed its solubility limit in the compound under study. Moreover, even if the potential does not exceed the decomposition potential, possible compound decomposition should be taken into consideration (Kröger, 1974).

In particular, compound decomposition may also be caused by its evaporation. As an example, we consider an A-type cell with the MX binary compound where X is a volatile component forming the X_2 gas. Let the cell be placed into a volume filled with the gas, the partial pressure being P_{X_2}. Upon reaching equilibrium between the sample and the gas phase, a potential difference V is applied to the cell, which changes the chemical potentials μ_M and μ_X in the sample.

As a result, at the SIM-blocking electrode interface, the equilibrium is upset with respect to component X in both the SIM and the gas phase. If the partial gas pressure around the blocking electrode is lower than at the interface, a current starts flowing through the cell. Such a situation is implemented under the condition of the constant pumping of an inert gas through the set-up with the sample. Because of electrolysis, a dc current, which is associated with the SIM decomposition (evaporation), flows in the A-type cell. In other words, the conditions for ion blocking are violated, and the above equations become invalid.

This effect was studied in detail experimentally (Mizusaki *et al.*, 1978, 1979, 1980, 1982). The authors have shown that the use of carbon (a porous material) as inert blocking electrodes requires the design of a special measuring cell so that the inert electrode would be completely isolated from the ambient medium. Then, at the MX/C interface, the equilibrium value of P_{X_2} corresponding to the applied voltage is attained.

On the whole, the type of cells used only slightly affects the slope of the CVCs. At the same time, the treatment of the obtained results, in accordance with Eqs. (66) and (67), may noticeably affect the values of the electronic-carrier characteristic parameters. Figure 13 shows the corresponding data for silver iodide. It is seen that the results obtained with the use of conventional cells do not obey Eq. (66) (the electronic current is independent of the sample size, curve 1), whereas the results obtained on an improved measuring cell give the expected dependence of i_{el} on L.

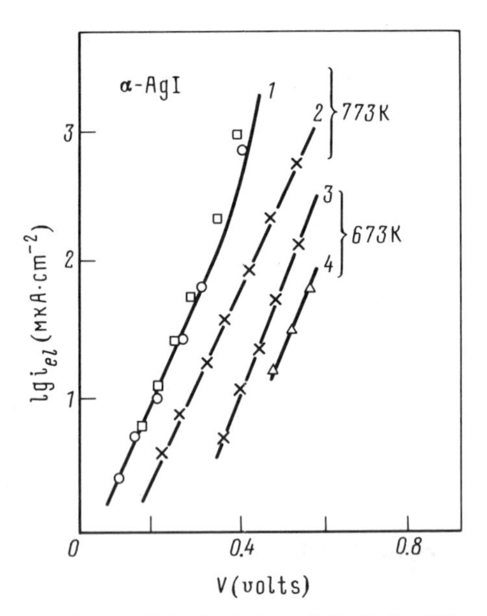

FIG. 13. Current-voltage characteristics for Ag|α-AgI|C cells for different sample thicknesses and temperatures: (1) conventional cell, $L = 2.7$ and $L = 6.0$ mm; (2), (3) improved cell, $L = 3.3$ mm; (4) improved cell, $L = 6.5$ mm. [From Mizusaki *et al.* (1978).]

In a number of experiments (Joshi and Liang, 1975; Lingras and Simkovich, 1978; Ivanov-Shits *et al.*, 1978; Schoomman *et al.*, 1983), qualitative deviations from the characteristics described by Eqs. (66) and (67) were observed, which may be explained by partial blocking of ion current at the electrodes. Brook *et al.* (1971) have shown that if the residual current i_{ir} flowing through the cell is small ($i_{ir} \ll i_{io}$, where i_{io} is the ionic current in a cell with reversible electrodes), then Eq. (66) acquires the following form for $\sigma_h(O) \gg \sigma_e(O)$:

$$i_{el} = i_h = \left(\frac{kT}{eL}\right)\sigma_h(O)\left\{\exp\left[\left(1 - \frac{i_{ir}}{i_{io}}\right)\frac{eV}{kT}\right] - 1\right\}. \tag{79}$$

Since $i_{io} \simeq V/R_{io}$, where R_{io} is ionic resistance of the SIM, the residual current i_{ir} at large potentials will affect only the intercept on the ordinate axis, not changing the slope of the $(\ln i_{el}, V)$ line. At the same time, in the majority of experiments, gently sloping CVCs were observed. In other words, the experimental results obtained may be explained only on the assumption that i_{ir} depends on the applied voltage.

Joshi and Liang (1975) have analyzed the applicability range of the model used in experiments with polarized cells. If Eq. (66) is to be valid, V should not exceed a certain value,

$$V < V_{max}, \qquad V_{max} \equiv -\left(\frac{kT}{e}\right)\ln\left[\frac{\sigma_h(O)}{\sigma_i}\right], \tag{80}$$

which means that the electronic component of conductivity should be small at every point of the crystal.

It is also clear that the electronic-carrier concentration depends on the cell type and the applied voltage—see Eqs. (66) and (67). Since the mobilities of ions and electrons (holes) are essentially different, Schoonman *et al.* (1983) suggested using less rigid limitations—the electronic-carrier concentration should be maintained at such a low level that the deviations from stoichiometry should be small for any voltage V.

Baklykov (1980) and Baklykov and Zakharov (1980) have analyzed the effect of point defects in the SIM bulk on CVCs for A-type cells. They came to the conclusion that the deviations of CVC behavior from the theoretical Eq. (66) may be explained by an increase of the hole concentration up to the value of the ionic-carrier concentration.

As an example of a comprehensive analysis of polarization measurement data, we cite the work of Ross and Schoonman (1984), who studied the $La_{1-x}Ba_xF_{2+x}$ solid solutions with high conductivity due to transport of interstitial fluorine ions. For a (−)La/SIM/Pt(+) cell, in which La is used as a reversible electrode, it has been found that electronic conductivity

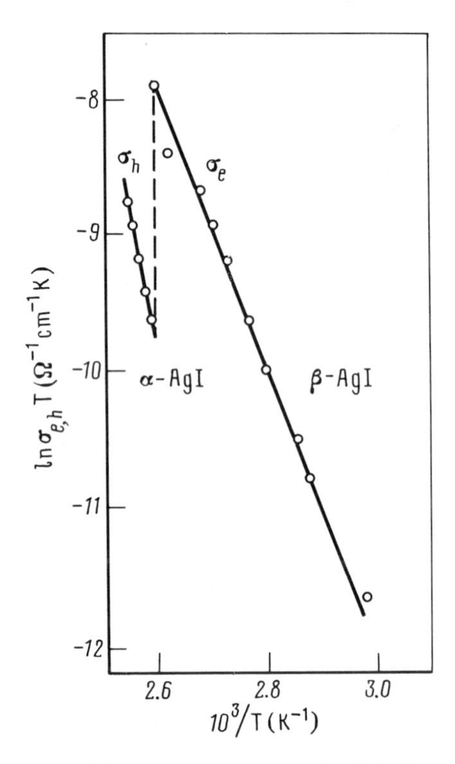

FIG. 14. Electronic conductivity versus temperature for AgI. Dotted line shows a drastic change in electronic conductivity at the temperature of the phase transition. [From Mazumdar *et al.* (1982).]

prevails in the temperature range 323–433 K, the measured value of σ_{el} being the maximum possible one, since it corresponds to the nonstoichiometric composition of the solid solution in the equilibrium with lanthanum.

Now, we should like to discuss the results obtained by Mazumdar *et al.* (1982), who studied electronic conductivity of the classical superionic material—silver iodide. They have revealed (Fig. 14) that upon the phase transition into the high-temperature superionic α-phase, the electronic component σ_{el} changes in a jumpwise manner, and—what is even more important—conductivity changes from the electron- to the hole-type. The authors believe that the observed effect is associated with the change in the band gap (from 2.8 to 2.5 eV in the β- and α-phases, respectively) and the position of the Fermi-level with respect to the top of the valence band (1.6 and 1.1 eV for the β- and α-phases). Thus, β-AgI should be a n-type semiconductor, and α-AgI a p-type one, which was, in fact, observed to be true in the experiments.

d. Kinetic Mode of Electronic Currents Flow

From the method used for the calculation of CVCs [Eqs. (63)–(67)] and the analogy with the expressions known in the theory of liquid electrolytes, it follows that these equations correspond to the model of diffusion kinetics. In other words, Eqs. (63)–(67) have been derived on the assumption that the rate of the resulting process is determined by the rate of electronic-carrier transport from the SIM bulk to the interface. On the other hand, under certain conditions, the limiting stage of the current flow may turn out not to be the transport of charged particles in the crystal, but the charge transfer across the interface.

Such a situation is often observed at the semiconductor-metal contact. It is essential that the current-voltage characteristic in this case be described by an equation similar to Eq. (66). Specifically, if we consider, for simplicity, only the electronic component in an A-type cell, the electronic current may be expressed (Strikha, 1974; Rhoderick, 1978) as

$$i_e = i'_{e,o}\left[1 - \exp\left(-\frac{eV}{kT}\right)\right]. \tag{81}$$

The value of the parameter $i'_{e,o}$ in Eq. (81) depends on the character of the charge-carrier motion in the vicinity of the interface. Physically, this parameter describes the maximum possible electronic current. The passage of electrons through the metal-semiconductor interface may be considered with the aid of the diode and diffusion models.

According to the diode model, the electronic current from the semiconductor is restricted by the emission current to the metal. If there is a "thin" barrier layer with dimensions smaller than the electron free path, then $i'_{e,o}$ coincides with the thermoemission current i_T [Eq. (27)]. The meaning of $i'_{e,o}$ here is analogous to the exchange current in electrochemical systems [see Eq. (61)], and not to the limiting current flowing from the bulk to the interface, as it takes places for i_{eo} in Eq. (66).

In accordance with the diffusion model, the partial current of electrons is limited by the rate of their transport through a depleted layer formed in the space-charge region at the interface. This model is used for a relatively "thick" barrier layer with dimensions essentially larger than the electron free path (although, of course, the dimensions are much smaller than the sample dimensions). Proceeding from the general equation for current density (28) in the form

$$i_e = en_e^o u_e \frac{d\tilde{\mu}_e}{dx}, \tag{82}$$

and using Expression (12) for concentration n_e^o, we arrive at

$$i_e = eu_e N_c \exp\left[-\frac{e(E_c - \tilde{\mu}_e)}{kT}\right] \frac{d\tilde{\mu}_e}{dx}. \tag{83}$$

Integrating Eq. (83) with the appropriate boundary conditions, we can calculate the CVC of the junction. For comparatively high voltages ($|V| \gg kT/e$), $i'_{e,o}$ is determined by the following expression:

$$i'_{e,o} = eu_e N_c \mathfrak{E} \exp\left(-\frac{eE_B}{kT}\right), \tag{84}$$

where $\mathfrak{E}$ is the maximum electric field at $x = 0$, $\mathfrak{E}$ being dependent on V.

Experimentally, the saturation current is seldom attained; i.e., i_e weakly increases for sufficiently high values of V. This may be associated, in particular, with a decrease of the potential-barrier height at the interface with an increase of the applied voltage. One of the general reasons for the barrier-height field dependence is the presence of an intermediate layer, e.g., of a thin oxide layer, between the metal and the semiconductor (Rhoderick, 1978).

We should like to indicate an important role of surface states (see Sec. 4), although the corresponding effects cannot yet be consistently taken into account. The main explanation of such a situation lies in the fact that, at present, there are no standard methods for the preparation of "good" SIM surfaces providing reproducible surface characteristics (Borovkov and Khachaturyan, 1976; Armstrong and Dickinson, 1976; Ukshe and Bukun, 1977).

Important information on the changes in the surface state with the variation of the external conditions (the composition of the gas atmosphere, tightness and quality of polished contacts, the electrode material, etc.) may, in principle, be obtained from the contact potential-difference measurements. Such measurements were carried out for a series of superionic materials (Malov *et al.*, 1979). At the same time, it should be noted that the characteristics of the metal-SIM interface differ from those of metal-electrolyte solution interface (Vershinin *et al.*, 1982) for some important reasons. The characteristics of metal-SIM interface is associated, first of all, with the specific character of the electrode reaction between two solids and, also, with the mechanical stresses arising in the specimen due to the incorporation (or removal) of ions into (from) the crystal lattice of the SIM.

For the kinetic mode of electronic current flow in an A-type cell, Eq. (81) may be used if the second contact is "ohmic" (formally, this corresponds to the case when the condition $i_{e,o} \to \infty$ is fulfilled). In particular, the ion-reversible contact may be considered as an "ohmic" one if its equilibrium is not upset during current flow. Otherwise, it is necessary to consider the processes occurring at both electrodes.

The maximum electronic current is determined either by the transport of species in the bulk (diffusion mode) or by the passage through the interface (kinetic mode), depending on which of two quantities, $i_{e,o}$ or $i'_{e,o}$, is smaller. Since the analytical forms of the $i_{el}(V)$ curves described by Eqs. (66) and (81) are the same, it follows from the above discussions that a unique determination of the model applicability to superionic materials under consideration cannot be drawn from the study of CVCs alone.

Choosing relatively thick samples (large L), it is possible to arrive at a situation in which the current is limited by electron transport in the SIM bulk and not by electron transfer through the interface. In the majority of cases, the assumption that the limiting characteristics of the transport are in the bulk seems to be justified. In some works this statement has been confirmed experimentally by studying dependence of electronic currents on the sample thickness. Thus, in studies of the conductivity of the α-phase of silver iodide (Mizusaki *et al.*, 1978), the dependence of the residual electronic current on the sample thickness was established (Fig. 13, curves 2–4).

At the same time, it should be noted that the reverse situation, when the key role is played by the phenomena occurring at the junction, cannot be *a priori* excluded from the consideration. Thus, Wagner (1976) and Wada and Wagner (1979) indicated that the CVC of the Cu/CuI/C system may be described within the framework of the model, taking account of the charge transfer through the interface. The same was also reported by Ivanov-Shits (1979) for the $Ag/Ag_4RbI_5/C$ system. These authors have established that the functional dependence of electronic currents on the potential at different temperatures is described as $i_{el} = i_o \exp(eV/kT)$ (Fig. 15) (see also Joshi and Liang, 1975; Kukoz *et al.*, 1977; Kennedy, 1977a). Therefore, proceeding from Eq. (61), the authors of the cited works come to a conclusion that the observed current is due to holes.

For an Ag_4RbI_5 single crystal, it has been found (Ivanov-Shits, 1979) that the partial conductivity of holes at 300 K is very small ($<10^{-17}\,\Omega^{-1}\,cm^{-1}$). The temperature dependence of i_o is described by the equation $i_o \sim \exp(-E_a/kT)$ where $E_a = 1.44$ eV. It may be assumed that, if the diffusion mechanism of the charge transport takes place, the main contribution to the temperature dependence of i_o comes from the carrier concentration. Then, the value of E_a should correspond either to the half-width of the forbidden band (similar to the case of intrinsic or compensated semiconductors) or to the position of the Fermi level F_s with respect to the top of the valence band. Since the band gap in Ag_4RbI_5 amounts to $\simeq 3.2$ eV (Table 2, Section 2), it is natural to assume that hole conductivity may be caused by the ionization of acceptor impurities associated with the excessive iodine.

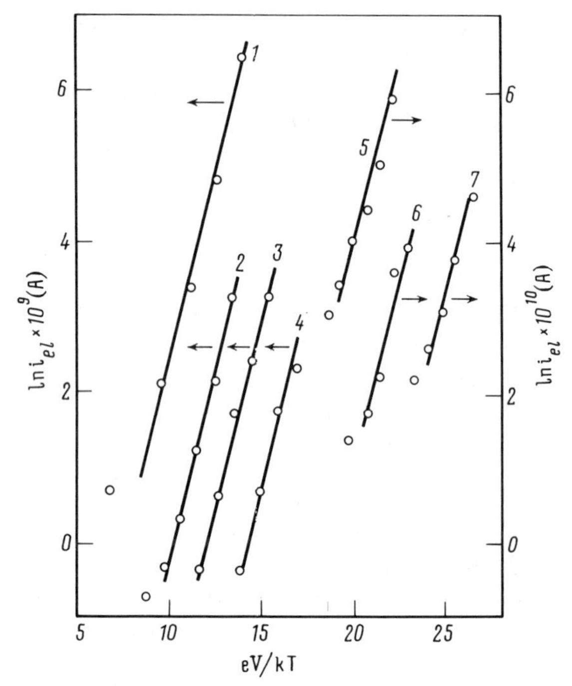

FIG. 15. Current-voltage characteristics for a Ag|Ag_4RbI_5 (single crystal)|C cell at different temperatures: (1) 438.6 K; (2) 395.3 K; (3) 378.8 K; (4) 353.4 K; (5) 334.1 K; (6) 315.8 K; (7) 297.0 K. [From Ivanov-Shits (1979).]

At last, i_o may be interpreted on the basis of the model that includes the electron transfer through a "barrier" layer at the Ag_4RbI_5/C contact. In this case, the potential barrier at the Ag_4RbI_5/C interface is $E_B \approx 1.47$ eV, which agrees quite well with the value of $E_a = 1.44$ eV, determined experimentally (it is assumed that E_B only weakly depends on temperature).

The analysis of the experimental results on the Ag/Ag_4RbI_5/C system has shown that the similar shape of electronic CVCs, $i_{el} = f(V)$, and the curves described by expression $i = i_o \exp(eV/kT)$ are not enough for drawing a conclusion on the diffusion model applicability, since similar equations may also be obtained using other approaches. For the unique choice of the model, additional experimental data should be invoked.

The polarization (steady-state technique) has been used for studying electronic conductivity of different SIMs, with both cationic and anionic conductivities. In principle, such experiments provide the determination of σ_{el} not only in binary, but also in multicomponent, compounds. It should also be emphasized that in the case of multicomponent compounds, chemical potentials of several components should be fixed in the general case in order to reduce the system to a quasi-binary one. Unfortunately, this fact has been overlooked in some works.

The results of electronic-conductivity measurements on different SIMs are given in the Appendix, where the type of polarization cells is also indicated. Note here that this table is based on numerous works on electronic conductivity measurements in SIMs and especially in cation-conducting materials for which the experimental conditions are of small importance (in particular, the partial gas pressure of the ambient atmosphere).

Two notes should be made in connection with the data listed in the Appendix.

Firstly, since the values of $\sigma_e(\sigma_h)$ are formally proportional to the local concentration, $\sigma_e(\sigma_h)$ varies along the sample. The tabulated values of $\sigma_e(\sigma_h)$ correspond to the electron (hole) concentrations at the ion-reversible electrode-SIM interface. Therefore, electronic conductivities of cells with different reversible electrodes may be quite different.

Secondly, there is a large scatter in the data for some materials, especially in the case of small values of $\sigma_e(\sigma_h)$. We have already discussed the possible reasons for such a scatter (see Sections 2,3,8).

9. Chemical (Ambipolar) Diffusion

In the previous sections, attention was focussed mainly on SIMs with relatively low electronic conductivity. At the same time, as has already been noted in the Introduction, there are compounds in which noticeable electronic conductivity is observed, along with high ionic conductivity. Such materials may be called mixed conducting superionic materials (MCSIM) or mixed ionic-electronic conductors (MIEC). We include in this type of ionic-electronic conductors, in particular, compounds with a large range of nonstoichiometry. The presence of a large number of delocalized electrons or holes in such compounds is due to "uncompensated" components of some elements.

The transport phenomena in nonstoichiometric compounds may possess a number of specific features physically related to the attainment of the local equilibrium between ions and electrons (holes) at every crystal point. We have briefly considered this problem in Section 8, where the concept of the local ionic-electronic equilibrium was used (Wagner, 1957) in the derivation of relationships describing the CVCs of electrochemical SIM-based cells. Now we shall consider in more detail the specific features of transport processes in materials with mixed ionic-electronic conductivity.

With the application of an electric field to a MCSIM, the gradients of ion and electron concentrations start developing. Due to the forces of Coulomb interaction, the diffusion flux of charged species gives rise to the flux of another species. If one species moves faster than another species, the arising electrical field makes the particles of the second species move together with the first species. Such "joint" diffusion of the charged particles is called ambipolar or chemical diffusion. In the process of such a diffusion, neutral atoms move along the sample, their flux being determined by the smaller current of ionic or electronic carriers.

Now consider a MCSIM that consists, for example, of monovalent mobile cations, M^+, and electrons. By virtue of the local equilibrium condition, $M \rightleftarrows M^+ + e^-$. Then, the densities of ionic- and electronic-carrier currents (i_i and i_e, respectively), in accordance with Eq. (28), are:

$$i_i = -\frac{\sigma_i}{e}\frac{d\tilde{\mu}_i}{dx}, \tag{85a}$$

$$i_e = \frac{\sigma_e}{e}\frac{d\tilde{\mu}_e}{dx} = \frac{\sigma_e}{e}\left(\frac{d\tilde{\mu}_M}{dx} - \frac{d\tilde{\mu}_i}{dx}\right). \tag{85b}$$

The form of Eq. (85) implies that the chemical potential of neutral atoms μ_M at each point is determined by the equilibrium condition

$$\mu_M = \mu_i + \mu_e = \tilde{\mu}_i + \tilde{\mu}_e. \tag{86}$$

In the case of multivalent mobile ions in MCSIM, Eq. (86) becomes more complicated and effects may arise that are associated with the formation of the intermediate products of the reaction corresponding to the local equilibrium.

The total current i in the system is the sum of the currents due to ionic and electronic carriers, $i = i_i + i_e$. Eliminating with the help of the above equation the derivatives $d\tilde{\mu}_i/dx$ and using the local equilibrium condition (86), it is possible to transform Eq. (85) to the form

$$i_i = \frac{\sigma_i}{\sigma_e + \sigma_i}i - \frac{\sigma_e\sigma_i}{(\sigma_e + \sigma_i)e}\frac{d\mu_M}{dx}, \tag{87a}$$

$$i_e = \frac{\sigma_e}{\sigma_e + \sigma_i}i + \frac{\sigma_e\sigma_i}{(\sigma_e + \sigma_i)e}\frac{d\mu_M}{dx}. \tag{87b}$$

The gradient of chemical potential, $d\mu_M/dx$, may be represented in the form

$$\frac{d\mu_M}{dx} = \frac{d\mu_M}{dn_M}\frac{dn_M}{dx}, \tag{88}$$

where n_M is the concentration of neutral atoms. The proportionality coefficient relating the partial currents of charged particles to the concentration gradient dn_M/dx of neutral particles is, by definition, the coefficient

$\tilde{D}_M$ of chemical diffusion. According to Eqs. (87) and (88), $\tilde{D}_M$ is given by the expression

$$\tilde{D}_M = \frac{\sigma_e \sigma_i}{(\sigma_e + \sigma_i)e^2} \frac{d\mu_M}{dn_M}. \tag{89}$$

From the definition of D_M, it follows that if the ionic component of conductivity prevails, the following approximate relationship is fulfilled:

$$\tilde{D}_M \approx \frac{\sigma_e}{e^2} \frac{d\mu_M}{dn_M} \equiv \tilde{D}_e \tag{90}$$

Let us consider in what way the coefficient of chemical diffusion introduced by Eq. (90), $\tilde{D}_e$, is related to the self-diffusion coefficient determined by the Einstein relation (31), D_e. With allowance for condition (72), Eq. (90) is transformed to the form

$$\tilde{D}_e = \frac{\sigma_e}{e^2} \frac{d\mu_M}{dn_M} \approx \frac{\sigma_e}{e^2} \frac{d\mu_e}{dn_e} = \frac{kT\sigma_e}{e^2} \frac{d \ln a_e}{dn_e} \tag{91}$$

It is seen from the latter equation that for "ideal" behavior of the electronic subsystem, when activity a_e is equal to concentration n_e, the coefficient $\tilde{D}_e$ should coincide with the coefficient of self-diffusion D_e.

Chemical diffusion is most often encountered, as will be shown later, in studies of nonstationary processes in MCSIMs.

10. Coulometric Titration

Electronic carriers may form in MCSIMs, due to ionization of incorporated atoms (or due to their deficit). Therefore, by varying the specimen composition, it is possible to vary the electron (hole) concentration. The MCSIM composition may be varied by different methods, e.g., by varying the synthesis conditions. At the same time, the presence in the MCSIM of ions with high mobility permits the use of another method for varying the composition. Transmitting an electric current through a sample, it is possible to inject or extract the definite number of atoms of one kind and to control the electronic-carrier concentration. Thus, a series of measurements may be carried out on one sample by varying the degree of its deviation from stoichiometry. It is also of great importance to be able to return to the initial MCSIM composition, or to create the conditions corresponding to the ideal stoichiometry of the compound under study.

Stoichiometry control of ion-conducting compounds by transmitting an electric current through an electrochemical cell consisting of the sample is called coulometric titration. This technique was suggested by Wagner (1953,

1957) and then developed by a number of scientists (Valverdi, 1970; Weiss, 1969a,b; Miyatani, 1955, 1958a,b, 1973; Rickert, 1982).

As an example, we shall consider, as we did earlier, a binary compound of the composition $M_{n+\delta}$ X with cationic (relative to M^+) and electronic conductivity. The quantity δ characterizes the degree of nonstoichiometry determined by Eq. (8). Let this compound be one of the constituents of an electrochemical cell of the type

$$B|M|SIM|MCSIM|B, \tag{C}$$

where SIM is the superionic conductor with pure cationic conductivity with respect to singly charged cations M^+; MCSIM is the ionic-electronic conductor $M_{n+\delta}X$; M is the cation-reversible electrode; and B denotes blocking electrodes with metallic (pure electronic) conductivity. Then the emf of such a partially equilibrium C-type cell may be written in the form (Wagner, 1957)

$$e\mathcal{E} = -(\mu_M - \mu_M^M) = -(\mu_e - \mu_e^M) - (\mu_i - \mu_i^M) \tag{92}$$

Here, μ_M is the chemical potential of M atoms in the nonstoichiometric compound, and μ_M^M is the chemical potential of the same M atoms in metal M. Equation (92) was derived on the assumption that, at the M/SIM and MCSIM/SIM interfaces, equilibrium is attained with respect to cations M^+, whereas at the B/M and MCSIM/B interfaces, it is reached with respect to electrons. Moreover, Eq. (92) implies that the chemical potential of a neutral atom may be represented as a sum of the chemical potentials of ion (μ_i) and electron (μ_e).

Since the high ionic conductivity of a MCSIM is due to a high concentration of mobile ions, it may be taken that, with the variation in δ, the concentration of cations M^+ will change only slightly (μ_i is virtually constant); i.e., by Eq. (71), $\mu_i = \mu_i^M$. Then Eq. (92) may be rewritten in the form

$$e\mathcal{E} = -(\mu_e - \mu_e^M). \tag{93}$$

For the ideal stoichiometric composition, $\delta = 0$, the electron and hole concentrations are equal. When the composition deviates from stoichiometry, additional neutral atoms with concentration n_M appear in the sample bulk. These atoms are fully dissociated into ions and electrons, in accordance with the scheme $M \rightleftarrows M^+ + e^-$. Then, since the crystal is still electrically neutral, the electron carrier concentration in the bulk varies in such a way that the following relationship is valid:

$$n_M = n_e^o - n_h^o. \tag{94}$$

The equilibrium concentrations of electrons and holes are determined by general equations (9).

Now, let us assume that the current is flowing in a C-type cell in a direction such that additional M-atoms are injected into the compound. The number of atoms thus injected is then determined by electric charge Q passed through the cell;

$$n_{\text{inj}} = \frac{Q}{\text{e}}. \tag{95}$$

It is convenient to choose, as a "zero" or initial state of the MCSIM, a state of the compound $M_{n+\delta}X$, which is in equilibrium with the non-metallic phase X (generally speaking, in this case $\delta \neq 0$). The equilibrium emf developed in the C-type cell is denoted by $\mathcal{E}^{C}$. Then

$$n_{\text{inj}} = 0, \qquad \text{if } n_{M} = n_{M}^{C}, \tag{96}$$

where n_{M}^{C} is the concentration of excessive M atoms in the C-type cell at $\mathcal{E} = \mathcal{E}^{C}$. Using Eqs. (94)–(96), we may readily see that $n_{\text{inj}} = n_{M} - n_{M}^{C}$, or, by considering Eq. (8), that

$$\delta = \frac{(n_{M}^{C} + Q/\text{e})V_{m}}{\mathfrak{N}_{A}} = n_{e}^{o} - n_{h}^{o}, \tag{97}$$

where V_{m} is the molar volume and $\mathfrak{N}_{A}$ is Avogadro's number.

Since $F - F_{c} = \mu_{e} - \mu_{eo}$, where μ_{eo} is the standard chemical potential, concentrations n_{e}^{o} and n_{h}^{o}, with allowance for Eq. (93), are functions of the emf of the C-type cell. Therefore Eq.(97) can be written in the following form:

$$\delta = \frac{(n_{M}^{C} + n_{\text{inj}})V_{m}}{\mathfrak{N}_{A}} = N_{c}\Phi_{1/2}(\zeta) - N_{v}\Phi_{1/2}(\xi), \tag{98}$$

where

$$\zeta \equiv \frac{\mu_{e} - \mu_{eo}}{kT} = \frac{\mu_{e}^{M} - \mu_{eo} - \text{e}\mathcal{E}}{kT};$$

$$\xi \equiv -\frac{\mu_{e} - \mu_{eo}}{kT} - \frac{E_{g}}{kT} = -\frac{\mu_{e}^{M} - \mu_{eo} - \text{e}\mathcal{E}}{kT} - \frac{E_{g}}{kT}.$$

Thus, Eq. (98) sets the relation between the deviation δ from the ideal stoichiometry of the sample (or, if it is more convenient, the number of injected atoms n_{inj}) and the emf of the C-type cell.

The graphic representation of the dependence described by Eq. (98) is a coulometric titration curve.

In some important instances, the general expression describing the coulometric titration curve may be simplified. Indeed, if the hole concentration is negligibly small (n-type material) and the deviation from the ideal stoichiometry is substantial, then, in accordance with Eq. (94), we have

$$n_{\text{inj}} \approx n_{\text{M}} \approx n_{\text{e}}.$$

Now Eq. (98), describing the titration curve, takes the form

$$n_{\text{inj}} = n_{\text{M}} = N_{\text{c}}\Phi_{1/2}(\zeta). \tag{99}$$

If the whole sample is in equilibrium with metal M (in other words, if phase M is precipitated on the right-hand blocking electrode of the C-type cell), the emf of the cell is $\mathcal{E} = 0$, and Eq. (99) is transformed to the form

$$n_{\text{M}} = n_{\text{M}}(\zeta_{\text{o}})\frac{\Phi_{1/2}(\zeta)}{\Phi_{1/2}(\zeta_{\text{o}})}, \tag{100}$$

where

$$n_{\text{M}}(\zeta_{\text{o}}) = N_{\text{c}}\Phi_{1/2}(\zeta_{\text{o}}); \qquad \zeta_{\text{o}} \equiv \zeta(\mathcal{E} = 0). \tag{101}$$

Differentiation of Eq. (100) yields

$$\left(\frac{dn_{\text{M}}}{d\mathcal{E}}\right)_{\mathcal{E}=0} = -n_{\text{M}}(\zeta_{\text{o}})\frac{\text{e}}{kT}\frac{\Phi'_{1/2}(\zeta_{\text{o}})}{\Phi_{1/2}(\zeta_{\text{o}})}. \tag{102}$$

Knowing experimental values of $(dn_{\text{M}}/d\mathcal{E})_{\mathcal{E}=0}$ and $n_{\text{M}}(\zeta_{\text{o}})$, and using the tabulated values of $\Phi_{1/2}$ and $\Phi'_{1/2}$, we can determine from Eq. (102) the dimensionless argument ζ_{o}. At last, using the ζ_0 value determined at a given temperature T, we may calculate $F - F_{\text{c}}$, i.e., the difference between the Fermi level and the bottom of the conduction band of the material. Since N_{c} is known, the effective mass of electrons, m_{c}, can be evaluated from Eq. (101).

If there is no degeneracy, the scheme of such a consideration is essentially simplified and the electron and hole concentrations are described by Eqs. (1). As a result, the equation of coulometric titration (98) may be written in the form (Wagner, 1953, 1957)

$$\delta = 2n_{\text{e}}^{\text{id}}\,\text{sh}\left[\frac{\text{e}(\mathcal{E} - \mathcal{E}^{\text{id}})}{kT}\right]. \tag{103}$$

Here, $n_{\text{e}}^{\text{id}} = n_{\text{h}}^{\text{id}}$ is the electronic-carrier concentration in an ideal stoichiometric sample ($\delta = 0$), and $\mathcal{E}^{\text{id}}$ is the emf of the cell with an ideal stoichiometric sample.

The coulometric titration technique has been successfully applied to study the properties of the electronic subsystem in silver and copper chalcogenides (Wieger, 1976; Ishikawa and Miyatani, 1977; Schmalzried, 1980; Horvatič and Yucic, 1984).

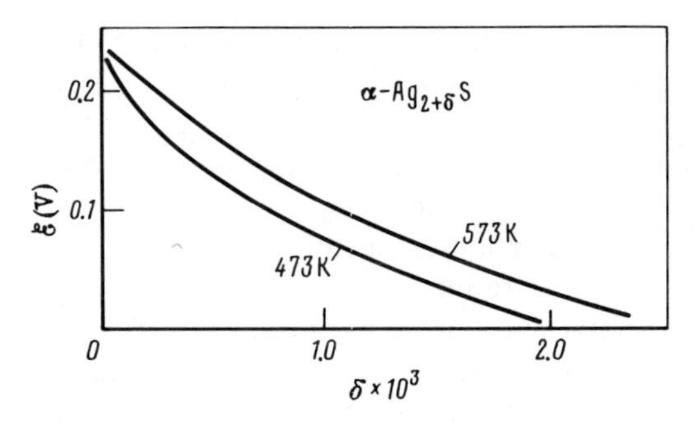

FIG. 16. Coulometric titration curves for $Ag_{2+\delta}S$ at different temperatures. [From Rickert *et al.* (1975).]

Let us consider in more detail the results obtained for the high-temperature α-phase of silver sulphide (Schmalzried, 1980; Rickert, 1982; Mostafa and Abd-Elreheem, 1985). The experimental curves of coulometric titration are shown in Fig. 16. It is seen that a crystal, which is in equilibrium with sulphur, has an almost ideal stoichiometric composition. The experimental data give the value $(dn_M/d\mathcal{E})_{\mathcal{E}=0} = 0$. Then, with the aid of the above described scheme, we may determine the values of $F - F_c$ and m_c.

Thus, Eq. (99) provides the calculation of electron concentations (and, if necessary, also of hole concentrations) as functions of the deviation from stoichiometry δ (Fig. 17).

As a result, it has been established that ideal stoichiometric silver sulphide is an intrinsic semiconductor. But even small deviations from stoichiometry

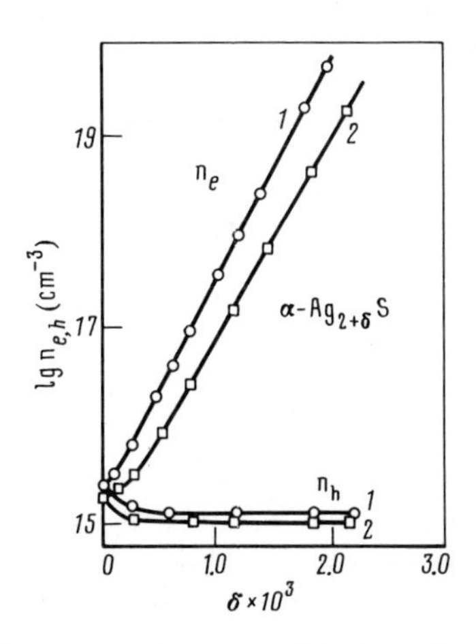

FIG. 17. Electron and hole concentrations versus δ for $Ag_{2+\delta}S$: (1) 498 K; (2) 598 K. [From Mostafa and Abd-Elreheem (1985).]

may affect conductivity, making such silver sulphide closer to the metallic type (for more detail see Section 19). If the value of δ is small, conductivity is strongly affected by thermal "throwing" of electrons from the valence to conduction band. However, with an increase in the amount of dissolved silver, the Fermi level passes into the conduction band, and σ_{el} becomes proportional to the deviation from stoichiometry δ.

Proceeding from the results of coulometric titration, it is possible to calculate the electronic-carrier concentrations. If independent data on electronic conductivity are known, it is easy to calculate mobilities of electronic carriers (electrons and holes). Thus, the u_e values for $Ag_{2+\delta}S$ have been found (Shukla and Schmalzried, 1979). It turns out that for samples with $\delta > 1.5 \times 10^{-4}$, mobility somewhat decreases with the temperature rise, whereas for samples with smaller deviations from stoichiometry, u_e increases with temperature. Such behavior seems to be explained by the transition from the semiconductor to metallic conductivity (for details see Section 19).

Also, electron and hole mobilities were obtained from Hall measurements in mixed conductors (Miyatani, 1955, 1958a,b; 1967, 1973; Sohége and Funke, 1984). Table 4 lists some characteristic parameters of electronic carriers for several MCSIMs.

TABLE IV

ELECTRONIC CARRIER CHARACTERISTICS OF SOME MCSIMs*

		MCSIM			
		α-$Ag_{2+\delta}S$	α-$Ag_{2+\delta}Se$	α-$Ag_{2+\delta}Te$	Ag_2Se-Ag_3PO_4
$\sigma_i, \Omega^{-1}\,cm^{-1}$		~2	~2-3	~1-2	~0.1
$\sigma_{el}, \Omega^{-1}\,cm^{-1}$		10^2-10^3	~10^3	10^2-10^3	~10^3
m_c/m_0		0.21-0.24	0.12-0.22	0.05-0.07	0.17-0.21
m_v/m_0			0.1-0.5	1-2	
$u_{e,h}$, $cm^2/V\,c$	u_e	~10^3	400-650	750-1440	250
	u_h			13-18	

These values are taken from the data collected by Miyatani (1955, 1958a,b), Miyatani *et al.* (1967), Takahashi and Yamamoto (1972), Shukla and Schmalzried (1979), Oehsen and Schmalzried (1981), Mostafa *et al.* (1982) and Sohége and Funke (1984).

* The given values were obtained at temperatures corresponding to the existence range of phases with high ionic conductivity.

11. Thermoelectric Phenomena

a. *Thermoelectric Power*

The physical nature of thermoelectric power is associated with the presence of the temperature gradient (∇T), which gives rise to the gradients of the chemical potential of particles. As a result, mobile carriers (both ionic and electronic) are redistributed in the SIM bulk. Since the particles are charged, such a redistribution is accompanied by the appearance of an electric field ($\nabla \varphi$) and voltage in the direction of the temperature gradient. The induced potential is infelicitly called a thermoelectromotive force or thermo-emf.

We shall consider the development of a thermo-emf in terms of thermodynamics of irreversible processes, restricting ourselves, for simplicity, to the one-dimensional case. The general equation for a charge flux (electric current) associated with the motion of certain species with charge ez may be written in the form (De Groot and Mazur, 1969; Gurov, 1978)

$$i = -\frac{\sigma}{(ez)}\left[\left(\frac{Q^*}{T} + S\right)\frac{dT}{dx} + \frac{d\tilde{\mu}}{d\bar{x}}\right]. \tag{104}$$

Here Q^* is the heat of transport. If we use Eq. (29), defining μ and expressions $\mu = H - TS$ and

$$\frac{d\tilde{\mu}}{dx} = \left(\frac{\partial\tilde{\mu}}{\partial x}\right)_T + \frac{\partial\tilde{\mu}}{\partial T}\frac{\partial T}{\partial x}; \quad \left(\frac{\partial\mu}{\partial x}\right)_T = \frac{\partial H}{\partial x} - T\frac{\partial S}{\partial x}; \quad \frac{\partial\mu}{\partial T} = -S,$$

where H and S are the partial enthalpy and entropy, respectively, per particle, we arrive at the following equation for the current:

$$i = -\frac{\sigma}{(ez)}\left[\frac{Q^*}{T}\frac{dT}{dx} + ez\frac{d\varphi}{dx} + \left(\frac{\partial\mu}{\partial x}\right)_T\right]. \tag{104'}$$

If there is a temperature gradient but the current is absent, a potential difference appears in the sample, which may be determined from Eq. (104). If there is only one species then, assuming that $i = 0$ in Eq. (104), we arrive at

$$ez\frac{d\varphi}{dx} = -\left(S + \frac{Q^*}{T}\right)\frac{dT}{dx} - \frac{d\mu}{dx}. \tag{105}$$

The homogeneous thermoelectric power is, by definition, $\theta_{\text{hom}} = d\varphi/dx$, whence, in accordance with Eq. (105), we have

$$\theta_{\text{hom}} = -\frac{1}{ez}\left[\left(S + \frac{Q^*}{T}\right) + \frac{d\mu}{dx}\right]. \tag{106}$$

A real cell for measuring the induced thermo-emf has electrodes contacting with the SIM. At the electrode-SIM interfaces, jumps in the contact potential steps are observed, their temperature dependence determining the so-called heterogeneous component of the thermoelectric power, θ_{het}. The value of θ_{het} may, in principle, be determined from the condition of equality of the electrochemical potentials due to corresponding current carriers in the SIM and electrode (Lidiard, 1957; Gurevich and Ivanov-Shits, 1982; Chebotin, 1984).

For example, we assume that there are two types of carriers in the SIM—monovalent cations and electrons. Now let us consider some types of thermoelectric cells with different electrode materials.

(1) Let electrodes R in a cell be cation-reversible, i.e., the equilibrium at the interface attainable only for one species. In this case, the current of each species is zero in the sample bulk,

$$i_{\mathrm{i}} = 0, \qquad i_{\mathrm{e}} = 0. \tag{107}$$

The substitution of Eq. (104) for currents into Eq. (107) yields the homogeneous thermoelectric power. The heterogeneous component may be found from the equality condition for electrochemical potentials of cations at the SIM/R interfaces. Thus, the total ionic thermoelectric power may be written in the form

$$\mathrm{e}\theta_{\mathrm{i}} = -S_{\mathrm{i}}^{*} - (S_{\mathrm{e}}^{*})^{\mathrm{R}} + S^{\mathrm{R}}, \tag{108}$$

where quantity

$$S^{*} \equiv S + \frac{Q^{*}}{T} \tag{109}$$

is called the transport entropy and S^{R} and $(S^{*})^{\mathrm{R}}$ are the entropy and transport entropy, respectively, in the electrode material.

(2) Let a cell for measuring thermo-emf have electrodes B possessing electronic conductivity alone. Then, at the SIM/B interface, equilibrium is attained only with respect to electrons. It may readily be shown that the electronic thermoelectric power is written in the form

$$\mathrm{e}\theta_{\mathrm{e}} = S_{\mathrm{e}}^{*} - (S_{\mathrm{e}}^{*})^{\mathrm{B}}. \tag{110}$$

We would like to emphasize that the equations for θ_{e} and θ_{i} have no transport numbers. Therefore, by choosing appropriate electrodes, we may study the processes of electronic transport against the background of strong ionic conductivity, and vice versa.

(3) Let the equilibrium relative to cations and electrons be attained simultaneously at both electrode/SIM interfaces. Such an electrode, M,

will be called fully reversible. Under the stationary conditions, the total current in any cross-section of the SIM in the open-circuit cell is zero,

$$i_i + i_e = 0. \tag{111}$$

We assume that currents i_i and i_e relate to one another as corresponding conductivities σ_i and σ_e, so that $t_{i,e} = t'_{i,e}$ (see Section 6). The substitution of Expression (104) for currents into Eq. (111), with due regard for the above discussion, yields

$$e\theta_{hom} = -t_i S_i^* + t_e S_e^* - t_i \frac{d\mu_i}{dT} + t_e \frac{d\mu_e}{dT}. \tag{112}$$

By virtue of the electrode properties, the electrochemical potentials of cations and anions at the interfaces become equal:

$$\begin{aligned} \tilde{\mu}_i &= \tilde{\mu}_i^M; \\ \tilde{\mu}_e &= \tilde{\mu}_e^M. \end{aligned} \tag{113}$$

Nevertheless, for finding the equilibrium jump in the potential we must determine the potential-inducing particle. Then, we may find the heterogeneous thermoelectric power and, finally, the total thermoelectric power.

The above considerations are based on the phenomenological approach and, therefore, the above relationships are of a general nature, independent of the type of mobile particles. Using various model representations, it is possible to derive more exact expressions for the thermoelectric power θ for electronic carriers.

Thus, for the model of quasi-free electrons (Smith, 1978), we have:

$$\theta_e = -\frac{k}{e}\left[\left(\frac{5}{2} - s\right) + \ln\frac{N_c}{n_e}\right] \tag{114}$$

$$\theta_h = \frac{k}{e}\left[\left(\frac{5}{2} - s'\right) + \ln\frac{N_v}{n_h}\right] \tag{115}$$

Here s and s′ are the constants characterizing scattering of electrons and holes in a SIM.

Since $(k/e) \approx 10^{-4}$ V/K, the values of the homogeneous thermoelectric power for semiconductor materials are $\approx 10^{-3}$ V/K (for comparison, an analogous value in metals is $\approx 10^{-6}$ V/K).

Equations similar to Eqs. (114) and (115) may readily be obtained for degenerate semiconductors.

The majority of experimental studies of thermoelectric power in SIM-based cells concerns the measurements of the ionic thermoelectric power of materials that are either cationic (Shahi, 1977) or anionic (Chebotin and Perfil'ev, 1978) conductors.

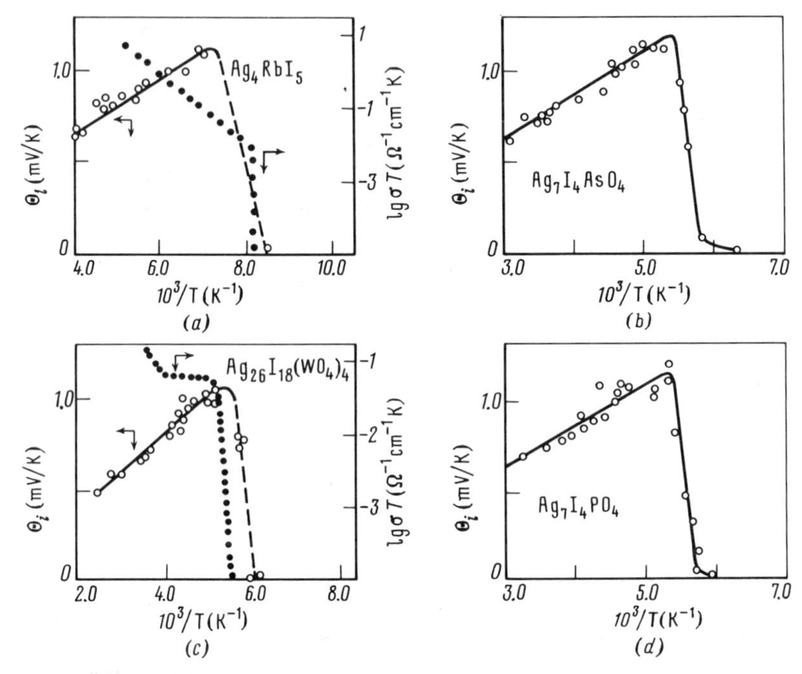

FIG. 18. Thermoelectric ionic power, Θ_i, versus temperature for some AgI-based SIMs (a–d). Temperature behavior of ionic conductivity is also shown (a, c). [From Ivanov-Shits *et al.* (1980).]

The conventional temperature curves of the thermoelectric power for AgI-based superionic materials are shown in Fig. 18 (Ivanov-Shits *et al.*, 1980). The slope of $(\theta, 1/T)$ lines permits one to determine, under simplifying assumptions, the heat of transport Q_i^* (Shahi, 1977). The results obtained indicate that the values of the heat transport for ions are very close to the activation energy of conduction. As a rule, $Q_i^* \lesssim E_a$.

The electronic component of thermoelectric power was measured for materials with a relatively high electronic conductivity, such as silver and copper chalcogenides (Miyatani, 1955, 1967, 1968) or intercalation compounds (Honders *et al.*, 1983a,b).

Thus, for Ag_xTiS_2 and Li_xTiS_2, the use of a cell with platinum electrodes (i.e., electron-reversible electrodes) provided the determination of the sign of θ_{el}, permitting the authors to draw the conclusion that, in this case, electronic carriers are mainly conduction-band electrons. It follows from Eq. (110) that, in this case,

$$e\theta_{el} = \frac{Q_e^*}{T} + S_e - S_{Pt,e}^*, \tag{116}$$

where $S^*_{Pt,e}$, which may be determined from independent experiments, is the entropy of electron transport in platinum.

Using Eq. (114) and measuring θ_e, we can determine the electronic carrier density n_e. The results are somewhat unexpected and contradict the conclusions drawn from a simple model in which each atom incorporated into the TiS_2 matrix emits an electron into the conduction band.

It turns out that, whereas the electron concentration in $Ag_{0.07}TiS_2$ should amount to $n_e = 1.23 \times 10^{21}\ cm^{-3}$, the experimentally-determined value is $n_e = 1.93 \times 10^{20}\ cm^{-3}$ (at 300 K). Honders *et al.* (1983) explain such a difference by noting incomplete ionization of intercalated metal atoms.

As has already been noted in Section 10, the composition of silver chalcogenides may be varied by transmitting current through a C-type cell containing the sample, and recording the resulting emf, $\mathcal{E}$. Then thermo-emf measurements are performed on a sample with the given nonstoichiometry. As an example, Fig. 19 shows the dependence of the dimensionless thermoelectric power $\kappa_T = \theta_e(e/k)$ on emf ($\mathcal{E}$) for the superionic phase α-Ag_2Se.

Thus, the thermo-emf measurements in SIMs yield important additional information on the characteristics of the electronic subsystem. Knowing the sign of thermoelectric power, θ_{el}, and using Eqs. (114)–(115), one may readily determine the type of electronic carriers (electrons or holes) and, by knowing the absolute value of θ_{el}, also their concentration.

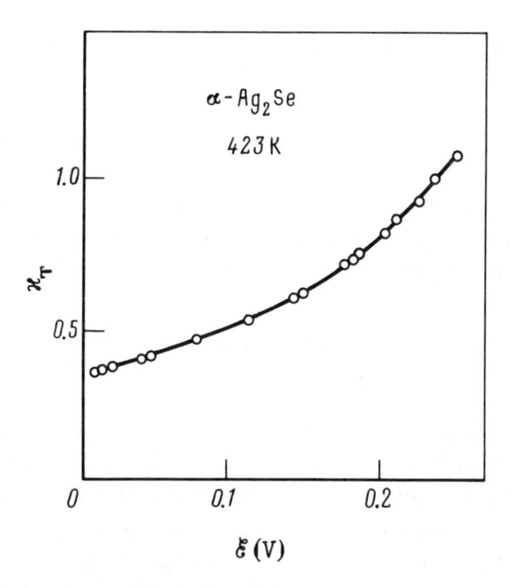

FIG. 19. Dimensionless thermoelectric electronic power of Ag_2Se, $\kappa_T \equiv \Theta_e e/k$, as a function of emf, $\mathcal{E}$, for C-type cell. [From Miyatani *et al.* (1967).]

b. Thermal Diffusion

As has already been noted in Section 9, the application of an electric field to a sample with mixed ionic-electronic conductivity may result in the chemical diffusion of the components. If there is also a temperature gradient, then the action of the temperature field results in the redistribution of carriers. Then, in accordance with Eqs. (85) and (104), the ionic and electronic currents may be written as

$$i_i = \frac{\sigma_i}{\sigma_i + \sigma_e} i - \frac{\sigma_i \sigma_e}{(\sigma_i + \sigma_e)e}\left(\frac{\partial \mu_M}{\partial x} + S_M^* \frac{\partial T}{\partial x}\right), \tag{117}$$

$$i_e = \frac{\sigma_e}{\sigma_i + \sigma_e} i + \frac{\sigma_i \sigma_e}{(\sigma_i + \sigma_e)e}\left(\frac{\partial \mu_M}{\partial x} + S_M^* \frac{\partial T}{\partial x}\right). \tag{118}$$

Here, $S_M^* = S_e^* + S_i^*$ is the transport entropy of M atoms.

It is easy to see that the flux of neutral atoms j_M is determined by the smaller of two currents—the ionic or electronic-carrier current. Therefore, taking Eq. (89) into account, we may write

$$j_M \approx \frac{i_i}{e} = \frac{\sigma_i}{\sigma_i + \sigma_e} \frac{i}{e} - \tilde{D}_M \frac{\partial n_M}{\partial x} - D_T \frac{\partial T}{\partial x}, \qquad i_e > i_i, \tag{119}$$

and

$$j_M \approx \frac{i_i}{-e} = -\frac{\sigma_e}{\sigma_i + \sigma_e} \frac{i}{e} - \tilde{D}_M \frac{\partial n_M}{\partial x} - D_T \frac{\partial T}{\partial x}, \qquad i_i > i_e. \tag{120}$$

Here, $D_T = [\sigma_e \sigma_i/(\sigma_i + \sigma_e)e^2](S_M^* - S_M)$ is the so-called thermal diffusion coefficient.

As has been shown by Ohachi and Taniguchi (1977, 1981), D_T may be expressed via the thermoelectric power of ionic (θ_i) and electronic (θ_e) carriers in a mixed conductor and the temperature derivative of emf, $\mathcal{E}$, for a galvanic C-type cell:

$$D_T = \frac{\sigma_e \sigma_i}{(\sigma_i + \sigma_e)e}\left(\theta_e - \theta_i - \frac{\partial \mathcal{E}}{\partial T}\right).$$

Thermal diffusion in nonstoichiometric silver and copper chalcognides has been studied by Miyatani (1968), Ohachi and Taniguchi (1977, 1981), and Konev *et al.* (1985). Figure 20 shows the coefficient D_T of thermal diffusion for α-$Ag_{2+\delta}Se$ single crystals versus their composition (or versus emf in a C-type cell). As is seen from Fig. 20, the sign of D_T is positive, indicating, according to Miyatami (1968), the diffusion of silver atoms from the hot sample end to the cold one.

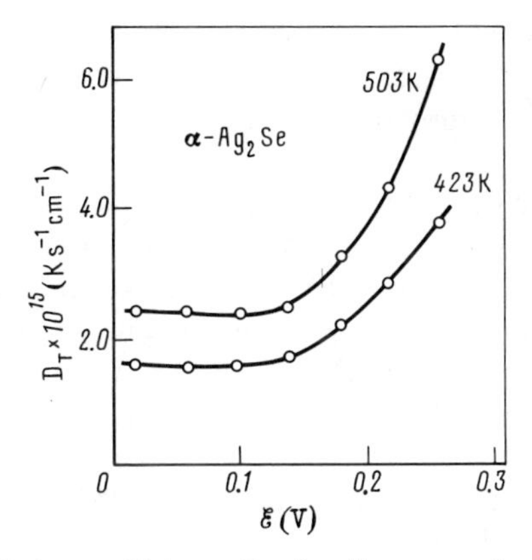

FIG. 20. Thermal diffusion coefficients, D_T, for silver atoms in α-Ag_2Se as a function of material composition at different temperatures. [From Ohachi and Taniguchi (1981).]

Summing up, we should like to note that the studies of stationary processes in SIMs permit us to obtain one of the most important characteristics of a material—electronic conductivity—and also to control its chemical composition.

V. Nonstationary Processes in Superionic Materials

The study of steady-state processes in systems with superionic materials allows one to determine electronic conductivity σ_{el}. Such studies, however, do not permit the separation of the contributions into σ_{el} that come from concentration and mobility of electronic carriers. This problem may be solved by methods based on nonstationary processes in SIMs. Knowledge of laws governing the relaxation processes in the ionic and electronic subsystems is also important for understanding the operating principles of numerous electrochemical devices and apparatus (see Part VIII).

Let us consider in more detail some nonstationary phenomena caused by electronic currents in superionic materials. When varying the polarizing voltage in an A- or B-type cell, the following processes may be observed:

(1) Charging of the double-layer capacitance at its interface with the blocking electrode (Raleigh, 1966, 1967, 1973; Hull and Pilla, 1971).

(2) Redistribution of charge carriers in the SIM bulk (Weiss, 1968, 1969a,b; Joshi, 1973; Joshi and Wagner, 1975).

(3) The penetration of mobile ions into the blocking electrode and their subsequent diffusion and/or migration in the electrode (Rickert and Steiner, 1966; Rickert and El-Miligy, 1968; Raleigh, 1967; Goldman and Wagner, 1974; Weppner and Huggins, 1977a,b).

The last process can be suppressed by an appropriate choice of the electrode material and, therefore, will not be considered here. We shall assume instead that ions moving in the SIM do not cross the SIM-electrode interface.

In many important cases, mathematical analysis of corresponding problems reduces to the solution of transient diffusion equations for concentrations of different species (electrons, holes, ions and vacancies). These equations have been studied in detail—in particular, in connection with investigations of nonstationary processes in liquid electrolytes (Damaskin and Petrii, 1983) and heat conduction problems (Carslow and Jaeger, 1959). Therefore, the considerable number of mathematical results obtained in such works (Crank, 1970; Raichenko, 1981; Mikhailov and Özişik, 1984) may be extended to the systems examined in this survey.

We will restrict ourselves to the consideration of mainly those nonstationary processes that have already been studied experimentally.

12. Statement of the Diffusion Problem and Its General Solution

As follows from what was stated in Section 9, the redistribution of charge carriers in the SIM bulk is associated with the chemical diffusion and migration of ionic and electronic carriers. The kinetics of such a joint transport is characterized by the chemical diffusion coefficient $\tilde{D}_{\mathrm{M}}$ of neutral atoms. The time characteristics describing the processes of attainment of the steady state depend on the initial and boundary conditions. The condition necessary for the application of diffusion equations to the description of such processes is that the relaxation times, $\tau_{\mathrm{r}} = L^2/\tilde{D}_{\mathrm{i,e}}$, should essentially exceed the time necessary for the attainment of electrical neutrality in the system (the Maxwell relaxation time, $\tau_\sigma \sim \sigma^{-1}$).

Quantity τ_σ is determined by the maximum conductivity for the components (usually, by σ_{i}); for $\sigma_{\mathrm{i}} = 10^{-1}$–$10^{-2}\ \Omega^{-1}\ \mathrm{cm}^{-1}$, this time is $\tau_\sigma \sim 10^{-10}$–$10^{-11}$ s.

The boundary conditions used for the diffusion equation also require the fulfillment of inequality $\tau_{\mathrm{r}} > \tau_{\mathrm{s}}$ where τ_{s} characterizes the times of transient processes at the interface. As a rule, $\tau_{\mathrm{s}} < \tau_\sigma$, and, therefore, the inequality $\tau_{\mathrm{r}} > \tau_{\mathrm{s}}$ does not impose any additional restrictions.

For chemical diffusion (see Section 9) the kinetics of the relaxation process is characterized by $\tilde{D}_{\mathrm{M}}$, and, therefore, $\tau_{\mathrm{r}} = L^2/\tilde{D}$. Along with the above-indicated conditions, the inequality $\tau_{\mathrm{r}} \gg \tau_{\mathrm{ie}}$ should also be fulfilled where τ_{ie} is the time necessary for the formation of a neutral component (atom) from the initial charged particles (ion and electron).

Note also that an important role is played by the joint motion of ions and electrons or holes (atomic motion) in nonstationary processes in SIMs.

This is clearly seen in materials with mixed ionic-electronic conductivity, e.g., strongly nonstoichiometric compounds. Some features of atomic motion in such compounds resemble those inherent in ambipolar diffusion in semiconductors.

In the case of chemical diffusion, a flux of neutral atoms, j_M, is determined by Eqs. (119)–(120), which, at $\nabla T = 0$, have the forms

$$j_M \approx \frac{i_e}{e} = \frac{\sigma_i}{\sigma_e + \sigma_i}\frac{i}{e} - \tilde{D}_M\frac{\partial n_M}{\partial x}, \qquad i_e > i_i, \tag{121}$$

and

$$j_M \approx \frac{i_e}{-e} = \frac{-\sigma_e}{\sigma_e + \sigma_i}\frac{i}{e} - \tilde{D}_M\frac{\partial n_M}{\partial x}, \qquad i_i > i_e. \tag{122}$$

The process of the attainment of the stationary state in a SIM is described by Fick's second law, which relates the atomic concentrational changes in time and space, and follows from Eqs. (121), (122):

$$\frac{\partial n_M}{\partial t} = -\frac{\partial j_M}{\partial x} = \tilde{D}_M\frac{\partial^2 n_M}{\partial x^2}. \tag{123}$$

In the derivation of Eq. (123), it was assumed that $\tilde{D}_M$ (and, therefore, σ_e, σ_i, $\partial\mu_M/\partial n_M$) is independent of coordinates. This assumption is justified if the currents flowing through the cell are sufficiently small. Otherwise, although the problem may be interpreted in the same terms, its mathematical solution becomes much more complicated (for details see Miyatani, 1981a).

The main methods of solving nonstationary diffusion equations are the separation of variables and methods using Laplace and Heaviside transformations (Raichenko, 1981), the latter being especially efficient for solving nonstationary problems in a half-space.

The general solution of Eq. (123) for the region of finite dimensions (in particular for a SIM sample of thickness L) when $\tilde{D}_M$ = const is described by a Fourier series (see, e.g., Raichenko, 1981):

$$n_M(x, t) = \frac{A_o}{2} + \sum_{k=1}^{\infty}[A(\nu_k)\cos\nu_k x + B(\nu_k)\sin\nu_k x]\exp(-\nu_k^2\tilde{D}_M t). \tag{124}$$

The spectrum of eigen-values (the set of quantities ν_k) and values of A_o, $A(\nu_k)$ and $B(\nu_k)$ is determined by the boundary and initial conditions.

Infinite series (124) converge very slowly, and, therefore, for small $\tilde{D}_M t$, it may be convenient to represent the solution via an error function.

The case of the galvanostatic mode often encountered and easily implemented experimentally was considered in detail by Yokota (1961). In this mode, a dc current flows in the circuit. In our case, this means that atoms pass through the electrodes of the electrochemical cell at a constant rate, i.e., the atomic flux is constant, and we have

$$\left.\frac{dn_{\mathrm{M}}}{dx}\right|_{x=0,L} = b. \tag{125}$$

If at the initial instant of time, the concentration in the sample is $n_{\mathrm{M}}(x, \mathrm{o})$, the solution of the boundary problem is given in the form

$$n_{\mathrm{M}}(x, t) - n_{\mathrm{M}}(x, \mathrm{o}) = bL\left[\frac{x}{L} - \frac{1}{2} + H\left(\frac{x}{L}, \frac{t}{\tau}\right)\right]. \tag{126}$$

Here

$$H\left(\frac{x}{L}, \frac{t}{\tau}\right) \equiv \frac{4}{\pi^2}\sum_{k=0}^{\infty}\frac{1}{(2k+1)^2}\exp\left[-(2k+1)^2\frac{t}{\tau}\right]\cos\left[(2k+1)\pi\frac{x}{L}\right], \tag{127}$$

where $\tau \equiv L^2/\pi^2\tilde{D}_{\mathrm{M}}$ is the characteristic time.

For relatively large times when $t \geq \tau$, it is possible to omit all but the first term of series (127). Then the approximate solution (the error not exceeding 1% already at $t = 0.3\tau$) has the form

$$n_{\mathrm{M}}(x, t) - n_{\mathrm{M}}(x, \mathrm{o}) = bL\left[\frac{x}{L} - \frac{1}{2} + \frac{4}{\pi^2}\exp\left(-\frac{t}{\tau}\right)\cos\left(\pi\frac{x}{L}\right)\right]. \tag{126'}$$

Lastly, for very large times ($t \to \infty$)—i.e., in the stationary mode—$H(x/L, \infty) = 0$. It follows from Eq. (126) that

$$n_{\mathrm{M}}(x, \infty) - n_{\mathrm{M}}(x, \mathrm{o}) = b\left(x - \frac{L}{2}\right). \tag{128}$$

In order to consider depolarization processes occurring in the cell after switching off the applied voltage, Eq. (123) should be solved with the initial conditions of Type (128). However, at the boundary, the conditions determined by Eq. (125) remain valid.

The transmission of a direct current through a SIM-based cell (and, hence, the creation of the constant concentration gradient at interfaces) may be implemented in two ways. Firstly, it is possible to use a cell with two "ionic" electrodes, i.e., electrodes which transmit the ionic current and completely block the electronic current. Secondly, a cell may be supplied with electronic electrodes whose action is the opposite with respect to that of ionic electrodes. In the former case, it follows from Eq. (87b), with allowance for condition $i_{\mathrm{e}} = 0$, that

$$\frac{\partial n_{\mathrm{M}}}{\partial x} = -\frac{ei}{\sigma_{\mathrm{i}}(\partial\mu_{\mathrm{M}}/\partial n_{\mathrm{M}})}. \tag{129}$$

In the latter case, taking account of condition $i_i = 0$, we obtain from Eq. (87a)

$$\frac{\partial n_M}{\partial x} = \frac{ei}{\sigma_e(\partial \mu_M/\partial n_M)}. \tag{130}$$

Quantity i in Eqs. (129)–(130) is the total current flowing through the cell.

It is possible to obtain the concentration (activity) changes for carriers by measuring the potential difference V at points 1 and 2 along the sample with the aid of specially-chosen electrodes (the so-called potential probes). Indeed, if we use electronic probes (i.e., probes reversible only with respect to electrons and holes), we obtain the following expression for the potential difference V_e:

$$V_e = -\frac{1}{e}\int_1^2 d\tilde{\mu}_e = -\int_1^2 \frac{i}{\sigma_i + \sigma_e}\,dx - \frac{1}{e}\int_1^2 \frac{\sigma_i}{\sigma_i + \sigma_e}\frac{\partial \mu_M}{\partial x}\,dx. \tag{131}$$

Here $\tilde{\mu}_e$ is determined by Eqs. (85) and (86), and the integration limits are the values of the corresponding quantities at points 1 and 2 of the probe location.

For ionic probes (i.e., probes reversible only with respect to ions) the potential difference V_i, with due regard for Eqs. (85) and (86), has the form

$$V_i = \frac{1}{e}\int_1^2 d\tilde{\mu}_i = -\int_1^2 \frac{i}{\sigma_i + \sigma_e}\,dx + \frac{1}{e}\int_1^2 \frac{\sigma_e}{\sigma_i + \sigma_e}\frac{\partial \mu_M}{\partial x}\,dx. \tag{132}$$

It may be assumed that conductivity σ_i is independent of the sample composition since n_i is large and, therefore, only weakly changes with the variation in the sample composition under the conditions of the direct current flow. At the same time, σ_e and n_e linearly change with n_M. Using the above-formulated assumptions, different cases of the times dependence of voltage V have been studied (Yokota, 1961; Miyatani, 1984, 1985; Ohachi, 1974; Ohachi and Taniguchi, 1977; Dudley and Steele, 1977, 1980). If an ionic current is switched on, the voltage relaxation at the ionic probes is described by the expression

$$V_i = -\frac{iL}{\sigma_i}\left[\frac{x}{L} - \frac{1}{2} + \frac{\sigma_e}{\sigma_i + \sigma_e} H\left(\frac{x}{L}, \frac{t}{\tau}\right)\right], \tag{133}$$

and, at the electronic probes, by the expression

$$V_e = \frac{iL}{\sigma_e} H\left(\frac{x}{L}, \frac{t}{\tau}\right), \tag{134}$$

where function H is given by Eq. (127).

Thus, by studying the voltage variations in time in a SIM-based cell with mixed conductivity, it is possible to determine the characteristic time τ and, hence, the chemical-diffusion coefficient $\tilde{D}_M$ for any combination of current and potential electrodes.

Fomenkov (1982) has suggested a simple experimental technique for determining transport numbers in materials with ionic-electronic type of conductivity. Depending on the prevalent conductivity, either electronic or ionic current is transmitted through a sample. Potential difference is measured in the cell in the stationary state during current flow and at the moment of its switching off. The analysis of the Eqs. (131)–(132) provides the derivation of simple expressions, which yield σ_i and σ_e.

Dudley and Steele (1980) gave the solution of the diffusion problem for the modified current Eqs. (121) and (122) with the cross-conductivities σ_{ie} and σ_{ei}, reflecting possible interaction between the ionic and electronic subsystems. It has also been shown that, in this case, time dependences similar to Eqs. (131) and (132), describing the system relaxation, are also valid.

13. Determination of Characteristic Parameters of Electronic Conductivity

This section will deal in more detail with the most widely used methods of determining the parameters of electronic conductivity in SIMs from data on transient processes.

a. Voltage Relaxation Methods

Weiss (1968, 1972), Wagner (1976) and Sasaki *et al.* (1981) have suggested methods for determining the mobility of electronic charge carriers based on the study of relaxation processes in A-type electrochemical cells. The studies were arrived out under conditions in which the effect of double-layer charging could be neglected.

If a polarizing voltage is applied to a cell, then, in a sufficiently long period of time ($t \gg \tau$), the stationary distribution of mobile particles is attained [in the simplest case, the distribution of electronic carriers is linear in accordance with Eq. (64)]. With the external voltage source switched off, the concentration gradient starts equalizing. Since the circuit is open, there is no electric current, and $i = 0$. Therefore, we may write the following expression for the electronic current [using Eq. (89)]·

$$i_e = \tilde{D}_e e \frac{\partial n_M}{\partial x}. \tag{135}$$

According to Eq. (135), the electron transport is provided by the chemical diffusion coefficient $\tilde{D}_e$ [notation $\tilde{D}_e$ is used here instead of $\tilde{D}_M$ in order to emphasize the electronic—in the narrow sense of the word—and not the hole character of the current—see Eq. (141)].

The voltage in a cell is determined by Eq. (65). Since, at the boundary, $x = 0$, the concentration (activity) n_e is constant and equals $n_e(o) = n_{e,s}$ (at this boundary, the electrode is reversible). The change in concentration at the second interface results, in accordance with Eq. (65), in the change of the voltage; i.e.,

$$V(t) = \frac{kT}{e} \ln \frac{n_{e,s}}{n_e(t)}. \tag{136}$$

The method of measuring the voltage relaxation of a cell with an external source switched off is called the open-circuit method.

The corresponding diffusion equation for electrons and boundary and initial conditions have the form

$$\frac{\partial n_e}{\partial t} = \tilde{D}_e \frac{\partial^2 n_e}{\partial x^2}; \tag{137}$$

$$\begin{gathered} \frac{n_e(x) - n_{e,s}}{n_e(L) - n_{e,s}} = \frac{x}{L}, \qquad t = 0, \qquad 0 < x \le L; \\ n_{e,s} = \text{const}, \qquad x = 0, \qquad t > 0; \\ \left(\frac{\partial n_e}{\partial x}\right)_{x=L} = 0, \qquad t > 0. \end{gathered} \tag{138}$$

[An analogous equation may also be written for ions (Weiss, 1968).] Condition (138) at $t = 0$ corresponds to the linear electron distribution in the SIM bulk at the initial moment of time; the same condition at $x = 0$ corresponds to the existence of a reversible electrode; and at $x = L$, there is no exchange at the blocking electrode either with respect to ionic or electronic carriers. The solution of the set of equations (137)–(138) for concentration $n_e(t)$ can be represented in the form

$$\frac{n_e(t) - n_{e,s}}{n_e(t = 0) - n_{e,s}} \approx \frac{8}{\pi^2} \exp\left(-\frac{t}{4\tau}\right), \qquad t > \tau. \tag{139}$$

With due regard for Eqs. (136) and (139), we arrive at

$$\tilde{D}_e t = -\frac{4L^2}{\pi^2} \ln\left[\frac{\pi^2}{8}\left\{\frac{1 - \exp(-eV/kT)}{1 - \exp(-eV_o/kT)}\right\}\right]. \tag{140}$$

As follows from the latter equation, by measuring $V(t)$, it is possible to determine $\tilde{D}_e$.

If a current is carried by holes, the corresponding expression may be derived from Eq. (140) by a mere substitution of e with −e.

Further simplifications are possible if $V(t)$, $V_o \gg kT/\mathrm{e}$. Then we have (Joshi, 1973; Joshi and Wagner, 1975)

$$\tilde{D}_\mathrm{h} t = -\frac{4L^2}{\pi^2}\left\{\frac{\mathrm{e}[V(t) - V_o]}{kT} + \ln\frac{\pi^2}{8}\right\}. \tag{141}$$

The above formulae, (140) and (141), have been observed experimentally in the study of transient currents in cells with CuCl (Joshi, 1973; Joshi and Wagner, 1975) and AgX (X = I, Br, Cl) (Mizusaki *et al.*, 1979, 1980, 1982) and also with a zirconium dioxide based SIM (Weppner, 1977a). Figure 21 shows the curve "$\log\{[n_\mathrm{e}(t) - n_{\mathrm{e,s}}]/[n_\mathrm{e}(t = 0) - n_{\mathrm{e,s}}]\}$ versus t/L^2" for an AgBr sample. As is seen, for large values of time, the experimental points form a straight line, the slope of which, in accordance with Eq. (139), permits one to determine the value of $\tilde{D}_\mathrm{e}$.

Temperature behavior of the quantities $\tilde{D}_\mathrm{e}$ and $\tilde{D}_\mathrm{h}$ obtained for AgCl crystals are shown in Fig. 22.

It is of interest to compare the values of $\tilde{D}_\mathrm{e}$ with D_e calculated by Einstein's relation. According to Mizusaki and Fueki (1982), the self-diffusion coefficient for $650 < T < 710$ K equals to 0.85 cm^2/s; i.e., $D_\mathrm{e}/\tilde{D}_\mathrm{e} \approx 10^4$. The authors believe that such a difference indicates that only a small amount of excessive silver dissolved in AgCl is ionized and forms conduction electrons.

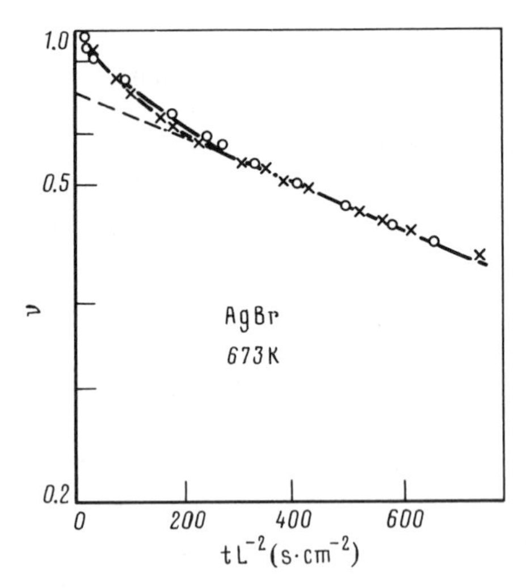

FIG. 21. Relative variation of electron concentration, $\nu = \lg\{[n_\mathrm{e}(t) - n_{\mathrm{e,s}}]/[n_\mathrm{e}(t = 0) - n_{\mathrm{e,s}}]\}$, as a function tL^{-2} for AgBr, with sample thickness, × − L = 0.84 cm; ○ − L = 0.51 cm. [From Sasaki *et al.* (1981).]

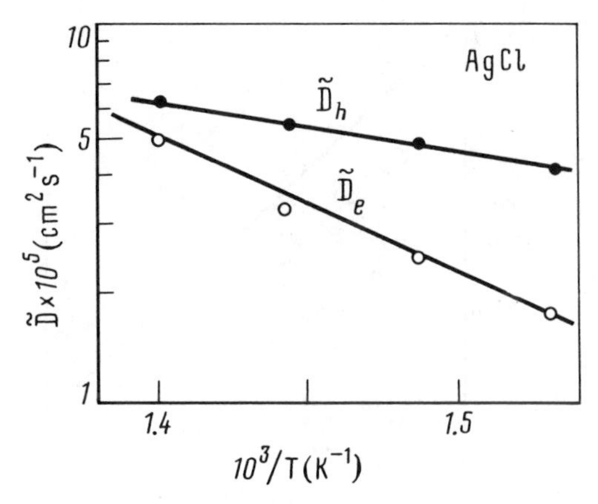

FIG. 22. Temperature dependence of chemical diffusion coefficient of electrons, $\tilde{D}_e$, and holes, $\tilde{D}_h$, for AgCl. [From Mizusaki and Fueki (1982).]

For small values of time ($t \ll \tau$), the solution of Eqs. (137)-(138) may be written in the form (Hartman *et al.*, 1976)

$$\frac{n_e(t) - n_{e,s}}{n_e(t = 0) - n_{e,s}} = \frac{2}{L}\sqrt{\frac{\tilde{D}_e t}{\pi}}. \tag{142}$$

The dependence of type (142) was experimentally observed for α-$Ag_{2+\delta}S$ at 470 K (Rickert, 1982) (Fig. 23). Such a dependence permits the determination of the diffusion coefficient.

The above consideration operates on the assumption that, in a SIM, one species of minority carriers—electrons or holes—is prevalent. If the concentrations of the species of minority carriers are comparable, the mathematical description of nonstationary processes becomes more complicated (Weppner, 1976, 1977a,b). As is shown by the corresponding calculations, if $\tilde{D}_e \gg \tilde{D}_h$, there are two characteristic times, and the relaxation process proceeds in two stages. First, the redistribution of electrons occurs, the holes almost not participating in restructuring. Then, at longer times, the changes in potential are determined by the establishment of an equilibrium hole concentration in the crystal.

If a voltage is applied to an A-type cell, then, as has already been noted, the capacitance of a double layer is charged, and diffusion proceeds in the bulk. If the effects associated with the charging of the double-layer capacitance may be neglected, then it is convenient to consider the behavior of the cell potential with the abrupt change in voltage from V_1 to V_2—the so-called voltage step method (Weiss, 1972; Wagner, 1976; Sasaki *et al.*, 1981). Then, diffusion of ionic and electronic carriers proceeds under the initial

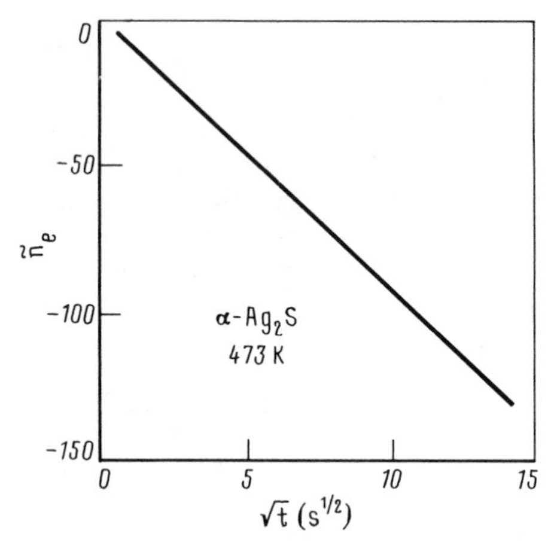

FIG. 23. Electronic concentration $\tilde{n}_e \equiv [n_e(t) - n_{e,s}] \times 10^8$ (in g atom/cm^3) as a function of $\sqrt{t}$ for α-Ag_2S. [From Rickert (1982).]

conditions of type (138) and boundary conditions, in the form

$$n_e(L, t) = n_{e,s} \exp\left(-\frac{eV_1}{kT}\right), \qquad t < 0\,, \tag{143}$$

$$n_e(L, t) = n_{e,s} \exp\left(-\frac{eV_2}{kT}\right), \qquad t \geq 0, \tag{144}$$

$$n_e(o, t) = n_{e,s}\,. \tag{145}$$

The solution of the above-formulated boundary problem for sufficiently large times is as follows:

$$\frac{n_e(L/2, \infty) - n_e(L/2, t)}{n_e(L/2, \infty) - n_e(L/2, o)} = \frac{4}{\pi} \exp\left(-\frac{\tilde{D}_e \pi^2 t}{L^2}\right), \qquad t > \tau. \tag{146}$$

Measuring $V(L/2, t)$ and using Eq. (146), it is possible to determine the chemical diffusion coefficient $\tilde{D}_e$ (or $\tilde{D}_h$). Such experiments were, in fact, performed for AgBr and α-AgI (Sasaki *et al.*, 1981; Mizusaki *et al.*, 1979, 1982) in the temperature range 600 to 800 K. The temperature behavior of the thus-found quantity $\tilde{D}_{e,h}$ is shown in Fig. 24.

If the electronic conductivity and the deviation from stoichiometry for the sample are known, it is possible to find mobilities $u_{e,h}$ of electronic carriers by the relationship following from Eqs. (8) and (2):

$$\sigma_{e,h} = \left(\frac{\mathfrak{N}_A \delta}{V_m}\right) u_{e,h} e. \tag{147}$$

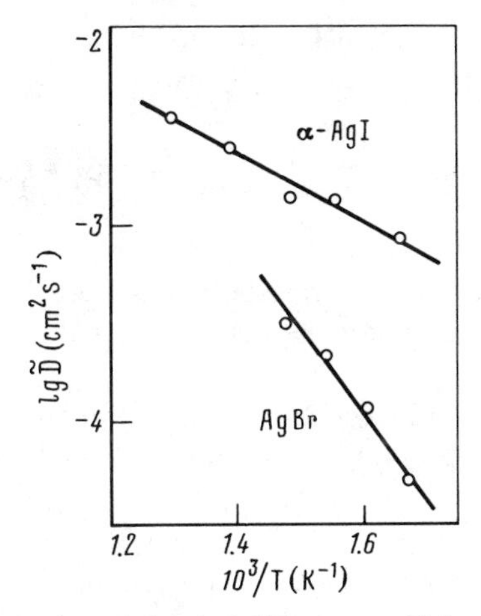

FIG. 24. Temperature behavior of chemical diffusion coefficient, $\tilde{D}$, for SIMs: (1) α-AgI: $\tilde{D}_h = 0.495 \times \exp(-0.34/kT)$, $cm^2\,s^{-1}$; (2) AgBr: $\tilde{D}_e = 710 \times \exp(-0.85/kT)$, $cm^2\,s^{-1}$. [From Mizusaki *et al.* (1979).]

Thus, it has been found that, at 670 K, mobility of electrons in AgBr is $u_e = 5.5 \times 10^3\ cm^2/V\,s$, and that of holes in α-AgI is $u_h = 2.5 \times 10^{-2}\ cm^2/V\,s$.

b. Charge Transfer Technique

In order to avoid the effect of charging of the double layer capacitance on transient currents, specially-designed cells with SIM samples of various lengths—but equal areas of the electrodes—may be used (Joshi, 1973; Joshi and Wagner, 1975, Mazumdar *et al.*, 1982). Assuming that the capacitances of double layers for all the samples are equal, it is possible to single out the component of the stationary current in the cell that is associated with the carrier redistribution in the crystal bulk. Such a double cell permits the determination of some other interesting characteristics, e.g., electronic carrier concentrations $n_e(o)$ and $n_h(o)$.

Let the superionic material under study be, for example, an n-type electronic conductor. Under stationary conditions, the concentration distribution of electrons is linear [Eq. (138)]. If we do not take the narrow regions, $L_D \ll L$, close to the interfaces into consideration, the total number of electrons, N_e, in the bulk of a sample of length L and cross-section S is

$$N_e = S\int_0^L n_e(o)\left[\frac{x}{L}\left(1 - \frac{n_e(L)}{n_e(o)}\right)\right] dx = n_e(o)\frac{LS}{2}\left[1 - \exp\left(-\frac{eV}{kT}\right)\right]. \tag{148}$$

The total change of the A-type cell charge with the variation of the polarizing voltage from V_1 to V_2, using Eq. (148), is

$$Q = Q_{el} + Q_{dl} = \frac{e}{2} SLn_e(o)\left[\exp\left(-\frac{eV_1}{kT}\right) - \exp\left(-\frac{eV_2}{kT}\right)\right] + S\int_{V_1}^{V_2} C_{de}\, dV. \quad (149)$$

Here, Q_{el} is the charge due to electron redistribution in the crystal, Q_{dl} is the charge of the double layer, and C_{dl} is the capacitance of the double-layer surface unit.

Comparing the results obtained on samples of different lengths (L_1 and L_2) but equal contact area S, we may exclude the contribution which comes to Q from the term Q_{dl}. From Eq. (149), we have ($L_1 < L_2$)

$$\Delta Q = \frac{eS}{2}(L_1 - L_2)n_e(o)\left[\exp\left(-\frac{eV_1}{kT}\right) - \exp\left(-\frac{eV_2}{kT}\right)\right]. \quad (150)$$

Here, $\Delta Q = Q_1 - Q_2$ is the experimentally-measured difference between quantities Q_1 and Q_2 related to cells of length L_1 and L_2, respectively.

An analogous expression may be obtained for a SIM which is a *p*-type semiconductor. Replacing e by −e, we obtain, from Eq. (150),

$$\Delta Q = \frac{eS}{2}(L_1 - L_2)n_h(o)\left[\exp\left(\frac{eV_2}{kT}\right) - \exp\left(\frac{eV_1}{kT}\right)\right]. \quad (151)$$

If $V_2 \gg V_1$, it follows from Eq. (151) that

$$\ln \Delta Q = \ln\left[\frac{eS}{2}(L_1 - L_2)n_h(o)\right] + \frac{eV_2}{kT}. \quad (152)$$

Thus, we may determine $n_{e,h}(o)$ from the ($\ln \Delta Q$, $V_{1,2}$) plot. Furthermore, if we also know $\sigma_{e,h}(o)$ (from, say, polarization measurements), we may also find the corresponding mobility $u_{e,h}$ and compare it with the values obtained by other relaxation techniques.

Such a comparison was, in fact, carried out by Mazumdar *et al.* (1982), in particular, for β-AgI, which is a *n*-type semiconductor. The authors experimentally obtained linear dependence ($\ln \Delta Q_e$, V_1), which yielded the mobility of electronic charges, $u_e \approx 5.1 \times 10^{-5}$ cm^2/V s (at 306 K). The described method was also used by Joshi (1973), who studied cuprous chloride. He obtained a value of $u_h = 10^{-1}$ cm^2/V s for the hole mobility at 553 K.

c. Current Relaxation Technique

One more relaxation method has been developed. It is used for determining the electronic-charge mobility and is based on measurements of a current

$i(t)$ flowing through a cell under conditions when $Q_{el} \gg Q_{dl}$ (Joshi, 1973; Wagner, 1976).

For example, consider a SIM which is a p-type semiconductor. The scheme of the experiment is the same as in the voltage step method, the only difference being that a relaxation current flowing through the cell is now measured. The diffusion equation should be solved only for short times $t < \tau$. In this approximation, the hole current at the blocking electrode is $i_h = -e\tilde{D}_h(\partial n_h/\partial x)_{x=0}$.

Joshi and Wagner (1975) have obtained

$$i_h(t) = \left(\frac{L}{\sqrt{\pi \tilde{D}_h t}}\right)[i_h(V_2) - i_h(V_1)] + i_h(V_1), \tag{153}$$

where

$$i_h(V) \equiv \left(\frac{kT}{eL}\right)\sigma_h(o)\left[\exp\left(\frac{eV}{kT}\right) - 1\right]. \tag{154}$$

As follows from Eq. (153), the time dependence of the relaxation current should be a straight line in the $i, t^{-1/2}$ coordinates. The coefficient of chemical diffusion, $\tilde{D}_h$, is determined from the slope of experimental straight lines. Such a treatment was carried out, in particular, for experimental data on cuprous chloride (Joshi and Wagner, 1975) (Fig. 25). The experimental data confirm the suggestions on the electronic-hole character of conductivity in CuCl.

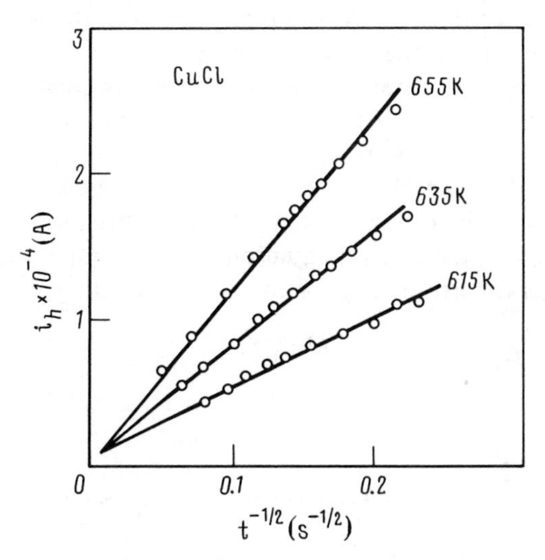

FIG. 25. Hole current, i_h, versus $t^{-1/2}$ for CuCl (sample thickness L = 1.672 cm) for a stepwise variation in voltage (at different temperatures). [From Joshi and Wagner (1975).]

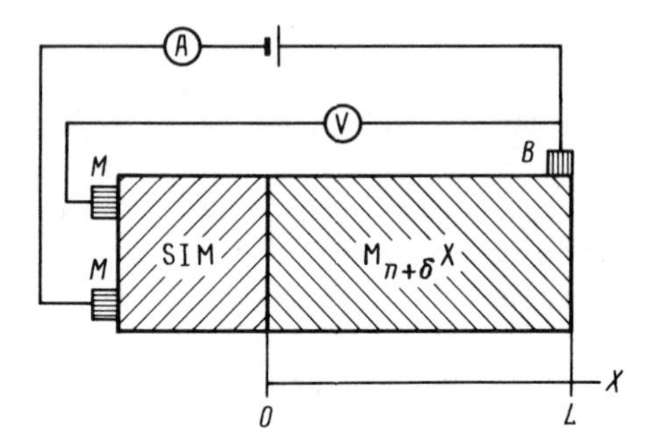

FIG. 26. An electrochemical cell for studying relaxation behavior of nonstoichiometric superionic materials $M_{n+\delta}X$: M—reversible with respect to M^+ ion metal electrode; B—electronic conductive ion blocking electrode; V—voltmeter; A—amperemeter for recording of relaxation current.

The current relaxation method was used for the determination of $\tilde{D}_M$ in SIMs with a high degree of nonstoichiometry, such as wüstite $Fe_{1-\delta}O$ and silver sulphide $Ag_{2+\delta}S$ (Rickert and Weppner, 1974; Rickert, 1982). An A-type cell in which the material under consideration $M_{n\pm\delta}X$ (Fig. 26) was placed at the site of a blocking electrode was used. In addition, in order to avoid the effect of electrode polarization on potential measurements, an auxiliary electrode was used.

With the application of a constant potential difference, the steady-state concentration distribution of ions $n_{i,o}$ is reached in the $M_{n\pm\delta}X$ electrode, corresponding to voltage V_0. At the moment $t = 0$, the voltage changes in a stepwise manner up to the value V_1; then, at the electrode/SIM interface ($x = 0$), a new concentration of mobile carriers is created in a jumpwise manner. The occurring diffusion results in equalization of particle concentrations in $M_{n\pm\delta}X$, with ions arriving from the superionic material and electrons from the second (blocking) electrode. As a result, an electric current, which may be recorded, starts flowing through the cell.

The solution of the diffusion equation with the boundary conditions of type (138) yields the expression for the relaxation current

$$i(t) = ze(n_{i,o} - n_{i,1})\left(\frac{\tilde{D}_M}{\pi t}\right)^{1/2}, \qquad t \ll \tau \tag{155}$$

and

$$i(t) = \frac{2ze}{L}(n_{i,o} - n_{i,1})\tilde{D}_M \exp\left(-\frac{\pi^2 \tilde{D}_M t}{4L^2}\right), \qquad t \gg \tau. \tag{156}$$

Here, $n_{i,1}$ is the concentration of mobile ions at the reversible electrode (V- and t-independent). Figure 27 shows the experimental data on current

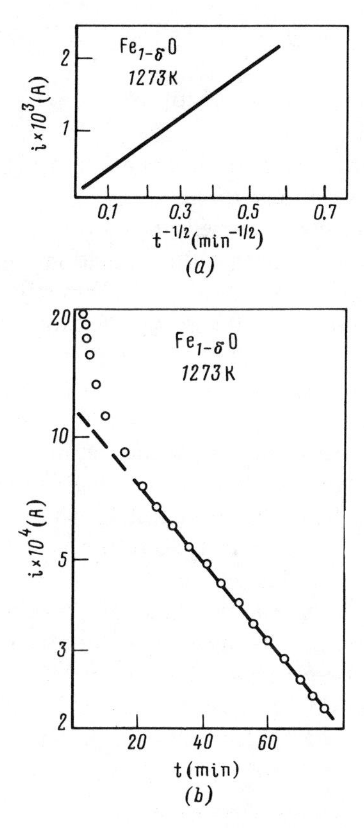

FIG. 27. Relaxation current versus time for $Fe_{1-\delta}O$ ($\delta = 0.106$). [From Rickert (1982).]

relaxation for the $Fe_{1-\delta}O$ SIM, which, at high temperatures, behaves as an oxygen–ion conductor. As is seen from Fig. 27a, for relatively short times, the experimental points are described quite well by a straight line corresponding to Eq. (155), and, for large times (Fig. 27b), to Eq. (156). The slope of these lines yields chemical diffusion coefficient, $\tilde{D}_M$, $\tilde{D}_M = 3.2 \times 10^{-6}\ cm^2\ s^{-1}$ at 1300 K.

Now, consider transient processes in which the double-layer capacitance plays an essential part.

d. Double-Layer Capacitance Charging

The effect of electronic carriers manifests itself in the charging and discharging of the double-layer capacitance (Raleigh, 1966, 1967, 1973; Hull and Pilla, 1971). If we assume that the capacitance C_{dl} is constant during the discharge process, we may readily obtain C_{dl} from the relationship describing the relaxation current in an A-type cell upon potential switching off:

$$i(t) = i_0 \exp\left(-\frac{t}{RC_{de}}\right), \tag{157}$$

where i_0 is the current flowing in the cell at an instant of time $t = 0$, and R is the resistance of a solid electrolyte.

Experimental studies were carried out on thin CuCl samples (Joshi and Wagner, 1975) in such a way that the relaxation current was only slightly affected by the charge-carrier redistribution in the specimen bulk. Moreover, to obtain small values of capacitance C_{dl}, the experiment was carried out at relatively low temperatures. The cell potential varied in a stepwise manner (from V_1 to V_2), with steps of 0.02 V. Then the relaxation (presumably, capacitance) current, $i(t)$ was measured with V_2 = const. In accordance with Eq. (157), the dependence of ln i on t in the region of relatively low potentials turned out to be linear (Fig. 28, curve 1). At higher potentials (Fig. 28, curve 2), the curve deviated from the straight line, which is explained (Joshi and Wagner, 1975) by the V-dependent electron-hole current.

We should like to emphasize that such studies are of great applied importance, since the superionic materials permit the design of high-capacitance electrolytic double-layer capacitors (Scott, 1972; Gailish *et al.*, 1975) (for more detail see Section 25).

Now let us consider self-discharge in specific SIM-based capacitors whose capacitance is determined by the double-layer capacitance at the interface.

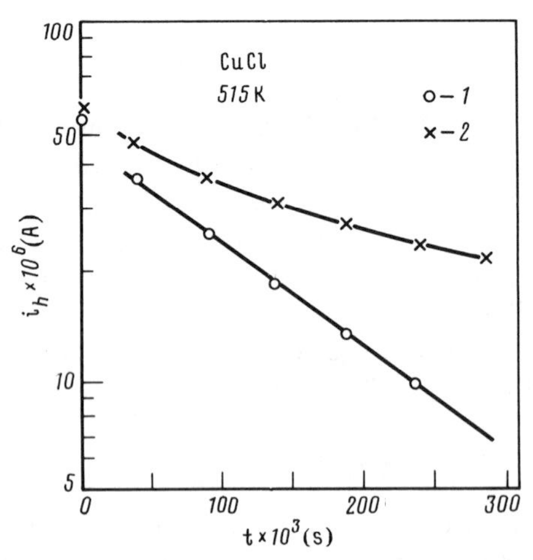

FIG. 28. Hole current, i_h, for a CuCl sample of thickness L = 0.085 cm versus time, t, at 515 K. Voltage steps (V_1 to V_2): (1) V_1 = 0.2476, V_2 = 0.2677; (2) V_1 = 0.4287, V_2 = 0.4489. [From Joshi and Wagner (1975).]

The depolarization process in an A-type cell, which models self-discharge, is determined by the effective leakage resistance tightly associated with electronic conductivity in the SIM. [The arguments in favor of the above statement have been discussed by Raleigh (1969) and Gurevich and Ivanov-Shits (1978).]

The process of self-discharge is rather slow, and, therefore, electronic currents in the capacitor are determined by Eqs. (66)–(67). The change dQ_{dl} in the charge occurring in time dt is $dQ_{dl} = i_{el} S\, dt$. Taking into account that $dQ_{dl} = C\, dt$, where C is the a priori measured cell capacitance, we obtain, for the case where conductivity is provided by electrons alone,

$$C\, dV = -i_{e,o} S \left[1 - \exp\left(-\frac{eV}{kT}\right)\right] dt. \tag{158}$$

For definiteness, we use Eq. (68) for $i_{e,o}$. Then, the factor before the square brackets in Eq. (158) may be written in the form $(kT/e)R_e^{-1}$, where $R_e \equiv L/\sigma_e(o)S$ is the effective capacitor resistance to the electron current.

In the most important potential range, $V \gg kT/e$, the functional dependence $C(V)$ is weak—$\sim V^{-1/2}$, and even slower (Gurevich and Kharkats, 1976). Therefore, taking into account only the strong exponential dependence of the current on potential and performing the integration in Eq. (158), we arrive at (Gurevich and Ivanov-Shits, 1978)

$$-\frac{e(V - V_o)}{kT} + \ln \frac{|1 - \exp(-eV_o/kT)|}{|1 - \exp(-eV/kT)|} = \frac{t}{R_e C}. \tag{159}$$

Here V_o is the initial potential difference in the cell at $t = 0$. For $V \gg kT/e$ we have

$$-\frac{e(V - V_o)}{kT} = \frac{t}{R_e C}. \tag{159'}$$

Thus, at the initial stage, the self-discharge is characterized by a linear decrease of V with time.

If the current is due to holes, the self-discharge, under the same assumptions, is described by the relationship

$$\ln \frac{|1 - \exp(-eV/kT)|}{|1 - \exp(-eV_o/kT)|} = -\frac{t}{R_h C}, \tag{160}$$

where $R_h \equiv L/\sigma_h(o)S$.

In accordance with Eq. (160), at sufficiently long times

$$t \gg R_h C \exp\left(-\frac{eV_o}{kT}\right)$$

and for potentials $|V| \gg kT/\mathrm{e}$, a logarithmic (i.e., very slow) decrease of V with time should be observed:

$$V \approx -\frac{kT}{\mathrm{e}} \ln\left(\frac{t}{R_{\mathrm{h}}C}\right). \tag{160'}$$

As follows from the above discussion the use of the standard formula valid for the discharge of a "conventional" capacitor,

$$\ln\left(\frac{V}{V_{\mathrm{o}}}\right) = -\frac{t}{RC}, \tag{161}$$

for the description of the self-discharge process is not quite correct in general. Formally, the latter equation follows from Eq. (158) at $|V| \lesssim kT/\mathrm{e}$, when $V = R \times i$. But the standard logarithmic dependence of V on time, Eq. (161), may be invalid even in the region of small potentials. The point is that, at small V, the dependence $C(V)$ in electrochemical systems is essential and should be taken into consideration when integrating Eq. (158).

Concluding, we should like to note that the data on self-discharge caused by electronic carriers in superionic materials may also be used to determine the effective electron-hole resistance $R_{\mathrm{e,h}}$ [see Eqs. (159)–(160)] and conductivities $\sigma_{\mathrm{e,h}}$.

e. Galvanostatic Intermittent Titration Technique

In order to determine the chemical diffusion coefficient, Weppner and Huggins (1977a,b, 1978) have suggested galvanostatic intermittent titration technique (GITT), which combines the advantages of both steady-state and transient methods. An experimental galvanostatic cell is shown in Fig. 26. The sample being studied (for example, the compound $M_{1+\delta}X$, in which mobile carriers are monovalent M^+ cations and electrons) is used as an electrode. The potential difference V measured between the SIM and the sample of $M_{1+\delta}X$ composition is determined by the Nernst relationship, Eq. (53), which, in our case, is written as

$$\mathrm{e}V = kT \ln\left(\frac{a_{\mathrm{M}}^{1}}{a_{\mathrm{M}}^{2}}\right), \tag{162}$$

where $a_{\mathrm{M}}^{1,2}$ is the activity M in the reference electrode (made of a SIM shown in Fig. 26) and in the compound $M_{1+\delta}X$ under study, respectively.

If a galvanostatic current pulse i_{o} of duration Δt is applied to the cell, then, at the interface ($x = 0$), a constant concentration gradient $(\partial n_{\mathrm{M}}/\partial x)$ is developed. For maintaining a gradient constant, the voltage applied to the cell should be varied (increased or decreased, depending on the sense of current i_{o}). Upon the passage of a pulse, a new stationary distribution of current carriers is established in the cell, corresponding to the changed composition of the initial $M_{1+\delta}X$ compound. The solution of the corresponding boundary diffusion problem for short times is given by

$$\left.\frac{dn_M}{d\sqrt{t}}\right|_{x=0} = \frac{2i_o}{e\pi\sqrt{\tilde{D}_M}}. \tag{163}$$

Equation (163) may conveniently be transformed to the form

$$\frac{dV}{d\sqrt{t}} = \frac{2i_o}{e\sqrt{\tilde{D}_M}\,\pi}\,\frac{dV}{dn_M}, \tag{164}$$

where dV/dn_M is, by definition, the slope of the titration curve. The titration curve, i.e., the dependence $V(n_M)$, may be calculated from the plot of the steady-state voltage (measured upon the passage of each pulse) as a function of the charge transmitted through the cell.

Thus, proceeding from Eq. (164), it is possible to determine $\tilde{D}_M$. Let the change in the steady-state voltage upon the pulse passage be sufficiently small (i.e., $dV/dn_M = \Delta V_s/\Delta n_M$, where the value of ΔV_s is determined in Fig. 29). The experimental dependence of V on Δt is linear (Fig. 29). Then it can be shown that the chemical diffusion coefficient is described by the equation (Weppner and Huggins, 1978)

$$\tilde{D}_M = \frac{4}{\pi\,\Delta t}\left(\frac{V_m}{N_X S}\right)^2\left(\frac{\Delta V_s}{\Delta V_t}\right)^2, \tag{165}$$

where V_m is the molar volume of the specimen and N_X is the number of X-atoms in it.

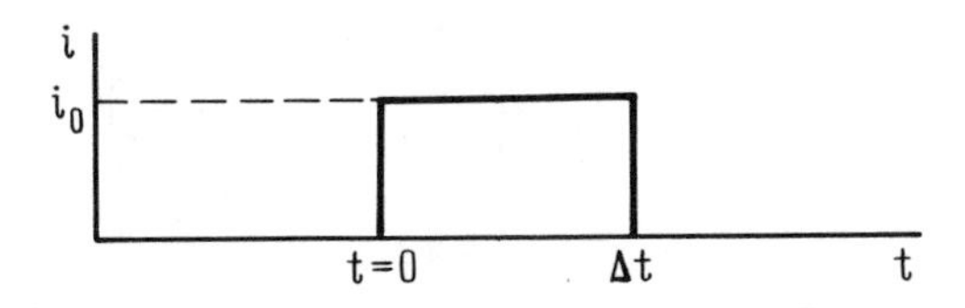

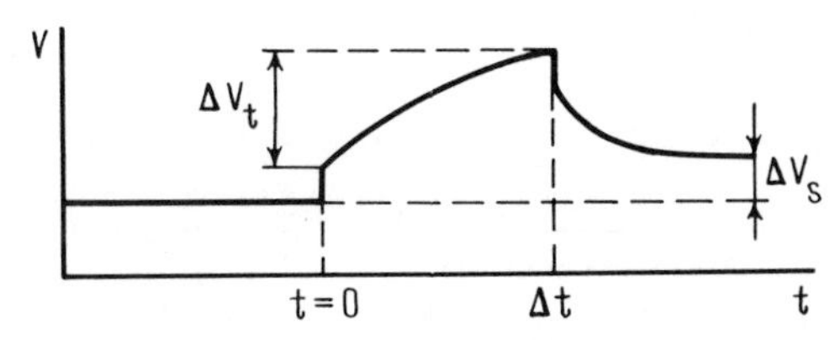

FIG. 29. Schematic diagram of recording current, $i(t)$, and voltage, $V(t)$, for a cell shown in Fig. 26.

The technique described allows also the determination of the specific conductivities of the components and their mobilities and self-diffusion coefficients. Thus, the self-diffusion coefficient D_M is determined by the relationship

$$D_M = -\frac{4kTV_m i_o \Delta V_s}{\pi n_M e^2 N_X (\Delta V_t)^2}, \qquad \Delta t \ll \tau. \tag{166}$$

The galvanostatic intermittent titration technique was successfully applied to the study of electrode materials with mixed conductivity in Li–A systems where A = Al, Sb, Bi (Weppner and Huggins, 1977a,b; Wen *et al.*, 1979) and in WO_3-based thin films (Raistrick *et al.*, 1981; Joo *et al.*, 1985) and Li-intercalated compounds with the general formula Li_xBO_n (B = V, W, Mo, Mn) (Dickens and Reynolds, 1981).

Slightly modified, this method was also applied to the investigation of the β-phase of silver sulphide with mixed ionic-electronic conductivity (Becker *et al.*, 1983) and to various TiS_2-, CoO_2- and NbS_2-based intercalates (Honders *et al.*, 1983a,b, 1985; Bouwmeester, 1985; Yamamoto *et al.*, 1985).

As an example, Fig. 30 shows chemical diffusion $\tilde{D}_{Ag}$ and self-diffusion D_{Ag} coefficients of silver in silver sulphide samples with different degrees of nonstoichiometry determined by the galvanostatic intermittent titration technique.

Concluding the discussion of transient currents in SIMs, we should like to stress again the limited character of concepts based on the chemical diffusion coefficient. We should keep in mind possible deviations from the local

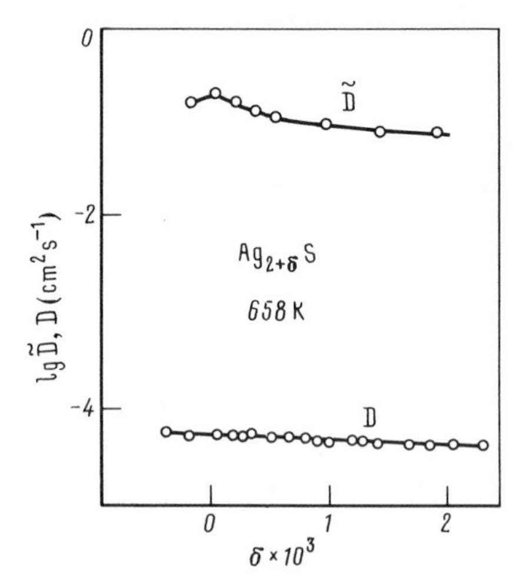

FIG. 30. Diffusion coefficient of silver, D, and chemical diffusion coefficient, $\tilde{D}$, of $Ag_{2+\delta}S$ as a function of the deviation from stoichiometry, δ. [From Weppner (1981).]

equilibrium between carriers, especially in nonstationary processes [see Eq. (86)]. It should also be remembered that the consideration of transient processes may be essentially complicated if the electronic current in an electrochemical cell is assumed to be of a non-diffusion nature, as determined by carrier transfer through the electrode-SIM interface (see, e.g., Couturier *et al.*, 1983).

VI. Photochemical Phenomena in Superionic Materials

Photochemical and photoelectrochemical processes in superionic materials became the subject of thorough investigations only recently. Such studies are closely related to the successful development of such new branches of physical chemistry as the study of photoelectrochemical processes at metal/liquid electrolyte and, especially, semiconductor/liquid electrolyte interfaces and photochemical processes in "conventional" ionic crystals (Pleskov and Gurevich, 1986). The results obtained in such works have already found some important applications—in particular, in connection with problems of light-energy conversion, information recording, electrochemical photosynthesis, etc. The participation of electronic carriers in such processes is of key importance. In this connection, it may be useful to dwell on some aspects of the photoelectrochemistry of superionic materials.

Special studies of the reflection and absorption spectra of light in the visible and adjoining ranges have shown that the optical properties of superionic materials are similar, on the whole, to those of conventional semiconductors.

Thus, the well-studied superionic conductor Ag_4RbI_5 shows, at room temperature, a sharp edge of fundamental absorption corresponding to an energy of $\hbar\omega_* = 3.2$ eV (Nimon *et al.*, 1977; see also Table 2). The analysis of electronic transitions shows that the Ag_4RbI_5 compound in the superionic state is a specific broadband semiconductor with the band gap equal to $E_g = \hbar\omega_*$.

For frequencies ω higher than ω_*, the energy of an absorbed radiation quantum is sufficient for photogeneration of electron-hole pairs ($\hbar\omega \rightarrow e^- + h^+$). The carrier redistribution caused by such photogeneration may drastically change not only the rate, but also the character, of the processes occurring at the interface.

14. Quasi-Thermodynamic Description

Consider first the quasi-equilibrium approach to the description of a photoexited state. Such an approach is especially useful for a qualitative understanding of specific features of the processes under consideration.

It is based on the concepts of Fermi quasi-levels introduced into consideration in the following way:

Let the material under study be exposed to light, resulting in additional, with respect to the equilibrium, generation of electronic carriers. In the steady-state mode, the dynamic equilibrium between photogeneration and recombination of electron-hole pairs is attained. As a result, certain stationary (but not equilibrium) values of the electron and hole concentrations, n_e^* and n_h^*, are reached. Let the lifetimes of the excited states be sufficiently long. Then the interaction of such states with the crystal lattice (in particular, with phonons) promotes the attainment of the equilibrium distribution in the electronic and hole subsystems, the temperature of such a distribution being equal to the lattice temperature. It is essential that the electron and hole gases, generally speaking, are not in equilibrium with one another. Under such conditions, both distributions (of electrons and of holes) are characterized by their own electrochemical potentials (Fermi levels), F_e and F_h, respectively. In distinction from the "complete" thermodynamic equilibrium, when $F_e = F_h = F$, the characteristic quantities F_e and F_h corresponding to a partial equilibrium are unequal. It is just the quantities F_e and F_h, that are called the Fermi quasi-levels.

Thus, the appearance of quasi-equilibrium electrons and holes in a superionic material, due to the illumination of this material with light, may be described as the "splitting" of the initial Fermi-level F into two quasi-levels, F_e and F_h. Now, instead of Eq. (1) describing the thermodynamic equilibrium, we have relationship describing the quasi-equilibrium,

$$n_e^* n_h^* = n_{\text{int}}^2 \exp\left\{\frac{(F_e - F_h)}{kT}\right\}. \tag{167}$$

Then, $n_e^* = n_e^o + \Delta n_e$; $n_h^* = n_h^o + \Delta n_h$. Since pairs are generated, outside the space-charge region, we have $\Delta n_e = \Delta n_h \equiv \Delta n$.

For example, consider a material in which the electronic (in the narrow sense of the word) conductivity prevails over the hole conductivity. In this case, F is given by the well-known equation

$$F = E_c + kT \ln\left(\frac{n_e^o}{N_c}\right), \tag{168}$$

whose structure is analogous to the general expression for chemical potential [see Eq. (29′)]. Substituting n_e^* for n_e^o in Eq. (168), and keeping in mind that $\Delta n \ll n_e^o$, we obtain $F_e = F$. Using Eq. (167), and substituting Expression (168) for F_e, we find that

$$F - F_h = kT \ln\left(\frac{n_h^o + \Delta n}{n_h^o}\right),$$

or

$$F - F_h \approx kT \ln \frac{\Delta n}{n_h^o}, \qquad \text{if } \Delta n \gg n_h^o. \tag{169}$$

Similarly, if, in a material with the predominant hole conductivity, $\Delta n \ll n_h^o$, then $F_h = F$, and

$$F_e - F = kT \ln\left(\frac{n_e^o + \Delta n}{n_e^o}\right),$$

or

$$F - F_e \approx kT \ln \frac{\Delta n}{n_e^o}, \qquad \text{if } \Delta n \gg n_e^o. \tag{170}$$

Thus, the shift of the electrochemical potential during photogeneration is observed mainly for minority carriers (Fig. 31). Note also that Eqs. (169) and (170) describe the shifts of the electrochemical potentials of electrons and holes caused by the change of their concentrations, similar to the case of doped semiconductors. In connection with this analogy, the shift of F_e and F_h due to light exposure may be called light doping.

If the illumination intensity J is not too high, Δn_e and Δn_h are proportional to J; therefore, the changes in $F - F_h$ and $F_e - F$ are proportional to the intensity logarithm. With an intensity increase, the difference between quasilevels remains almost constant, due to recombination processes.

Let a superionic material under study be in contact with a metal or an electrolyte solution that includes an oxidation-reduction system. If electron exchange proceeds through the interface, then, in the absence of illumination—i.e., in the state of thermodynamic equilibrium—the SIM is characterized by a definite equilibrium value of the electrode potential φ_{ed}^o (measured from a certain fixed reference electrode).

If the potential is φ_{ed}^o, the total current in the circuit is zero, with electronic and hole currents being also zero. Illumination increases the electron and hole concentrations, thus changing the currents due to electrons and holes that pass through the interface. Then the quasi-equilibrium potentials φ_h^o and

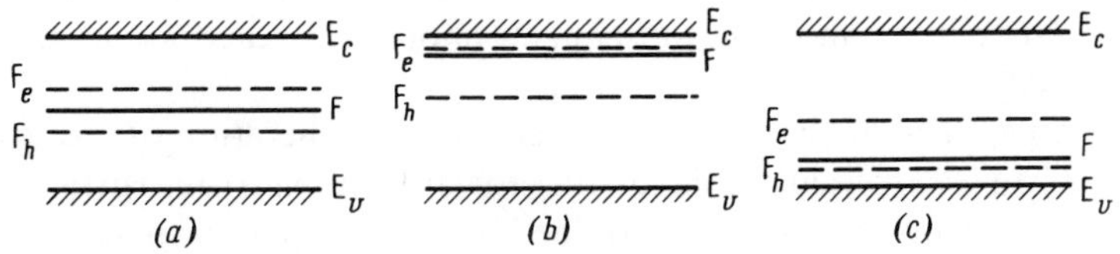

FIG. 31. Positions of quasi-Fermi levels of electrons, F_e, and holes, F_h, relative to the Fermi level, F, and the band edges, E_c and E_v, in materials with different electronic conductivity: (a) intrinsic, (b) n-type, (c) p-type semiconductors.

φ_e^o, for which the hole and electron currents become zero, are not equal either to one another or to the equilibrium potential φ_{ed}^o. In other words, the illuminated electrode seems to be simultaneously under two different electrode potentials, one of which determines the rate of the process occurring with the participation of holes, and the other determining the rate of the process with the participation of conduction-band electrons.

Using the concept of Fermi quasilevels $F_{e,h}$ and proceeding from the general thermodynamic principles, we obtain the following equations for the deviations of the quasi-equilibrium potentials $\varphi_{e,h}^o$ from the equilibrium value φ_{ed}^o:

$$e(\varphi_{e,h}^o - \varphi_{ed}^o) = F - F_{e,h}. \tag{171}$$

As follows from Eq. (171), in extrinsic semiconductor materials (when $n_e^o \gg n_h^o$ or $n_h^o \gg n_e^o$), the sample illumination changes the potential of minority carriers alone. This change may be quite large, so that $e|\varphi_e^o - \varphi_{ed}^o|/kT \gg 1$ or $e|\varphi_h^o - \varphi_{ed}^o|/kT \gg 1$. Therefore, the illumination drastically changes the current in the system. It is seen from Eq. (171) that an electrochemical reaction may proceed at the illuminated semiconductor electrode even if its potential is lower than the equilibrium dark potential. This fact is of great practical importance.

Using equations relating currents to electrode potentials [in particular, Eq. (56) or other similar equations], it is possible, from Eq. (171), to determine how the current depends on the illumination intensity, and to analyze a number of other effects caused by illumination. Examples illustrating such an approach to photoelectric processes are considered in a monograph by Pleskov and Gurevich (1986).

15. Photolytic Processes

A detailed study of photoprocesses in the Ag_4RbI_5 SIM with the participation of electronic carriers was carried out by Nimon *et al.* (1977, 1979, 1980, 1981). Photocurrents were induced by illumination of an Ag_4RbI_5 crystal with a metal (Pt) electrode blocking the ionic current. The appearance of photocurrents is associated with the generation of electronic carriers in the SIM bulk and the subsequent photolytic reactions at the metal electrode.

The first stage of the resulting process is the generation of electron-hole pairs. Then electrons and holes diffuse towards the electrode surface. There, holes may easily be trapped by iodine ions I^- with the formation of iodine atoms, in accordance with the reaction $I^- + h^+ \rightarrow I$. Similar to the case of photogenerated holes in other I-containing compounds (Friedenberg, 1982), holes move mainly by hopping between I^- ions. Therefore, such motion resembles that of neutral I atoms. Lastly, photolytic iodine is reduced at the metal electrode ($I + e^- \rightarrow I^-$), thus providing the cathodic photocurrent. The anodic photocurrent is caused by electrons appearing in the conduction band, due to photogeneration. Some of these electrons return to the electrode.

In the simplest case, the following equation for the excess electron Δn_e and hole Δn_h concentrations is obtained:

$$\frac{\partial \Delta n_{e,h}}{\partial t} = D_{e,h} \frac{\partial^2 \Delta n_{e,h}}{\partial x^2} + Y\alpha J \exp(-\alpha x). \tag{172}$$

Equation (172) follows from general expressions (33) and (35), if the migration component in i_e and i_h is neglected. The last term in Eq. (172) describes source q generating electrons and holes; J is the intensity of photoexciting light (the number of light quanta transmitted through a unit surface of the superionic material per unit time); α is the coefficient of light absorption; and Y is the yield of photogeneration of electron-hole pairs.

If the reactions of photolytic particles at the electrode proceed at a sufficient rate, then, at the electrode ($x = 0$), $\Delta n_{e,h}(0, t) = 0$. If we also take into account that in the material bulk, $\Delta n_{e,h}(\infty, t) = 0$, and that the initial conditions are $\Delta n_{e,h}(x, 0) = 0$, then Eq. (172) yields the corresponding photocurrents $i_{e,h} = -eD_{e,h}(\partial \Delta n_{e,h}/\partial x)\big|_{x=0}$:

$$i_{e,h} = eYJ[1 - \exp(\alpha^2 D_{e,h} t)\operatorname{erfc}(\alpha\sqrt{D_{e,h} t})]. \tag{173}$$

The experimentally-measured photocurrent i_{ph} (the difference between the cathodic and anodic photocurrents), in accordance with Eq. (173), is

$$i_{ph} = \begin{cases} \dfrac{2eY\alpha J}{\sqrt{\pi}}(\sqrt{D_h} - \sqrt{D_e})\sqrt{t}, & \alpha^2 D_{e,h} t \ll 1, \\[2ex] \dfrac{eYJ}{\alpha\sqrt{\pi}}\left(\dfrac{1}{\sqrt{D_h}} - \dfrac{1}{\sqrt{D_e}}\right)\dfrac{1}{\sqrt{t}}, & \alpha^2 D_{e,h} t \gg 1. \end{cases} \tag{174}$$

As is seen from Eq. (174), shortly after the light is switched on, the absolute value and the sign of the photocurrent are determined by the diffusion coefficient of the more mobile particles. For longer times after the light is switched on, the diffusion coefficient of less mobile particles determines the photocurrent.

The experimental data obtained for sufficiently long periods of time after the light is switched on yield straight lines (in the i_{ph}, $t^{-1/2}$ coordinates), a fact which agrees with the theory (Fig. 32). The above analysis also shows that $D_h > D_e$.

Absorption of photoactive light in Ag_4RbI_5 provides a convenient and controllable (with respect to intensity) generation of halogen in the region close to the electrode. The study of the photoeffect characteristics at the

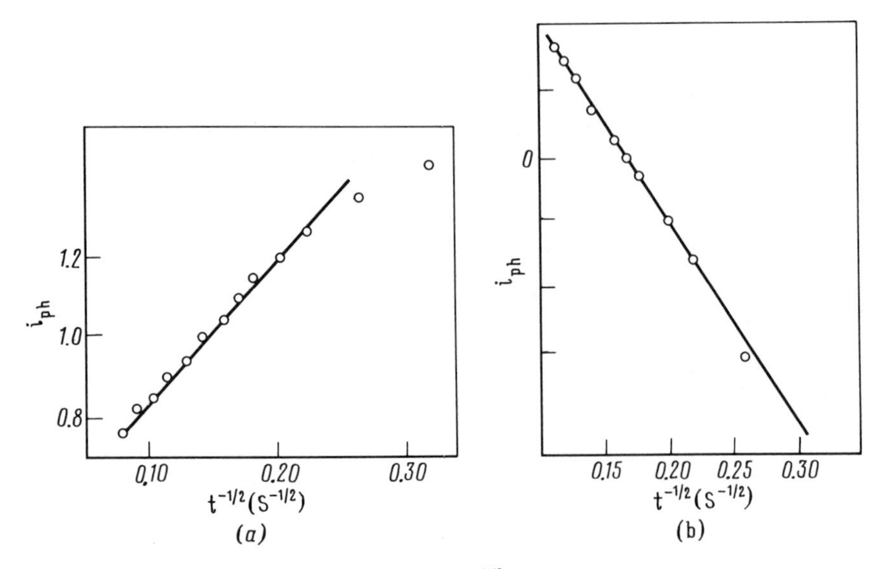

FIG. 32. Photocurrent, i_{ph} (arb.un.), versus $t^{-1/2}$: (a) upon illumination switching on, (b) upon illumination switching off. [From Nimon *et al.* (1979).]

Ag_4RbI_5 boundary yields information on the mechanism and kinetics of dark reactions of iodine electroreduction and silver anodic oxidation.

Note also that in Ag^+-conductive superionic materials (silver halides), specific photoprocesses may occur. Similar to the situation with ordinary silver halides, trapping of photogenerated electrons by Ag^+-ions may result in the appearance of silver centers and, in the final analysis, in the formation of coloidal silver particles.

This phenomenon in Ag_4RbI_5 solid electrolytes was studied by Nimon *et al.* (1980) by the optical spectroscopy technique. They established, in particular, the mean radius of photolytic silver particles forming in this compound during exposure to ultraviolet radiation. With an increase of exposure time, the particle size first rapidly increases, up to a mean radius of about 40 nm, and then the process drastically decelerates.

The long-wavelength irradiation of a solid electrolyte containing silver centers results in their specific clean-out. This effect is analogous to the Herschel effect in silver halides, where the optical density of a light-sensitive layer exposed to light, but not developed, decreases if this layer is illuminated with red or infrared light.

The microscopic mechanism of the Herschel effect is related to the electron photoemission from photolytic silver particles into a solid electrolyte (Gurevich and Barshchevskii, 1970). The emission in a relatively low-frequency range ($\hbar\omega = 1.1$ eV) indicates an essential decrease of the work function for electrons passed from a silver to a superionic material, in comparison with the work function for electrons passed to vacuum. The data obtained permits one to find the work function for electrons passed from Ag to Ag_4RbI_5 and, then, using Eq. (27), to estimate the thermoemission current from the metal to the SIM. It turns out that this current is of the order of 10^{-12}–10^{-11} A/cm^2 for the potential of the silver-electrode.

The above value approaches, in the order of magnitude, the values of electron currents in the Ag|Ag_4RbI_5|B system, where B is the blocking electrode. Such a correspondence favors mechanism 3 of electronic-carrier generation in the given compound (see Section 4).

16. Optical Electronic Transitions

The study of optical transitions in superionic materials yields important additional information on the properties of the electronic subsystem of a crystal. In particular, as has already been noted, by studying the fundamental absorption edge, it is possible to estimate the band gap E_g. During the so-called direct (vertical) transitions occurring with the participation of an electron and a radiation photon, the electronic momentum (quasi-momentum) is conserved. This is explained by the fact that the photon momentum $\hbar\omega/c$ (c is the velocity of light) is negligibly small. The value of the quantum threshold energy $\hbar\omega_o$ does not coincide with the thermodynamic E_g value, since it does not correspond to the minimum interband energy. If, however, light absorption is accompanied by the emission or absorption of one or several phonons (indirect transitions), then the threshold energy is determined by the condition $\hbar\omega \approx E_g$. The E_g values indicated in Table 3 were determined from optical measurements with regard to this type of electronic transitions.

An electron and a hole arising due to photoexcitation may interact and form a two-particle bound state–exciton. The total energy of an exciton is smaller than the band gap. Therefore, in the optical spectra of conventional semiconductors, exciton lines are located lower than the fundamental absorption edge, corresponding to the interband transitions.

The picture in a SIM is more complicated, since disorder of the ionic subsystem may influence the exciton states that are "sensitive" to inner crystal fields. It is natural, therefore, that most attention has been focussed on studying the effect of phase transitions on the superionic state on the optical characteristics of SIMs, e.g., of AgI-based SIMs.

Studying Ag_4RbI_5 luminescence, Afanas'ev *et al.* (1982, 1983) came to the conclusion that the narrow lines of the observed luminescence spectrum from the γ-phase of Ag_4RbI_5 correspond to the transitions of Ag^+-ions from the sublevels of the $4d^9 5s^1$ excited state to the ground $4d^{10}$-level (Radhakrishna *et al.*, 1979; Afanas'ev *et al.*, 1983). The explanation suggested is confirmed by the absence of temperature shifts of luminescence lines, since the electron-hole excitation localized at an ion is almost insensitive to the changes occurring in the crystal lattice with the temperature rise (Afanas'ev *et al.*, 1983).

At the same time, lines of the luminescence spectrum in the range 2.95–3.24 eV were attributed to impurity-band transitions (Akopyan *et al.*, 1983, 1984). The direct exciton transition in γ-Ag_4MI_5 (M = Rb, K, NH_4) corresponded to the reflection spectrum in the range 3.2–3.4 eV. The authors believe that the absence of luminescence in the exciton range is explained by a high impurity content and by the radiationless recombination of carriers on crystal defects.

The transition of Ag_4MI_5 crystals to the superionic state (disordered phase) is accompanied by essential changes in the halfwidth, spectral absorption, and intensity of the exciton reflection bands. Similar effects were also observed (Findley *et al.*, 1983) in fluoride conducting PbF_2, which experiences the diffuse phase transition to the superionic state. Figure 33 shows the temperature dependences of the exciton-band halfwidth in Ag_4RbI_5 and PbF_2.

The complex character of the above discussed phenomena and the insufficient experimental data in this field permit only qualitative explanation of the observed phenomena. It is believed that the exciton band broadening during the superionic transition may be caused by the spreading of the valence band edge, which is attributed to the fluctuations in Ag–I bond lengths [Akopyan *et al.* (1984)], and by the screening of electron-hole interaction by mobile ions. The frequency changes observed for the exciton structure in the reflection spectra during the phase transition seem to be

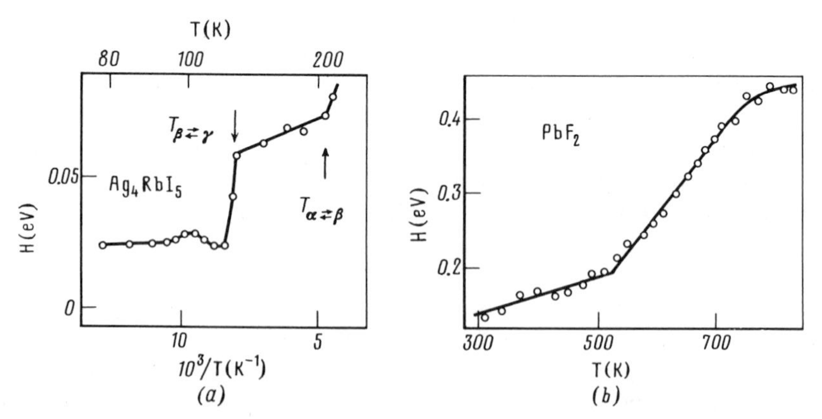

FIG. 33. Exciton linewidth, H, versus temperature for different SIMs: (a) Ag_4RbI_5 [from Akopyan *et al.* (1984)], (b) PbF_2 [from Findley *et al.* (1983)].

caused by the change of the exciton in the binding energy with the band gap variation.

Analogous features of optical and luminescence spectra were also observed for copper-conductive SIMs—$RbCu_4Cl_3I_2$ (Afanas'ev *et al.*, 1984) and Cu_6PS_5X (X = I, Br) (Studenyakin *et al.*, 1984).

Electronic optical transitions may have a significant effect on the behavior of ionic carriers in a SIM. Thus, according to Kristofel (1985a), photoelectrons due to illumination may be located at the cations of the disordered sublattice. As a result, the cations are transformed into neutral atoms and, therefore, do not participate any more in ionic current. Moreover, such a localization affects the interaction between the cations and surrounding particles. In other words, photoexcitation of the electronic subsystem in a superionic conductor may, firstly, change ionic conductivity and, secondly, influence the interactions in the ionic subsystem, thus changing the mechanism of the superionic transition.

Concluding this section, we should like to dwell on a peculiar mechanism of ionic-carrier generation associated with photoexcitation of electrons. As has been shown in a series of special studies, the defect formation in dielectric ionic crystals under the action of intense beams of neutrons and γ-rays is similar to the defect formation caused by ultraviolet irradiation. Lushchik *et al.* (1977), Hersch (1971) and Sibley and Poolex (1974) have assumed that, in both cases, the defect formation has the same mechanism—the decay of electronic excitations gives rise to structural Frenkel defects (for electron and proton beams, the main mechanism of the point-defect formation consists of direct collisions of beam particles and the nuclei of atoms forming the crystal).

Further investigations have completely confirmed this assumption and shown that the efficiency of the conversion of excitons into point defects depends on the excitons' origin. In particular, the efficiency of this process is maximal for direct optical generation of excitons.

Different mechanisms of defect formation upon the exciton decay have been suggested. Without going into details (see, e.g., Lushchik *et al.*, 1977, or Kristofel, 1985b), we should like to consider one of the most important mechanisms of defect formation associated with the decay of autolocalized excitons.

An autolocalized exciton may be regarded as an electrically neutral, excited quasi-molecule. Structural defects appear as a result of the dissociation of such a molecule. For conventional di- and multi-atomic molecules, such processes have been studied in detail. Figure 34 show the adiabatic potentials (terms) of a diatomic molecule as a function of the distance between its nuclei. The process of photoexcitation ($1 \rightarrow 2$ transition) is accompanied by the formation of a state with substantial excess of vibrational energy.

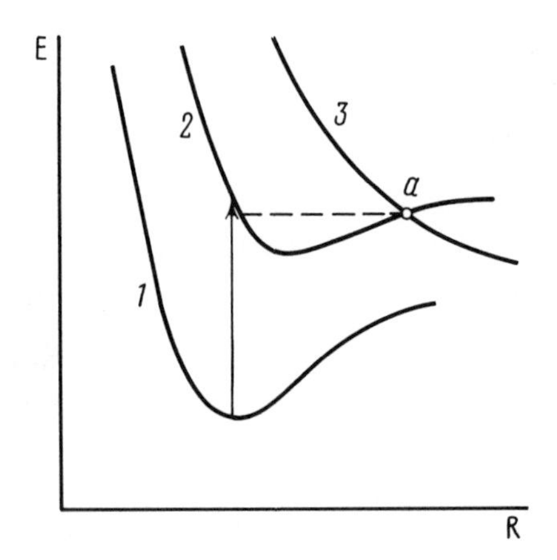

FIG. 34. Schematic diagram of diatomic molecule terms as a function of a internuclear distance: (1) Initial term, (2) Activated-molecule term, (3) Dissociated-molecule term. Terms 2 and 3 cross at point "a."

If a molecule is in state 3 having no minimum, then, during vibrational relaxation, the radiationless transitions $2 \rightarrow 3$ are possible in the vicinity of point *a*, and the molecule may dissociate into atoms (or larger fragments).

A diatomic photoexcited molecule "immersed" in a crystal lattice is, of course, a more complicated subject, but, in this case as well, there is sufficiently high probability of molecule dissociation.

Such transitions have been studied in detail for haloid crystals. The decay of di-haloid excitons in these compounds may be accompanied by the transition of a neutral halogen atom into an interstitial, with the formation of a vacant anion site V^+ and a bound electron. Thus-obtained, the $[V^+ + e^-]$ pair is called the color- or F-center. Similar processes seem to also be possible in superionic materials.

We have not mentioned here numerous works on studies of SIMs by infrared and Raman spectroscopies. These works have been reviewed by Delaney and Ushioda (1979).

VII. Selected Aspects of Electronic Phenomena in Superionic Materials

This part deals with some trends in the studies of semiconductor properties of SIMs that, for some reason or another, have not so far been widely developed, although they are of significant scientific interest. Such methods and the major results obtained in the corresponding studies are considered.

17. Ionic-Electronic Cooperative Disordering

In considering the specific features of superionic conductivity in SIMs (Part 1), we mentioned that there is a series of SIMs in which the state of high-ionic conductivity is reached in a stepwise manner at a certain temperature and is explained by a first-order phase transition. It is of interest to follow the relation between the electronic and ionic subsystems under such conditions. In particular, the electronic-ionic interaction should necessary result in a jumpwise change in the concentration of electronic carriers at the temperature of the phase-transition in the ionic subsystem. And, vice versa, the electronic subsystem may produce an essential effect on the critical conditions of abrupt disordering in the ionic subsystem.

The first attempts at qualitative description of the effect of ionic disordering on the concentration of electronic carriers in superionic materials were made by Ramasesha (1982). Later, Gurevich and Kharkats (1985), using the earlier-developed thermodynamical approach (Gurevish and Kharkats, 1978, 1986d), carried out a thorough analysis of the mutual influence of the electronic and ionic subsystems in the process of temperature disordering.

The behavior of a disordering ionic subsystem in the state of thermodynamic equilibrium is characterized by the free energy

$$\mathbf{F}_{\mathrm{i}} = \mathbf{F}_{\mathrm{i}}^{(\mathrm{o})} + \mathbf{E}_{\mathrm{i}}(n_{\mathrm{i}}) - TS, \tag{175}$$

where $\mathbf{F}_{\mathrm{i}}^{(\mathrm{o})}$ is the free energy in the ordered state, $\mathbf{E}_{\mathrm{i}}(n_{\mathrm{i}})$ is the energy of the subsystem of disordered ions, which depends on their concentration n_{i}, and S is the entropy due to disorder. Quantity $\mathbf{E}_{\mathrm{i}}$, with allowance for collective interactions in the ionic subsystem, may be written in the form

$$\mathbf{E}_{\mathrm{i}} = wn_{\mathrm{i}} - \frac{\lambda_{\mathrm{i}} n_{\mathrm{i}}^2}{2N_{\mathrm{i}}}. \tag{176}$$

Here, $w > 0$ is the energy required to remove an ion from a lattice to an interstitial site, in the absence of interaction between point defects (i.e., between ions at interstitials and the vacancies they have left); λ_{i} is a phenomenological constant, the sign and the value of which are determined by the sum of interactions (possibly indirect ones) in the defect ensemble in the crystal; $\lambda_{\mathrm{i}} > 0$ corresponds to the effective attraction and $\lambda_{\mathrm{i}} < 0$ to the effective repulsion; and N_{i} is the concentration of the initial sites.

Proceeding from Eqs. (175) and (176), we may show that function $n_{\mathrm{i}}(T)$ is ambiguous in a certain range of parameters. The necessary condition for such ambiguity is $\lambda_{\mathrm{i}} > 0$. In the ambiguity range of $n_{\mathrm{i}}(T)$, a jumpwise transition from one branch of the solution to another occurs under certain conditions and at temperature $T = T_{\mathrm{tr}}$. In other words, a first-order phase

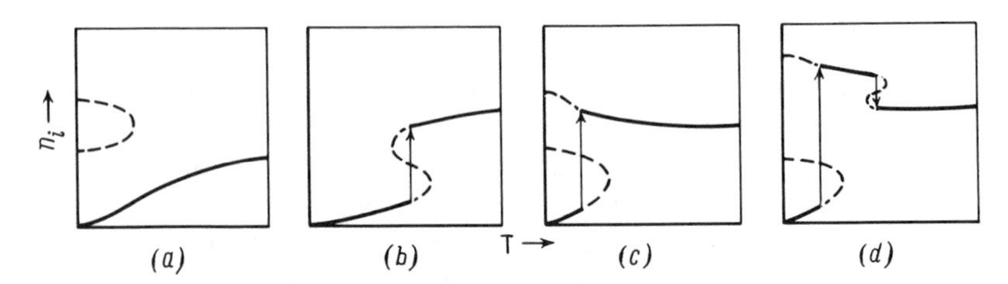

FIG. 35. Schematic $n_i(T)$ dependence for different cases. Solid lines are the physically-allowed solutions. Arrows indicate the transitions between different branches of the solutions. [From Gurevich and Kharkats (1985).]

transition occurs in the ionic subsystem. As a result, the concentration of mobile ions in interstitial positions changes in a jumpwise manner.

Figure 35 shows typical temperature curves of concentration n_i and possible schemes of temperature phase transitions under conditions when the effect of electrons on ionic disordering may be neglected. The curves (a–d) are not equivalent, since the parameters of the ionic subsystems are different (for details see Gurevich and Kharkats, 1985).

The electronic subsystem is characterized by free energy $\mathbf{F}_e$ which may be represented in a form similar to Eq. (175), but with parameter w acquiring here a different physical meaning. The main idea of the cited works was to express the free energy $\mathbf{F}_{ie}$ of an electron-ion ensemble is the form

$$\mathbf{F}_{ie} = \mathbf{F}_i + \mathbf{F}_e - \frac{\lambda_{ie} n_e n_i}{N_i}. \tag{177}$$

The last term in Eq. (177) is proportional to the concentrations of "quasi-free" ions, n_i, and electrons, n_e, and takes into account the cross interaction between the ionic and electronic subsystems. Quantity λ_{ie} in Eq. (177) is the phenomenological constant characterizing the energy of such an interaction (attraction if $\lambda_{ie} > 0$). Independent of the sign of the resulting interaction, a jumpwise change of the particle concentration in one subsystems is necessarily followed by a jumpwise change of the concentration in the other subsystem.

If inequality $n_e \ll n_i$ is valid, then, despite the small effect of the electronic subsystem on the ionic one, the reverse effect may be substantial. The temperature curves of electronic concentration $n_e(T)$ for such a case are given in Fig. 36, four of which correspond to $\lambda_{ie} > 0$ and four to $\lambda_{ie} < 0$ (with four variants of $n_i(T)$ in Fig. 35). Depending on the sign of λ_{ie}, the jumps in n_e may be either symbatic (when $\lambda_{ie} > 0$) or antibatic (when $\lambda_{ie} < 0$) to the jumps in n_i.

In the case of a strong superionic transition in a material that is an intrinsic semiconductor ($n_e^0 = n_h^0 = n_{int}$), the concentration ratio at the jump

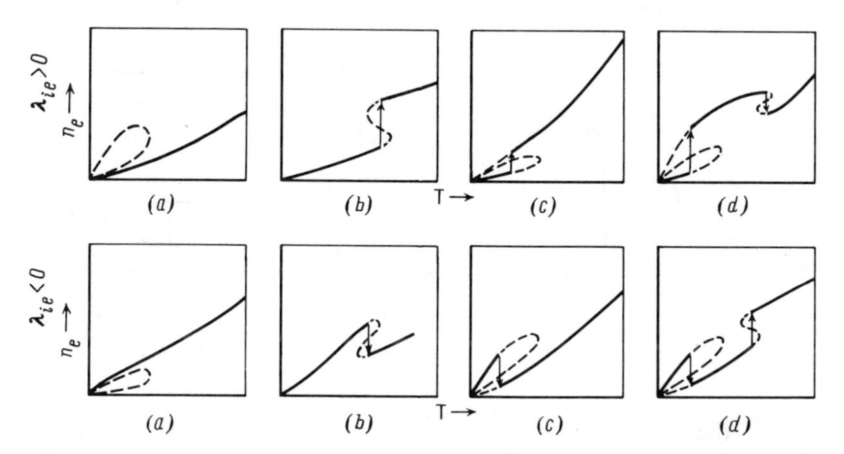

FIG. 36. Schematic $n_e(T)$ dependence for different cases. Solid lines are the physically-allowed solutions. Arrows indicate the transitions between different branches of the solutions. [From Gurevich and Kharkats (1985).]

temperature $T = T_{tr}$ is

$$\frac{n_e^l(T_{tr})}{n_e^r(T_{tr})} = \exp\left(\frac{\lambda_{ie}}{kT}\right). \tag{178}$$

Here, subscripts l and r characterize the left and right branches of the $n_e(T)$ curve, between which the jumpwise transition occurs. It is seen from Eq. (178) that the direction of the jump depends on the sign of λ_{ie}, and its magnitude becomes noticeable even for relatively small (of the order of kT) value of electronic-ionic interaction.

Consider the behavior of concentration $n_e(T)$ in a doped conductor under conditions where the temperature of the superionic transition T_{tr} differs from temperature T_e characterizing the boundary of complete ionization of impurities (donors, for example).

As follows from Eq. (7a,b), T_e is determined by the condition $n_1 = N_d$, or $kT_e = (E_c - E_d)/\ln(N_c/N_d)$. Let $T_{tr} > T_e$, so that for $T = T_{tr}$, we have $n_e \simeq N_d$. Then, if $\lambda_{ie} > 0$, n_e remains almost constant, whereas n_i changes in a jumpwise manner (electrons are "insensitive" to restructuring of the ionic subsystem). On the other hand, if $\lambda_{ie} < 0$, concentration n_e drops in a stepwise manner at $T = T_{tr}$, and then, with the further rise in T, it gradually approaches the value of N_d. This is illustrated by Fig. 37a, topologically equivalent to Fig. 36b ($\lambda_{ie} < 0$) of the general scheme.

Now, let $T_{tr} < T_e$. Then, if $\lambda_{ie} < 0$, a jump in n_i does not affect the value of n_e, which remains negligibly small. On the contrary, if $\lambda_{ie} > 0$, then at $T = T_{tr}$, n_e may reach, in a jumpwise manner, a value of N_d (see Fig. 37b topologically equivalent to Fig. 36b for $\lambda_{ie} > 0$).

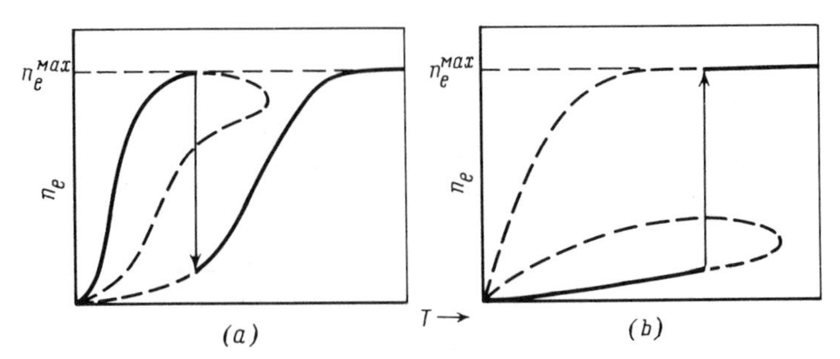

FIG. 37. Possible temperature curves $n_e(T)$ for doped semiconductors: (a) $-\lambda_{ie} < 0$, $T_{tr} > T_e$; (b) $-\lambda_{ie} > 0$, $T_{tr} < T_e$. [From Gurevich and Kharkats (1985).]

Thus, the behavior of $n_e(T)$ at $T = T_{tr}$ may be interpreted as an effective jumpwise change of the band gap in intrinsic semiconductors or the location depths of donor-acceptor levels in doped semiconductors. In general, this change is associated with a partial disturbance of the periodic structure of a superionic material (see Section 1).

The effect of the electronic subsystem on ionic disordering may be interpreted as the renormalization of the energy of formation for an isolated point defect in the ionic subsystem. Such a renormalization results in the change of the temperature T_{tr} of jumpwise ionic disordering, the direction of the temperature change being determined by the sign of the cross interaction energy (T_{tr} decreases if $\lambda_{ie} > 0$). It is natural to expect that the sign of λ_{ie} will be different for n- and p-type materials since the signs of electronic carriers in such materials are different. Thus, by changing the dopant type, it is possible to control, within a certain range, the temperature of the superionic transition.

18. Generation of Space-Charge Peak in Mixed Ionic-Electronic Conductors

It is interesting to consider the density distribution of the space charge in ionic-electronic conductors with a flowing electric current. It turns out that, in distinction from, e.g., pure ionic liquid or solid conductors or pure n- or p-type electronic conductors, the electric current in the mixed conductors may give rise to unexpected effects in the distribution of the space-charge density (Kharkats, 1984; Gurevich and Kharkats, 1987).

As has already been noted in Section 6, the calculation of migration-diffuse currents is carried out on the assumption of the local electroneutrality of the system beyond the limits of a double electric layer [Eq. (40)]. The validity of such an approximation is determined by the small value of the ratio of the Debye length L_D to the system size L.

If no current is flowing, we may obtain the distributions of the concentrations n_k of the charged components and of electric potential from the self-consistent solution of the Poisson equation (36). Under these conditions, the space charge density $\rho(x)$ decreases with the distance from the electrode according to the exponential law: $\rho(x) \sim \exp(-x/L_D)$ for $x \gtrsim L_D$. Current flowing may "extend" the region of deviation from electroneutrality.

This problem has been studied in detail for systems with two types of carriers having the opposite signs, e.g., for the solution of a binary electrolyte (Levich, 1959). This calculation reduces to the following: Using, in the zeroth approximation, the equations for electroneutrality, (40) [instead of Eq. (36)] and Eq. (33), we arrive at the set of equations (for simplicity, we take that $z_1 = -z_2 = 1$),

$$D_1\left(\frac{dn_1}{dx} + n_1\frac{d\psi}{dx}\right) = -\frac{i_1}{e},$$

$$D_2\left(\frac{dn_2}{dx} - n_2\frac{d\psi}{dx}\right) = 0, \tag{179}$$

$$n_1 = n_2.$$

Here, $\psi = e\varphi/kT$ is the dimensionless electrical potential, and subscripts 1 and 2 relate to cations and anions, respectively.

Equations (179) may readily be integrated. As a result, we obtain the potential distribution $\psi(x)$. Substituting the function $\psi(x)$ into the Poisson equation in the form $\rho = [kT\varepsilon_o\varepsilon/e]d^2\psi/dx^2$, we arrive at the unknown distribution of charge density, $\rho = e(n_1 - n_2)$. In particular, on the assumption that the values of concentrations ($n_1 = n_2 = n_o$) and potential ($\psi = 0$) are known at $x = L$, we obtain, from the second equation (179), that $n_2 = n_o\exp(\psi)$, and, from the third equation, that $n_1 = n_2 = n_o\exp(\psi)$. Substitution of this value (n_1) into the first equation yields

$$2D_1 n_o \exp(\psi)\frac{d\psi}{dx} = -\frac{i_1}{e}. \tag{180}$$

Integrating Eq. (180), we obtain $\psi(x)$ and then $\rho(x)$. Finally,

$$\rho(x) = 2en_o\left(\frac{L_D}{L}\right)^2\left(\frac{i}{2i_D}\right)^2\left[1 + \frac{i}{2i_D}\left(\frac{x}{L} - 1\right)\right]^{-2}, \tag{181}$$

where $i_D \equiv en_oD_1/L$. Quantity $2i_D$ has the meaning of the maximum possible (limiting) stationary current in a binary monovalent electrolyte. [In solutions with the excessive foreign ions not discharging at the electrode, the limiting current coincides with i_D.]

Note that the disturbance of electroneutrality is of the order of $\Delta = (L_D/L)^2$. This justifies the use of the calculation scheme according to which $\rho(x)$ is obtained by differentiation of function $\psi(x)$ which is determined on the assumption that $n_1 = n_2$. As is seen from Eq. (181), in distinction from the exponential decrease of $\rho(x)$—typical for the equilibrium ion distribution, in the case of flowing current $\tilde{i} \equiv i/i_D$, value $\rho(x)$ decreases rather slowly (Fig. 38a).

Unlike binary solutions, in systems with one mobile species and fixed charges of the opposite signs (e.g., in the doped n- or p-type semiconductors or superionic conductors with pure ionic unipolar conduction), the dependence $\rho(x)$ is of somewhat different character; the flowing current does not change the exponential character of the charge-density decrease with the distance from the interface towards the crystal bulk.

When the current flowing in the systems contains, along with stationary charges, mobile carriers of different signs, the situation drastically changes. It turns out that the density distribution $\rho(x)$ of the space charge may form a peak over a wide range of parameters characterizing the system. The position, height and shape of the peak depend on the applied current (Kharkats, 1984). [Note also that, in contrast to binary systems, the value of i is not limited here by the value of $2i_D$.]

For example, consider a material in which mobile carriers are cations and electrons (denoted by subscripts 1 and 2) and that has "background" of fixed

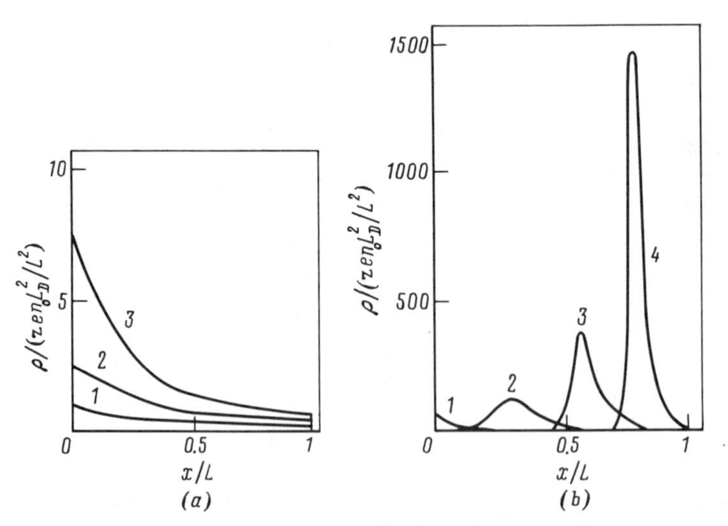

FIG. 38. Current-induced charge density distribution: (a) A binary system: 1 – $\tilde{i} = 1$; 2 – 1.25; 3 – 1.5. (b) A system with fixed charges, $\gamma = 0.9$: 1 – $\tilde{i} = 2$; 2–3; 3–5; 4–10. [From Kharkats (1984).]

charges (created, in this case, by the anion sublattice and the ionized donors of electrons). Assuming, as earlier, that the electrode is blocked, we arrive at the following set of transport equations:

$$D_1\left(\frac{dn_1}{dx} + n_1\frac{d\psi}{dx}\right) = -\frac{i_1}{e},$$

$$D_2\left(\frac{dn_2}{dx} - n_2\frac{d\psi}{dx}\right) = 0, \tag{182}$$

$$n_1 - n_2 = n_3,$$

where n_3 is the concentration of fixed charges. Assuming that n_3 is constant, and integrating Eqs. (182) with the boundary conditions $n_1(L) = n_0$, $n_2(L) = \gamma n_0$ and $\psi(L) = 0$ (then $n_3 = \text{const} = (1 - \gamma)n_o$), we have

$$\rho = 2en_o\left(\frac{L_D}{L}\right)^2\left(\frac{i}{i_D}\right)^2 \frac{2\gamma\exp(\psi)}{[2\gamma\exp(\psi) + 1 - \gamma]^3}, \tag{183}$$

$$2\gamma[\exp(\psi) - 1] + \psi(1 - \gamma) = \frac{i}{i_D}\left(\frac{x}{L} - 1\right). \tag{184}$$

Equations (183) and (184) describe the coordinate distribution of the space charge density. The parameter $\gamma = (0 \leq \gamma \leq 1)$ characterizes the relative contribution from the background of fixed charges. In the $\gamma = 1$ limit, this background is absent; in the $\gamma = 0$ limit, there are no mobile anions, or, in other words, the material is a unipolar cation conductor.

Figure 38b shows the $\rho(x)$ curves calculated by Eqs. (183) and (184) for different values of dimensionless current $\tilde{i} = i/i_D$. For relatively small current values, the function $\rho(x)$ monotonically decreases with x—a decrease that is qualitatively similar to its behavior in binary solutions. But with an increase in the current, the maximum of $\rho(x)$ starts shifting towards larger x values. The peak height increases, and its width decreases with the current. The "deformation" of the $\rho(x)$ curve occurs in such a way that the integral charge (the area under the curve) also increases with the current, reaching the following asymptotic form for $i \gg i_D$:

$$Q_\rho = 4en_oL\left(\frac{L_D}{L}\right)^2\left(\frac{i}{i_D}\right)^2 \frac{\gamma}{1 - \gamma^2}. \tag{185}$$

The potential corresponding to the maximum of the $\rho(x)$ distribution depends on γ alone and is given by the relationship $\psi_{max} = \ln[(1 - \gamma)/4\gamma]$. Substituting ψ_{max} into Eq. (183), we may readily determine the value of ρ_{max}.

Note also that the space-charge peak due to the current is essentially asymmetric—its right-hand wing decreases rather smoothly, as in binary systems, whereas the left-hand wing decreases in a more drastic way, as in unipolar conductors.

This situation where the concentration of fixed charges in a SIM is constant has been analyzed. At the same time, n_3 may depend on the conditions of the sample formation (especially for films), sample history, and other factors, and, in general, n_3 is a function of coordinate x. Additional details of the $\rho(x)$ distribution, in the case in which n_3 is a linear function of x, have been considered by Gurevich and Kharkats (1987). As discussed earlier, $\rho(x)$ is proportional to the parameter $(L_D/L)^2$, but depends on current i in a more complicated way. Now, $\rho(x)$ also depends on the sign of current i and parameters characterizing the $n_3(x)$ distribution.

The most interesting result is that current-induced space-charge density $\rho(x)$ may change its sign inside the sample. At sufficiently large currents, $\rho(x)$ shows a maximum in the region of positive $\rho(x)$ values. In the region of negative $\rho(x)$ values, $\rho(x) < 0$, the function has a relatively flat minimum.

19. Transition to Metallic-Type Conductivity in Mixed Ionic-Electronic Conductors

As indicated in Section 10, SIMs with large deviations from stoichiometry often have comparable values of σ_i and σ_{el} and may be regarded as mixed ionic-electronic conductors or as superionic materials with mixed conductivity (MCSIMs). Using coulometric titration, it is possible to determine a number of characteristic parameters for the electronic subsystem of such materials, e.g., electronic carrier concentration or the position of the Fermi level. The possibilities of the coulometric titration technique are illustrated by Fig. 39, which shows the dependence of the Fermi level F on δ for silver sulphide. It is seen that degeneracy occurs for values of $\delta \gtrsim 5 \times 10^{-4}$, so that F is located, similar to metals, in the conduction band ($F > E_c$).

Note also that the characteristic feature of semiconductors is an increase of electronic conductivity with temperature.. For metals, the situation is reverse—the temperature rise is accompanied by a decrease in conductivity. Thus, by studying function $\sigma_{el}(T)$ over a wide range of temperatures, it is possible to determine the character of electronic conductivity in the compound under consideration. Detailed measurements of σ_e were carried out by Shukla and Schmalzried (1979) on silver sulphide. Figure 40 shows the temperature dependence of electronic conductivity σ_e. Let us dwell on the results of σ_e measurements in the superionic α-phase of silver sulphide.

It has been found that for a sample with $\delta < 1.5 \times 10^{-4}$, conductivity is of the semiconductor nature, so that the temperature coefficient of σ_e is

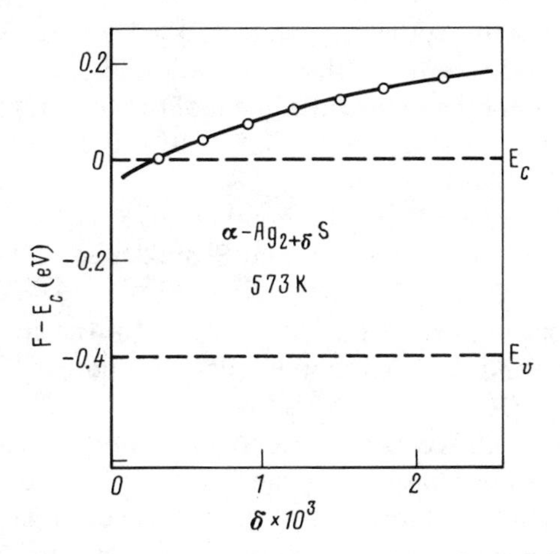

FIG. 39. Fermi level, F, in $Ag_{2+\delta}S$ as a function of the deviation from the ideal stoichiometry. [From Rickert *et al.* (1975).]

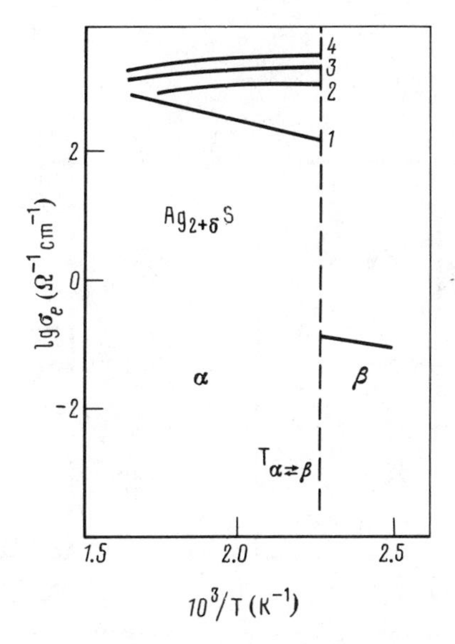

FIG. 40. Electronic conductivity, σ_e, versus reciprocal temperature for various stoichiometric compositions of silver sulphide, $Ag_{2+\delta}S$. $\delta \times 10^3$: 1 – 0.063; 2 – 0.69; 3 – 1.45; 4 – 2.16. $T_{\alpha\to\beta}$ is the temperature of $\alpha \to \beta$ phase transition in Ag_2S. [From Shukla and Schmalzried (1979).]

positive ($d\sigma_e/dT > 0$). At the same time, if $\delta > 1.5 \times 10^{-4}$ [or $\delta > 5 \times 10^{-4}$, according to Junod *et al.* (1977)], metallization of electronic conductivity is observed, corresponding to the fulfillment of the condition $d\sigma_e/dT < 0$. The latter inequality is valid up to $\delta \approx 2.5 \times 10^{-3}$ (i.e., for silver sulphide in equilibrium with silver).

Similar results were obtained for other silver chalcogenides (Miyatani, 1973; Junod *et al.*, 1977; Oehsen and Schmalzried, 1981; Mostafa *et al.*, 1982; Sohége and Funke, 1984).

Note one more important fact associated with the study of temperature behavior of ionic and electronic conductivity components for the compounds of the Ag_2X type (X = S, Se). As is known (see, e.g., the review by Funke, 1976), these compounds experience a polymorphous temperature transition, the high-temperature phase being characterized by structural disorder (see Introduction). This polymorphous transition is accompanied by a considerable jump in ionic conductivity with rise in temperature, but σ_i only weakly depends on sample nonstoichiometry (Fig. 41). Electronic conductivity in low-temperature β-phase is of a semiconductor nature, and in the superionic α-phase is either of semiconductor or metallic character, depending on δ. [The hatched regions in Fig. 41 show the limits of conductivity variations in the samples of different compositions determined by the limits of δ-variation.]

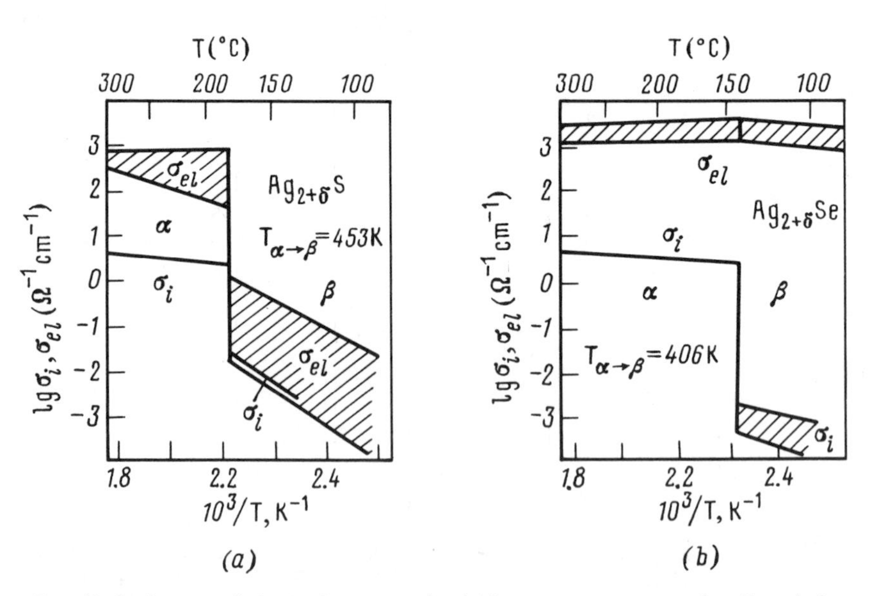

FIG. 41. Ionic, σ_i, and electronic, σ_{el}, conductivities versus temperature for silver chalcogenides. Dashed regions correspond to values of σ_i and σ_{el} for different δ.

The transition from nonmetallic to metallic conductivity may be interpreted in terms of the mechanism of ionic disordering (in particular, of silver cations in the α-phase of Ag_2X) (Junod *et al.*, 1977). Specifically, the distances between disordered ions in the structure and the occupancy of Ag^+-positions may vary, a variation which is qualitatively equivalent to the formation of the impurity band due to randomly distributed donor levels.

Such an interpretation is similar to that used in the description of the metallization of electronic conductivity in semiconductors due to a rise in impurity concentration. The mechanism of a drastic increase in the electronic conductivity of a material (dielectric-metal transition) may be understood by using the model theories of disorder (Ziman, 1979). If there is no noticeable chemical trend of impurity segregation or of cluster formation during crystallization, then impurities are distributed randomly and, on the average, uniformly over the whole sample bulk. Let the mean concentration of impurities be n_{im}; then the characteristic length r_s corresponding to such a concentration may be determined by the equation $n_{im}^{-1} = (4\pi/3)r_s^3$.

Now assume that the electronic wave function corresponding to an impurity that "fills" a sphere of radius r_a around the impurity center, and that the electronic states at different impurities may be considered as independent, unless the corresponding spheres touch one another or overlap. Thus, an electron located in the impurity center may reach only the region of the space that is occupied by a bound cluster of mutually overlapped spheres.

Within such an assembly-of-spheres model, the description of the electronic conductivity is associated with the percolation problem. As to the microscopic properties, the system under consideration behaves as a dielectric until the moment at which the density of the "gas" of spheres becomes so high that an infinite cluster of overlapping spheres is formed. It is inside such a cluster that an electron may move freely. In other words, the attainment of the percolation threshold in the given system manifests itself as a transition (at some critical concentration of impurities) to metallic conductivity.

The analytical solution of this problem has not been found yet, but numerical calculations show that the critical percolation threshold (radius), r_{cr}, is given by the expression

$$r_{cr} = 0.70 r_s. \tag{186}$$

This equation indicates that, at a given concentration n_{im} of the centers, the radius r_a characterizing each sphere should exceed r_{cr}. Equation (186) may be rewritten in the form

$$\eta_{cr} = \left(\frac{r_{cr}}{r_s}\right)^3 \approx 0.35. \tag{187}$$

Taking into account the physical meaning of quantities r_a and r_s, η_{cr} is the minimum volume of the total material bulk that should be occupied by "conducting" impurity spheres in order to provide the transition to metallic conductivity. It is worth noting that η_{cr} is close to the packing density in a regular diamond-type lattice. We should like to emphasize that the above given values were obtained not as a result of an exact solution, but by the way of numerical calculations. Depending on the details of the chosen model and the applied mathematical methods, the numerical coefficient in Eq. (186) may slightly vary, and, therefore, the values obtained for η_{cr} may differ from the above. Specifically, Kirkpatrick (1973) indicates the range of η_{cr} values as 0.25–0.30.

According to Eq. (187), the critical volume necessary for percolation is not too small. Nevertheless, it may readily be reached for the systems under consideration. The point is that the size of the impurity-center r_a, for electrons is, in fact, the effective Bohr radius, which corresponds to the lowest impurity level. As is shown by estimates, this radius is tens, or even hundreds, of times larger than the interatomic distances in the initial materials. Therefore, an impurity concentration of 10^{-4}–10^{-2} at. % should be considered, in our case, as very high and sufficient for the fulfilment of the inequality $r_a > r_{cr}$.

Note also that there are reasons to believe (although the fact has not been proved as yet) that Eq. (187) is valid not only for the assembly-of-spheres model. In more general cases, η_{cr} seems an appropriate description of the critical part of the total volume that should be occupied by a medium possessing different properties in order to provide the formation of a single cluster. These conditions broaden the applicability limits of the described model.

20. Interaction of Ionic and Electronic Currents

In the previous sections, we assumed, when considering ionic and electronic currents in superionic materials, that ionic and electronic carriers move independently—in other words, that each current i_k is due to the gradient of its own electrochemical potential $\tilde{\mu}_k$ [see Eqs. (85), (87)].

At the same time, it is natural to assume that, in mixed ionic-electronic conductors (MCSIMs) with prevalent electronic conductivity (e.g., in silver chalcogenides, Fig. 42) caused by a high electron (hole) concentration (up to 10^{21} cm^{-3}), ions may be dragged by electrons; i.e., the gradient μ_e may affect ionic current i_i.

Miyatani (1981) has considered the problem of two currents (ionic and electronic) flowing through a MCSIM sample with the assumption that the cross terms in the phenomenological transport equations cannot be neglected.

In this case, Eqs. (85a) and (85b) should be rewritten in the form

$$i_i = -\frac{\sigma_{ie}}{e}\frac{\partial\tilde{\mu}_e}{\partial x} - \frac{\sigma_i}{e}\frac{\partial\tilde{\mu}_i}{\partial x}, \tag{188}$$

$$i_e = \frac{\sigma_e}{e}\frac{\partial\mu_e}{\partial x} - \frac{\sigma_{ei}}{e}\frac{\partial\mu_i}{\partial x}. \tag{189}$$

Here, σ_{ie} and σ_{ei} are some phenomenological coefficients characterizing the cross conductivities. In virtue of the Onsager reciprocal relation, according to which the cross transference coefficients should be equal (see, e.g., De Groot and Mazur, 1969), the following condition should be fulfilled: $\sigma_{ie} = \sigma_{ei}$.

Let only electronic current be flowing through a sample under the stationary conditions (i.e., only the so-called electronic electrodes are applied—see Section 13). Using ionic and electronic probes, it is possible to measure the potential differences V_i and V_e between any points of the sample. These quantities are determined by Eqs. (131) and (132). According to Eq. (188), if $i_i = 0$, then

$$\sigma_{ie} = \frac{V_i}{V_e}\sigma_i. \tag{190}$$

If only ionic current is transmitted through the sample ($i_e = 0$, if the ionic electrodes are appropriately chosen), then we may readily obtain, from Eq. (188), that

$$\sigma_{ei} = \frac{V_e}{V_i}\sigma_e. \tag{191}$$

Thus, measuring the voltage drop in a sample with appropriately chosen electrodes and potential probes, we may check the validity of Eqs. (188) and (189).

Such experimental studies were carried out by Miyatani (1981) for silver selenide and copper sulfide. Figure 42 shows conductivity measured for α-$Ag_{2+\delta}Se$ samples with varying composition. First of all, the sample composition, as was to be expected, only slightly affects ionic conductivity, whereas σ_e changes by more than a factor of three.

But the main result here is, of course, the establishment of cross interaction, which manifests itself in a nonzero value of cross conductivity, $\sigma_{ie} = \sigma_{ei}$. Specifically, it was obtained that $\sigma_{ie}/\sigma_i \approx 10^{-2}$–$10^{-3}$.

If the electronic component of the current noticeably affects the ionic component, it may be expected that the value of the measured effective charge of ions in a MCSIM would change, due to transmission of strong electronic current through the sample. Attempts at experimental detection of

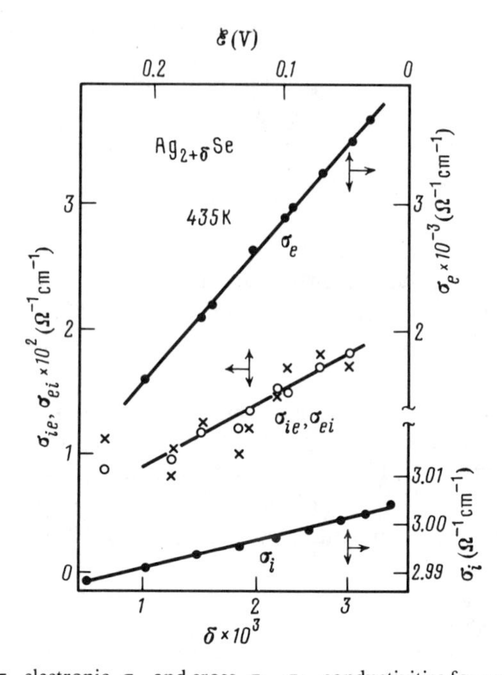

FIG. 42. Ionic, σ_i, electronic, σ_e, and cross, σ_{ie}, σ_{ei}, conductivities for various stoichiometric compositions of silver selenide, $Ag_{+\delta}Se$. O—σ_{ie}; ×—σ_{ei}. [From Miyatani (1981).]

this effect were made by Fomenkov (1982). Fomenkov studied copper chalogenides $Cu_{2-\delta}X$ (X = S, Se) and ascertained the effective copper charge to be equal to $(0.97 \pm 0.04) \cdot e$, which, the authors believe, indicates the independent character of ion and electron motion in $Cu_{2-\delta}X$.

At the same time, the accuracy of the effective charge determination in the above-mentioned experiments was relatively low (~5%); therefore, in the case of a weak drag of the different species on one another, the expected effect cannot be detected. This assumption is favored by the results obtained by Miyatani (1981).

Thus, the possible (in principle) effect of the cross interaction turns out to be rather small ($\sigma_{ie}/\sigma_i \ll 1$). The obtained results give grounds to believe that the conventional approach, in which the cross interaction is neglected, is well justified.

VIII. Apparatus and Devices Based on Superionic Materials

Unusual physical and chemical properties of superionic materials make them very promising for use in various applied fields and, especially, in electronics.

Modern microelectronics is characterized by an increasing sophistication and a miniaturization of apparatus, characteristics closely related to the widening scope of problems to be solved in electronics. Therefore, the interest in devices based on nontraditional physical principles is quite understandable.

Electrochemical devices (ECD) are the devices whose functioning is based on principles of electrochemistry. They are intended, first and foremost, for storage, conversion, and transmission of information. Sometimes, devices for conversion and storage of electrical energy are also considered as ECDs, and these have been considered in detail by Bagotskii and Skundin (1981). Since ECDs are the constituent part of electronic blocks, their electrical and design parameters should be consistent with other elements of such moduli. It is quite understandable that a substantial reduction in the size an weight of ECDs is possible only with new technological solutions and materials. Difficulties arising due to the use of liquid electrolytes may be reduced or even completely eliminated by designing new all-solid state systems.

Now we can discuss a new, promising branch of electronic instrument manufacturing—microionics. Microionics does not oppose microelectronics; on the contrary, it focusses its attention on seldom-used phenomena of solid-state electronics associated with the ionic transport.

The main functional element of all SIM-based devices is an electrochemical cell consisting of a system of electrodes and a SIM. Functioning of such a cell is provided mainly by the processes at the electrode-SIM interfaces (heterojunctions). Two main types of heterojunctions are used: completely or partly reversible and blocking ones. Depending on the value of the electronic component of conductivity in a SIM, the properties of heterojunctions may be essentially different, providing the design of various functional elements.

Below, we shall consider devices based on SIMs with noticeable electronic conductivity and devices in which electronic conductivity may deteriorate operational parameters.

21. Integrators

A special place among superionic-based devices is held by integrators or coulometers (Kennedy and Chen, 1969; Kennedy, 1972; Kennedy *et al.*, 1973) of a very simple design.

A discrete integrator may be an A-type cell (see Section 8),

$$M \,|\, \mathrm{SIM} \,|\, B, \qquad \text{(D)}$$

consisting of reversible (usually metallic) (M) and blocking (B) electrodes divided by a SIM layer with mobile cations M^+. Blocking electrode B, which is also called an indicator electrode, possesses electronic conductivity.

Such a discrete integrator operates in two modes—charging (recording) and discharging (reading). Its functioning is based on a reversible redox reaction of the dissolution (deposition) of metal occurring at the SIM/electrode interface. During integrator charging, i.e., when the current is transmitted through the cell and the reversible electrode acts as a source of ions and is dissolved, the substance mass is transported from the reversible electrode and deposited on the inert one. In the discharge mode, the direction of the current changes, and the substance deposited on the indicator electrode is dissolved in the SIM.

In the charging process, the integrator-coulometer is a system with two reversible electrodes. The voltage drop on a cell is determined only by the ohmic voltage drop in the SIM bulk and the sum of the polarization voltages on the heterogeneous juctions. During discharge, when the so-called read-out current is transmitted through the cell, the indicator electrode, upon the removal of the deposited mass from it, becomes a blocking electrode, and the voltage drop on the cell drastically increases. This jump in voltage indicates the completion of the reading process. It may also be used for activating the corresponding functional circuit of, e.g., the signal device.

As a rule, the read-out current is switched off automatically upon the attainment in the cell of a definite cut off voltage V_{co}, given *a priori*, which is smaller than the decomposition voltage of the SIM.

The schematic time dependence of the voltage at the cell during integration is shown in Fig. 43. In the ideal case, i.e., for proper functioning of a cell, the following relationship should be fulfilled:

$$i_{ch} \cdot t_{ch} = i_{dis} \cdot t_{dis} = Q = \text{const},$$

where i_{ch}, t_{ch} are the current and time of cell charging, and i_{dis}, t_{dis} are the current and time of its discharging.

Now it is clear what requirements should be met by a SIM used in an integrator: it should possess a high ionic conductivity in order to decrease ohmic losses, and it should have the minimum electronic conductivity, which

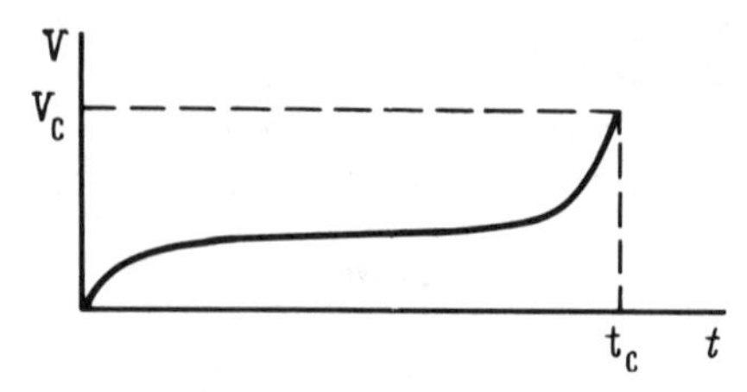

FIG. 43. Schematic time dependence of voltage for a discrete integrator during dc reading. V_c and t_c are operation voltage and time, respectively.

introduces additional errors into the element operation. The requirement of small σ_{el} is of special importance if a small charge is integrated.

Depending on the properties of the materials used and on the design, discrete devices may be used for integrating currents in the range from 10^{-6} to 10^{-2} A, within time intervals from fractions of a second to several years (Kennedy, 1972; Treier, 1978). In addition, integrators with discrete reading of information may find applications as timers in delay circuits, indicators of maintenance time, etc.

The use of these devices as memory elements is very advantageous, since they require no additional energy for the storage of the recorded information.

In addition to discrete integrators, in which the information is output only upon the completion of recording or reading, it is possible to create functional SIM-based elements providing the output of information at any given moment of time. Such devices received the name of analog integrators (Kennedy, 1972; Valverde, 1981). In distinction from discrete integrators, the functioning of analog elements is based on the use of SIMs with mixed ionic-electronic conductivity. An analog integrator is based on a C-type cell (see Section 10):

$$M \mid \mathrm{SIM} \mid \mathrm{MCSIM} \mid B. \qquad \text{(C)}$$

Consider the processes occurring in such a cell due to the application of an electric field. Depending on the polarity of the applied voltage, either the ions from the MCSIM are extracted into the SIM, or they are injected from the SIM into the MCSIM. The change in the electronic carrier concentration in the material with mixed conductivity is accompanied by a change in the electrical potential of the cell (for more detail see Section 10). With the external source switched off, the attained potential difference is stored. When metal cations are injected into the MCSIM, the cell discharges (Fig. 44).

In order to improve the characteristics of an analog integrator, a three-electrode system may be used, having an additional reversible reference

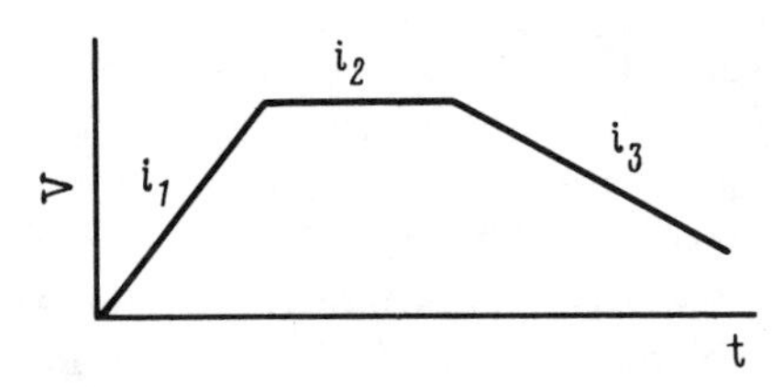

FIG. 44. Schematic the charge-discharge characteristic of analog integrator: i_1 = const; $i_2 = 0$; i_3 = const. (currents i_1 and i_2 have opposite directions).

electrode (2) in comparison with a two-electrode system:

$$\begin{array}{c}(1)\text{-}M \,|\, \text{SIM} \,|\, \text{MCSIM} \,|\, \text{SIM} \,|\, M\text{-}(2).\\ | \\ B\end{array}$$

Information reading is accomplished via measuring the potential difference between *M*-(2) and *B* electrodes, whereas the complex *M*-(1)/SIM electrode is used as a "reservoir" of ions. Such a design permits one to avoid errors in the process of integration that are associated with the polarization of the auxiliary electrode (1). It is important to note that the above-described elements can provide data without erasing the data.

Materials with ionic-electronic conductivity used in the above-described devices are solid solutions based on silver sulphides, selenides and tellurides, e.g., on the binary compounds of compositions $(a\text{Ag}_2\text{Se}) \times (b\text{Ag}_3\text{PO}_4)$ (coefficients a and b satisfy the relationship $a + b = 1$), or ternary compounds of compositions $(a\text{Ag}_2\text{S}) \times (b\text{Ag}_{1.7}\text{Te}) \times (c\Phi)$, where $\Phi = \text{Ag}_3\text{PO}_4$ or $\text{Ag}_4\text{P}_2\text{O}_7$ (and $a + b + c = 1$). Ionic conductivity in such compounds is $\sigma_i \approx 10^{-1}\ \Omega^{-1}\ \text{cm}^{-1}$ and electronic conductivity is $\sigma_{el} \approx 10^2\text{–}10^3\ \Omega^{-1}\ \text{cm}^{-1}$ (Ikeda *et al.*, 1976; Yushina, 1979; Yushina *et al.*, 1979; Ikeda and Tada, 1980).

An example of an analog integrator cell is the following system suggested by Yushina *et al.* (1979):

$$\text{Ag} \,|\, \text{Ag}_4\text{RbI}_5 \,|\, 0.915\text{Ag}_2\text{Se} \times 0.085\text{Ag}_3\text{PO}_4 \,|\, \text{C}$$

The corresponding cell, with a specific capacitance up to 80 F/cm^3, operates in the range from 0 to 0.24 V, the cell voltage being a linear function of charge.

22. Diode-Rectifier

One more very promising functional SIM-based element is an electrochemical diode-rectifier. The design of this element is very simple: if the electrode [for example, (1)-*M*] of a two-electrode cell,

$$(1)\text{-}M \,|\, \text{MCSIM} \,|\, M\text{-}(2),$$

has a much larger area than another one [microelectrode (2)], then the current flowing through the cell is dependent on its direction. Physically, it is explained by the fact that the dissolution and deposition reactions at the electrode are of an "asymmetric" character. If the current is flowing from electrode (1) to (2) (forward direction), then the cell resistance is determined by the ionic-electronic conductivity of the MCSIM and only slightly varies with the applied voltage. For the reverse current, the resistance of the material bulk adjacent to microelectrode (2) drastically increases because of the strong electrode polarization. As a result, the current flowing in the opposite direction is limited by the microelectrode transmittance, i.e., by its area. Thus, the system is characterized by an asymmetric current-voltage curve.

Also, if "backward" voltage exceeds the decomposition voltage for the MCSIM, then the cell resistance increases even more, due to additional blocking of microelectrode (2) by the products of MCSIM decomposition (but this may disturb the reversibility of element functioning).

Note also that if a MCSIM possesses cation conductivity, and rectifying occurs at its interface with microelectrode (2), then the above-considered diode rectifies the current flowing in the direction reverse with respect to the conventional point diode based on the metal/*n*-type semiconductor junction (Miyatani, 1968).

23. Double-Layer Capacitors-Ionistors with Superhigh Capacitance

The operation of electrochemical capacitors is based on the charging effect of the double-layer capacitance at the SIM/blocking electrode interface. When an A-type cell is charged (see Section 21) (the blocking electrode functions as an anode), mobile ions (cations, in our case) migrate under the action of the electrical field to the reversible electrode where the metal may deposit. At the SIM/blocking electrode interface, a negative space-charge layer is formed; cations which go into the SIM bulk "open" the anionic framework. At the same time, at the anode (i.e., in the metal), electrons attracted to the bulk of the electrode "open" the positively-charged cation framework of the metal. In other words, charge separation at the SIM-metal interface leads to the formation of a double electric layer, which plays the part of a specific capacitor with two oppositely-charged plates of different materials.

Since the gap between the plates of such electrochemical capacitor is of the order of interatomic distances in the condensed medium, i.e., of several angstroms, its specific capacitance is rather high, being $1\text{–}10^2\ \mu F/cm^2$. Thus, for heterojunction $C \mid Ag_4RbI_5$, it is $C_{dl} \approx 20\text{–}40\ \mu F/cm^2$ (Armstrong and Dickinson, 1976), and for $Pt \mid Na\text{-}\beta\text{-}Al_2O_3$, it is $C_{dl} \approx 0.03\text{–}1.00\ \mu F/cm^2$ at 300 K and $C_{dl} \approx 20\ \mu F/cm^2$ at 450 K (Bukun *et al.*, 1973; Armstrong and Dickinson, 1976; Hooper, 1977).

Using blocking electrodes of porous materials, it is possible to drastically increase the interface area and, thus, fabricate a system with a total capacitance up to tens of Farads per cubic centimeter. Such elements are called double-layer capacitors (Scott, 1972; Owens *et al.*, 1977; Sekido and Nimomiya, 1981; Xue and Chen, 1986; Pham-Thi *et al.*, 1986) or ionistors (in the Soviet scientific literature) (Gailish *et al.*, 1975; Ivanov-Shits *et al.*, 1978a).

The operating voltage of electrochemical capacitors cannot exceed the decompositon potential of the SIM, since, otherwise, a stationary current starts flowing through the cell due to an electrochemical reaction proceeding at earlier blocked anode. In other words, the decomposition potential of an electrolyte is the potential of the ionistor breakdown. Note also that effective leakage resistance R_e of a capacitor is determined by the electronic conductivity of a SIM.

Another specific feature of the electrochemical capacitors under study is the noticeable dependence of their capacitance on the frequency of the charging current. Physically, the point is that the variation in the charge of the double-layer in electrolytes is associated with ion redistribution and, therefore, proceeds at a rather low rate. For a frequency of several tens of hertz, the effective capacitance of the elements decreases by approximately two orders of magnitude (Owens *et al.*, 1977; Nigmatullin *et al.*, 1984).

Electrolytic solid-state capacitors have unique properties. They not only possess high energy storage capability, but are also highly reliable and may store the charge for a very long time. Therefore, they may be used as emergency power sources. High capacitance, small geometric sizes, and, hence, small inductance provide unique possiblities for the fabrication of electric circuits operating within the very low frequency range (Nigmatullin *et al.*, 1984). This is of great importance for, e.g., studies of seismic oscillations or circuits with the operational amplifiers in special function generators (Tanase and Takahashi, 1985).

Electronic currents deteriorate the ionistor characteristics. Thus, they may have negative effect on ionistor charging if electronic currents are sufficiently strong, but potentials are much lower than the potential of the SIM decomposition. Figure 45 shows the charge curves for an A-type element with the Ag_4RbI_5 superionic conductor. As is seen from Fig. 45, the voltage of the cell increases up to a certain value V_{cr}, which depends on the charging current. It may be assumed that the critical potential V_{cr} corresponds to the condition in which the electronic current in the electrochemical cell is approximately equal to the ionic one. This assumption is confirmed by experimental results showing that charging and electronic currents are almost equal (for potentials equal to V_{cr}).

The above effect is clearly seen in thin-film ionistors (Ivanov-Shits *et al.*, 1979a). Despite the fact that electronic conductivity is lower than ionic conductivity, the electronic currents for thin-film cells are noticeable at large potentials. Thus, for a 10 μ thick Ag_4RbI_5 sample, an electronic current exists up to 10^{-5}–10^{-6} A at 350 mV and 10^{-4}–10^{-5} A at 450 mV, affecting element charging (Ivanov-Shits *et al.*, 1979b).

Thus, if electronic conductivity is high, the range of working voltages of an electrochemical double-layer capacitor (ionistor) decreases.

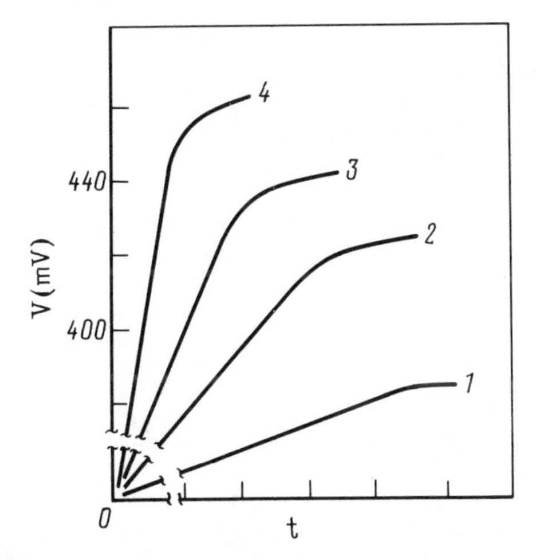

FIG. 45. Potential of a Ag | Ag_4RbI_5 | Au cell versus time during dc charging: 1—$i = 0.5\ \mu A$; 2—5; 3—10; 4—20. [From Ivanov-Shits *et al.* (1978).]

Electronic current also determines the process of ionistor self-discharge. The corresponding calculation is given in Section 13. We note here only that the real leakage current, e.g., for a double-layer capacitor with a capacitance of 1.8 F (with a SIM based on complex silver salts), is 3×10^{-9} A (Tanase and Takahashi, 1985); for a cell with $\approx$4 F, the capacitance (with a SIM of carbon and Na-β''-Al_2O_3 mixture) current is 2.5×10^{-3} A (Xue and Chen, 1986).

24. INJECTION SWITCHES

Injection switches are, in their essence, sophisticate analog integrators. The function of injection switches is based on the effect of drastic changes in the resistance of the material due to the injection of ions or electrons. Such a switch is an electrochemical cell of the type (Di Domenico *et al.*, 1979; Singh *et al.*, 1980; Ukshe *et al.*, 1982)

$$\begin{array}{c} \qquad\qquad\qquad\qquad B \\ \qquad\qquad\qquad\qquad | \\ (1)\text{-}M \,|\, \text{SIM} \,|\, \text{MCSIM} \,|\, B\text{-}(2), \\ \qquad\qquad\qquad\qquad | \\ \qquad\qquad\qquad\qquad B \end{array}$$

where *B* denotes ion-blocking electronically-conductive electrodes. A simplified structure of such a switch is shown in Fig. 46. It consists of control and load circuits. The former, in turn, consists of electrodes 1 and 2, a MCSIM and a SIM, and supplies the so-called active material (MCSIM) with ions and electrons. The loading circuit is built by *B* electrodes and a MCSIM.

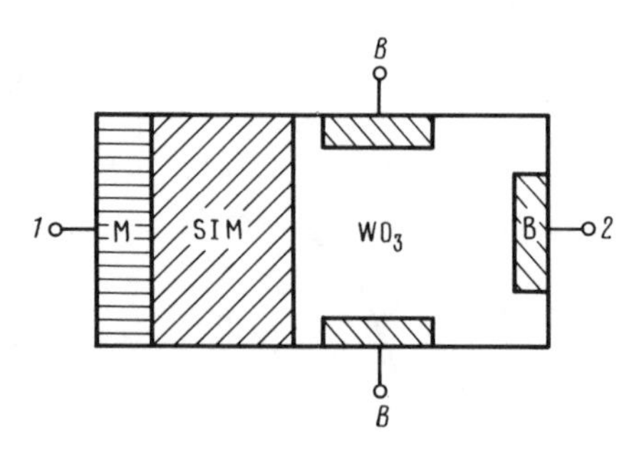

FIG. 46. Schematic diagram of an injection switch: (1) electronic conductive electrode reversible with respect to M^+ ion, (2) electronic conductive ion blocking electrode.

As an active material, tungsten trioxide WO_3 is often used, the structure and properties of which have already been discussed in Part I.

Tungsten trioxide nonactivated with impurities is a dielectric, and, therefore, the loading circuit B-WO_3-B possesses high resistance in the transitional state. A current flowing through the control circuit induces injection of ions from the SIM layer and electrons from the metallic electrode $B(2)$ into WO_3. As a result, tungsten bronze is formed. Therefore, the resistance in the load circuit drastically decreases, which may be recorded, in particular, by bringing a measuring device into the circuit.

Another switch schematically shown in Fig. 47 was designed by Ikeda and Tada (1980). The load circuit, (1)-MCSIM-(2), also includes a compound with mixed conductivity based on the complex silver salts. If a MCSIM is saturated with metal M, the solid solution formed has a low resistance, and a strong current may flow through the load. During discharge in the (1)-SIM-(3) circuit, the MCSIM is depleted of the charge carriers, and electronic conductivity drops.

For large potentials in the load circuits, a partial decomposition of the active material may occur, and one of B electrodes (2 or 3) is blocked by decomposition products, resulting, in the final analysis, in circuit cutoff.

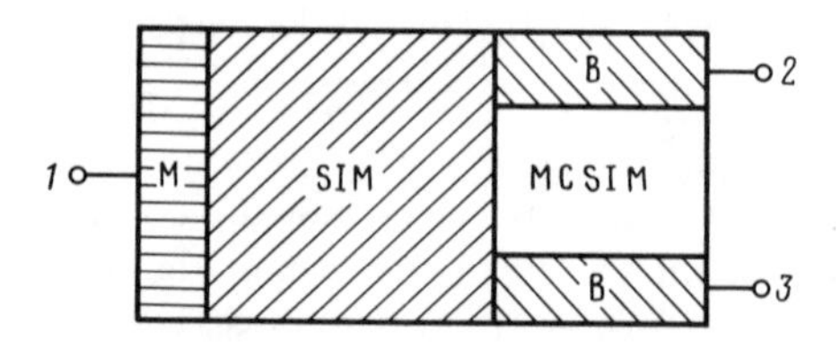

FIG. 47. Schematic diagram of an injection switch-memistor: (1) electrode reversible with respect to M^+ ion, (2) electronic conductive ion blocking electrode.

This effect manifests itself in a sharp increase of the voltage between (1) and (3), indicating the completion of the element discharge.

The above-described cells may also be used as memory or logic elements in which the load circuit is simultaneously the reading circuit. The advantages of such devices arise from the fact that the recording-reading cycle may be repeated many times without changing the device characteristics. Also, these devices have high operating voltage (indicating the cutoff of the load circuit) and consume a small amount of energy.

Possible changes in MCSIM resistance during injection (extraction) of electrons have been used in the design of a controllable resistor, the so-called memistor (Takahashi *et al.*, 1973; Ikeda and Tada, 1980). In this case, a three-electrode cell, analogous to that depicted in Fig. 47, is used.

If a direct current i is transmitted through a (1)–(2) circuit, the change observed in the conductivity of the MCSIM layer obeys the following law:

$$\sigma = \sigma_0 - a \int_0^t i \, dt, \tag{192}$$

where a is the unit cell parameter associated with its geometry, concentration and mobility of electronic carriers in the MCSIM. The resistance of a layer with the mixed conductivity is recorded in the reading circuit (2)–(3). As follows from Eq. (192), the total conductivity of the MCSIM linearly depends on the charging current of the element

$$\frac{1}{R} = \Gamma(\sigma_0 - ait), \tag{193}$$

Γ being dependent on the electrode geometry.

Just such a linear dependence was in fact, observed for memistors with a SIM and MCSIM based on the silver salts (Takahashi *et al.*, 1973),

$$Ag|Ag + Ag_6I_4WO_4|(Ag_2S)_{0.69} \times \underset{\substack{| \\ Au}}{\overset{\substack{Au \\ |}}{(Ag_{1.7}Te)_{0.285}}} \times (Ag_4P_2O_7)_{0.025}.$$

Note also that, if an alternating current passes through the reading circuit (2)–(3), it is possible to record the voltage changes instead of resistance variations.

25. Electrochromic Devices

Superionic materials have found wide applications in various optical devices based on recently-discovered phenomenon, the so-called electrochromic effect (see, e.g., the review by Chang, 1976; Faughan and Grandall,

1981; Lusis, 1981). The nature of the electrochromic effect seems to be associated with the appearance of the so-called color centers, forming upon ion and electron injection into SIM and changing the material color. The electrochromic effect, in the broad sense of the word, is understood as a change in any optical property of a crystal by an applied electric field, not necessarily associated with a change in color, and independent of whether an electrical current is flowing or not.

All-solid-state electrochromic cells are used for the electric-to-optic signal transduction on the basis of the injection effect. Superionic materials are used in electrochromic cells, firstly, as solid electrolytes, which are ion reservoirs, and, secondly, as electrochromic materials proper, which provide fast ionic and electronic transport. Now consider the operation principles of a four-layer electrochromic element, schematically shown in Fig. 48. Two external plates are made of a metal or semiconductor. They function as the electrodes to which the voltage is applied and may also play the part of an electron source or sink. The left plate contacts a layer of an electrochromic material, e.g., of tungsten trioxide, and the right-hand plate contacts a layer of a superionic conductor not conducting electrons. The solid electrolyte layer performs a double function: on the one hand, ions arrive to the electrochromic material through the conductor-electrochromic material interface, and, on the other hand, the layer serves as an electronic insulator that does not allow electrons from the right-hand metal electrode to penetrate into the electrochromic material.

Let an electrode neighboring the solid electrolyte play the part of an anode. The application of an electric field results in the double injection into the electrochromic material—that of electrons from the cathode and that of cations from the electrolyte. As a result, the electrochromic material acquires a certain color. Thus, injection of Ag^+ ions colors tungsten trioxide blue.

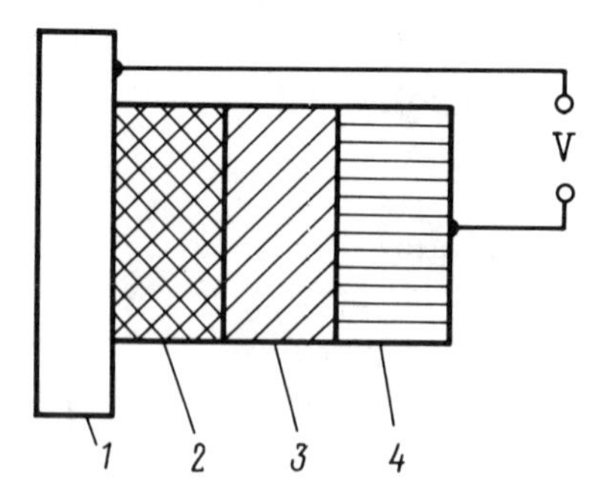

FIG. 48. Schematic diagram of a simple electrochromic cell: (1) optically transparent electronic conductive electrode, (2) electrochromic material, (3) SIM, (4) electrode reversible with respect to ions.

Coloration is proportional to the concentration of injected ions and, therefore, may be controlled. If a cell is switched on in the reverse direction, carrier extraction occurs, and WO_3 looses the color.

The nature of color centers is not quite clear as yet. A certain role is played here by variation in the valence of transition-metal ions in electrochromic oxide, due to trapping of injected electrons. Water molecules (or, more precisely, their hydrogen) present in the material bulk seem also to be of importance here.

At present, the number of known electrochromic materials is quite large—these are transition metal oxides (WO_3, MoO_3, Nb_2O_5, V_2O_5, solid electrochromic materials) and gels and solutions of organic complexes (liquid electrochromic materials). In addition to their technological advantages, the use of solid electrolytes also increases the operating temperature range of electrochromic elements and excludes undesirable side reactions, such as gas evolution.

Hydrogen-, silver-, sodium- and lithium-ion conductors are often used as superionic materials. Recently, great attention has been focused on a new electrochromic material—anodic irridium oxide films (AIROFs) (Dautremont-Smith *et al.*, 1979). In distinction from tungsten and similar bronzes, irridium films are colored due to incorporation of anionic carriers (with the simultaneous removal of electrons). Therefore, in fabrication of all-solid-state cells, anionic SIMs are used. Rice and Bridenbaugh (1981) carried out the investigations of an electrochromic cell with a fluoride ion conductor $PbSnF_4$.

The properties of electrochromic material affect the operating time of the device and the degree of "color reversibility." Thus, elements with AIROFs have a higher speed of response than devices based on "bronzes," but the contrast and graduation produced by AIROFs is much worse. At present, optimal combinations of materials used in electrochromic devices are searched for. There are some devices providing more than ten million coloring-bleaching cycles, the response time not exceeding several hundredths of a second (for more details see the collected papers edited by Pankove, 1980). The advantages of electrochromic devices are high image quality and low operating voltages. They are shock- and vibration-resistant. It is also of great importance that storage of the obtained optical image does not require a continuous energy supply.

26. Thin-Film Elements

In connection with the previously-mentioned, important problem of the miniaturization of electronic components in modern electronic devices, one should consider the possibility of growing superionic thin films (Kennedy,

1977b; Ivanov-Shits *et al.*, 1983, 1985). The use of SIM thin films essentially broadens the technological methods for designing miniature SIM-based functional elements.

One of the most widespread methods for the preparation of thin-film samples is vacuum deposition. It is necessary to determine the optimum deposition mode for each substance (evaporator and substrate temperatures, substrate type, evaporation rate, etc.). Depending on deposition conditions, thin superionic films may have different crystal orientation and structure (if a compound has several polymorphic modifications). Such films can be epitaxial, amorphous or polycrystalline, and possess different physical-chemical properties.

SIM films should meet the following requirements: possess maximum possible ionic conductivity, contain no pores, and be stable to the ambient atmosphere.

Superionic materials of complicated composition are often obtained by the evaporation of a mechanical mixture of initial components. Thus, films of the superionic Ag_4RbI_5 conductor (Ivanov-Shits *et al.*, 1976) were obtained by vacuum deposition of the AgI + RbI mixture containing 85–88 mole % of AgI (instead of 80 mole % of AgI, as is required by the chemical formula). The component ratio was adjusted empirically using X-ray and DTA data and the fact that simple binary salts were evaporated independently at different rates, and the compound was synthesized during the deposition of these salts onto the substrate.

Figure 49 shows the concentration dependence of ionic conductivity σ_i in the AgI–RbI film system (Borovkov *et al.*, 1975). It is seen that the maximum conductivity corresponds to the films containing 80 mole % of AgI (or 85–88 mole % of AgI in the initial mixture). Samples of other compositions are $AgRb_2I_3$ + Ag_4RbI_5 or Ag_4RbI_5 + AgI mixtures.

Exposed to air for a long time, the films decompose. As a result, the total resistance of a "sandwich" film cell, $Ag|Ag_4RbI_5|Ag$, increases within a few days from several Ohms up to $\sim 10^3\ \Omega$ (Kennedy *et al.*, 1973).

The electronic conductivity of Ag_4RbI_5 films was studied by Ivanov-Shits *et al.* (1978b) with the use of the Wagner polarization method. It has been established that conductivities are low and, for different samples, vary within the limits $\sigma_e(o) \sim 10^{-9}$–$10^{-11}\ \Omega^{-1}\ cm^{-1}$ and $\sigma_h(o) \sim 10^{-13}$–$10^{-14}\ \Omega^{-1}\ cm^{-1}$ at 300 K. A relatively large scatter in the experimental data is explained by the fact that the electronic component of conductivity, as was already indicated (Section II), is very sensitive to various impurities. In particular, the composition of a superionic film may differ from the ideal stoichiometry, which markedly affects the value of σ_{el}.

The deposition of complex fluorides was carried out from two evaporators of the initial components, AF_x and BiF_y (A = Pb, Ag, Rb, Tl). This method

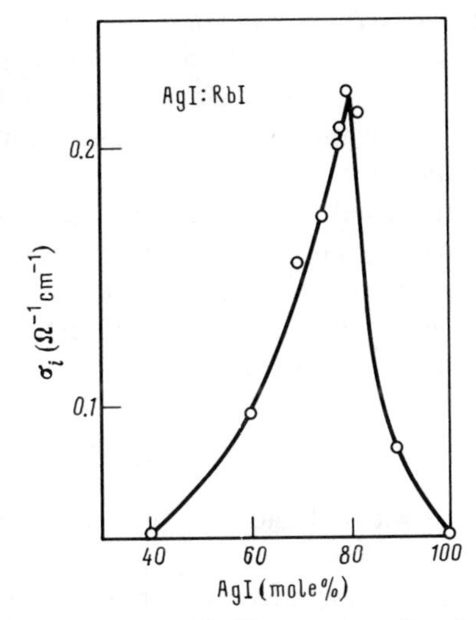

FIG. 49. Ionic conductivity of AgI : RbI thin film system as a function of composition. [From Borovkov *et al.* (1975).]

provides the formation of high-quality solid pore-free layers with high ionic conductivity (Couturier *et al.*, 1979; Murin *et al.*, 1982).

The majority of the known sodium- and lithium-superionic materials (e.g., Na-β-alumina and LISICON) are high refractory substances; therefore, the corresponding films were obtained by rf sputtering (Schnell *et al.*, 1981; Ohtsuka and Yamoij, 1983), a method which is widely used in semiconductor technology.

Along with the vacuum deposition techniques, considerable attention has been given to the chemical synthesis of superionic materials. Thus, in order to obtain single-crystal and polycrystalline dielectric and protective films for various electronic and optical devices, the so-called "hydrolysis-polycondensation" technology is used. Despite the fact that the SIM synthesis from the metal-organic compounds is quite a new method, it has already been successfully used for the synthesis and study of the thin films of such promising compounds as NASICON ($Na_3Zr_2Si_2PO_{12}$) and LISICON-$Li_{14}Zn(GeO_4)_4$ (Perthuis *et al.*, 1984; Perthuis and Colomban, 1985).

At present, one may encounter in the literature descriptions of solid-film functional elements with superionic materials—batteries, galvanic and fuel cells, coulometers, timers, ionistors, electrochromic devices, and gas sensors. For more detail about thin-film SIMs, we send the reader to the review by Kennedy (1977), Ivanov-Shits *et al.* (1985), and Gouet (1986). Currently, thin film devices are still in the state of exploratory development, since the problems associated with the fabrication of stable SIM films with high conductivity and design of multilayer elements are still far from being solved.

Thus, superionic materials are widely used in various devices and apparatus, and their application range increases rapidly. The compounds with pure ionic conductivity successfully compete with liquid electrolytes, and they have almost ousted them from, e.g., galvanic cells and electrochemical detectors. And, what is more important, at the current stage of microelectronics development, superionic materials will play a decisive role in the design of a new generation of functional solid devices, including those which have not had analogues as yet.

IX. Conclusion

As follows from the above discussion, there are a number of interesting properties of superionic materials that are important for practical applications, and certain properties inherent in the substances with superionic conductivity result in specific electronic characteristics of these compounds.

A large number of problems related to the electrical properties of superionic materials may be interpreted in terms of the phenomenological approach similar to that used for the description of semiconductor materials. The fruitfulness of such an approach is explained by the general character of the principles it is based on and by similar properties possessed by electronic carriers in semiconductors and superionic materials. The possibilities of the phenomenological approach are far from exhausted and may be used for investigating unsolved problems—for example, dendritic growth and the related problem of the stable functioning of SIM-based cells, the problem of thermodynamic instability of the SIM-metal interface under conditions of ionic-electronic conductivity, the problem of more detailed description of electrochemical phenomena at an interface between two solids, and the problem of the dependence of the phenomenologic transference coefficients on concentrations and electric fields.

Far from well understood is the important problem of consistent description of electron behavior in superionic materials, with allowance for disordering and electronic interactions, in particular, taking lattice vibrations into account. It should be noted that the ionic-electronic processes under conditions of partial ionic disorder play an important role, not only in SIMs, but also in films and layers in general. Therefore, this problem falls outside the limits of SIM studies, in the narrow sense of the world, as the problem is very important for many other branches.

At the same time, experimental data on fine details of electronic processes in superionic materials are still insufficient, although the knowledge of such processes is the premise for the creation of a microscopic theory. The most reliable way of obtaining complete and reliable results here is the simultaneous application of different experimental techniques, primarily those used in the physics of semiconductors. Thus, a wealth of information on electronic properties in SIMs may be obtained from Hall measurements, including those under the conditions of specimen illumination. An important role for studying the electronic energy spectra of SIMs and their electronic properties is played by optical experiments. The study of various temperature effects, including thermostimulated currents, may provide quantitative data on electronic traps in crystals and the characteristic parameters of such traps.

A wide spectrum of experimental techniques fruitfully used in physical-chemical studies of "traditional" solids—infrared and Mössbauer-spectroscopies, different methods of magnetic resonance, coherent and incoherent scattering of low-energy neutrons and gamma radiation, the effect of superhigh static pressures and ultrasound, etc.—also play an important role. Their wide applications to the study of SIM properties seem to be quite natural, as are the use of some detailed theoretical concepts (theories of critical phenomena, disordered media, percolations, polarons, excitons, radiationless transitions).

The rapid development of scientific research in solid-state physics and chemistry, in general, and in superionic materials, in particular, is a powerful impetus to further theoretical and experimental studies of electronic processes in superionic materials.

Acknowledgments

The authors are grateful to Dr. L. I. Man for translating the manuscript into English.

References

Afanas'ev, M. M., Goffman, V. G., and Kompan, M. E. (1982). *Fiz. Tverdogo Tela* **24**, 1540.

Afanas'ev, M. M., Goffman, V. G., and Kompan, M. E. (1983). *Zh. Eksper. Teoret. Fiz.* **84**, 1310.

Afanas'ev, M. M., Venus, G. B., Gromov, O. G., Kompan, M. E., and Kuz'min, A. P. (1984). *Fiz. Tverdogo Tela* **26**, 2956.

Akopyan, I. Kh., Monov, A. E., and Novikov, B. V. (1983). *Pis'ma v ZhETF* **37**, 459.

Akopyan, I. Kh., Gromov, D. N., Mishchenko, A. V., Monov, A. E., Novikov, B. V., and Yaufman, M. D. (1984). *Fiz. Tverdogo Tela.* **26**, 2628.

Albert, J. P., Jouanin, C., and Gout, C. (1977). *Phys. Rev. B* **16**, 4619.

Armstrong, R. D., and Dickinson, T. (1976). *In* "Superionic Conductors" (G. D. Mahan and W. L. Roth, eds.), p. 65. Plenum Press, New York.

Bagotskii, V. S., and Skundin, A. M. (1981). *Khimicheskie Istochniki Toka.* Energoizdat, Moscow.

Balescu, R. (1975). "Equilibrium and Nonequilibrium Statistical Mechanics." J. Wiley and Sons, Inc., New York.

Baklykov, S. P. (1980). *In* "Khimiya Tverdogo Sostoyaniya" (Yu. A. Zakharov, ed.), p. 14. Kemerovskii Gos. Univ., Kemerovo.

Baklykov, S. P., and Zakharov, Yu. A. (1980). *In* "Khimiya Tverdogo Sostoyaniya" (Yu. A. Zakharov, ed.), p. 3. Kemerovskii Gos. Univ., Kemerovo.
Bates, J. B., and Farrington, G. C., eds. (1981). "Fast Ion Transport in Solids" (Proc. Int. Conf.), *Solid State Ionics.* **5**.
Bauer, R. S., and Huberman, B. A. (1976). *Phys. Rev. B* **13**, 3344.
Becker, K. D., Schmalzried, H., and von Wurmb, V. (1983). *Solid State Ionics.* **11**, 213.
Blaka, P., Redinger, J., and Schwarz, K. (1984). *Z. Phys. B.* **57**, 273.
Bonch-Bruevich, V. L., and Kalashnikov, S. G. (1977). "Fizika Poluprovodnikov", Nauka, Moscow.
Bonch-Bruevich, V. L., Zvyagin, I. P., Kayner, R., Mironov, A. G., Enderlayn, R., and Esser, B. (1981). "Elektronnaya Teoriya Neuporyadochennykh Poluprovodnikov." Nauka, Moscow.
Bonch-Bruevich, V. L. (1983). *Uspekhi Fiz. Nauk.* **140**, 583.
Borovkov, V. S., Ivanov-Shits, A. K., and Tsvetnova, L. A. (1975). *Elektrokhimiya* **11**, 664.
Borovkov, V. S., and Khachaturyan, N. A. (1976). *Elektrokhimiya* **12**, 798.
Bouwmeester, H. J. M. (1985). *Solid State Ionics* **16**, 163.
Boyce, J. B., De Jonghe, L. C., and Huggins, R. A., eds. (1986). "Solid State Ionics" (Proc. 5th Int. Conf.), *Solid State Ionics.* **18/19**.
Bradley, J. N., and Green, P. D. (1967). *Trans. Faraday Soc.* **63**, 424.
Brook, R. J., Pelzmann, W. L., and Kröger, F. A. (1971). *J. Electrochem. Soc.* **118**, 185.
Burley, G. (1967). *Acta Crystallogr.* **23**, 1.
Burker, L., Rickert, H., and Steiner, R. (1971). *Z. Phys. Chemie.* **74**, 146.
Carslow, H. S., and Jaeger, J. E. (1959). "Conduction of Heat in Solids." Oxford Univ. Press, London.
Chandra, S. (1981). "Superionic Solids." North-Holland, Amsterdam.
Chandra, S. (1984). *Mat. Science Forum.* **1**, 153.
Chang, I. F. (1976). *In* "Non Emissive Electrooptic Displays" (J. Bruinink, I. F. Chang, and M. R. Zeller, eds.). Plenum Press, London.
Chebotin, V. N. (1982). "Khimiya Tverdogo Tela." Khimiya, Moscow.
Chebotin, V. N. (1984). *Elektrokhimiya.* **20**, 55.
Chebotin, V. N. (1985). *In* "Defekty i Massoperenos v Tverdofaznykh Soedineniyakh Perekhodnykh elementov," p. 3. Ural. Nauch. Tsentr Akad. Nauk SSSR, Sverdlovsk.
Chebotin, V. N. (1986). *Uspekhi Khimii.* **55**, 914.
Chebotin, V. N., and Perfil'ev, M. V. (1978). "Electrochemistry of Solid Electrolytes." Technical Information Center, U.S. Department of Energy.
Chu, H. F., Rickert, H., and Weppner, W. (1973). *In* "Fast Ion Transport in Solids" (W. van Gool, ed.), p. 181. North-Holland, Amsterdam.
Couturier, G., Danto, Y., and Pistre, J. (1979). *In* "Fast Ion Transport in Solids" (P. Vashishta, J. N. Mundy, and G. K. Shenoy, eds.), p. 181. Elsevier, North-Holland, Amsterdam.
Couturier, G., Salardenne, J., Sribi, C., and Rosso, N. (1983). *Solid State Ionics.* **9/10**, 699.
Crank, J. (1970). "The Mathematics of Diffusion." Oxford Univ. Press, London.
Damaskin, B. B., and Petrii, O. A. (1983). "Vvedenie v Elektro-Khimicheskuyu Kinetiku." Vysshaya Shkola, Moscow.
Davison, S. G., and Levine, J. D. (1970). *In* "Solid State Physics" (H. Ehrenreich, F. Seitz, and D. Turnbull, eds.), Vol. 25. Academic Press, New York.
Dautremout-Smith, W. C., Beni, G., Schiavone, L. M., and Shay, J. L. (1979). *Appl. Phys. Lett.* **35**, 565.
Dickens, P. G., and Reynolds, H. J. (1981). *Solid State Ionics* **5**, 331.
De Groot, S. R., and Mazur, P. (1969). "Non-Equilibrium Thermodynamics. North-Holland, Amsterdam.

Di Domenico, Jr., M., Singh, S., and Van Uitert, L. G. (1979). US Patent 4163982.
Dunn, B. (1986). *In* "37th Meeting ISE, Extended Abstracts," Vol. 1, p. 131. Vilnius, USSR.
Dudley, G. J., and Steele, B. C. H. (1977). *J. Solid State Chem.* **21,** 1.
Dudley, G. J., and Steele, B. C. H. (1980). *J. Solid State Chem.* **31,** 233.
Evarestov, R. A., Murin, I. V., and Petrov, A. V. (1984). *Fiz. Tverdogo Tela.* **26,** 2579.
Fabry, P., and Kleitz, M. (1976). *In* "Electrode Processes in Solid State Ionics" (M. Kleitz and J. Dupuy, eds.), p. 331. Reidel, Dordrecht-Holland.
Faraday, M. (1834). *Ann. Physik.* **107,** 241.
Faughan, B. W., and Crandall, R. S. (1980). *In* "Display Devices," *Topics in Applied Physics* (J. I. Pankove, ed.), Vol. 40, p. 181. Springer-Verlag, Berlin.
Findley, P. R., Wu, Z., and Walker, W. C. (1983). *Phys. Rev. B* **28,** 4761.
Friedenborg, A., and Shapiro, Y. (1982). *Surface Sci.* **115,** 606.
Fomenkov, S. A. (1982). Thesis. Uralskii Gos. Universitet, Sverdlovsk.
Friauf, R. J. (1972). *In* "Physics of Electrolytes" (J. Hladik, ed.), Vol. 2, p. 1103. Academic Press, New York.
Fujimoto, H. H., and Tuller, H. L. (1979). *In* "Fast Ion transport in Solids" (P. Vashishta, J. N. Mundy, and G. K. Shenoy, eds.), p. 649. Elsevier, North-Holland, Amsterdam.
Funke, K. (1976). *Progr. Solid State Chem.* **11,** 345.
Gailish, E. A., D'yakonov, M. N., Kuznetsov, V. P., and Kharitonov, E. V. (1975). *Elektronnaya promyshlennost.* **N 8,** 42.
Geller, S., ed. (1977). "Solid Electrolytes." Springer-Verlag, Berlin.
Goldman, J., and Wagner, J. B., Jr. (1974). *J. Electrochem. Soc.* **121,** 1318.
Goodenough, J. B., Jensen, J., and Potier, A., eds. (1985). "Solid Protonic Conductors, III, for Fuel Cells and Sensors." Odense Univ. Press, Denmark.
Gouet, M. (1986). *J. de Physique, Colloque Cl.* **47,** Cl-119.
Gurevich, Yu. Ya. (1975). *Dokl. Akad. Nauk SSSR.* **222,** 143.
Gurevich, Yu. Ya. (1980). *Elektrokhimiya* **16,** 1077.
Gurevich, Yu. Ya., and Barshtevskii, B. U. (1970). *Dokl. Akad. Nauk SSSR* **191,** 115.
Gurevich, Yu. Ya., and Kharkats, Yu. I. (1977). *Zh. Eksp. Teor. Fis.* **72,** 1845.
Gurevich, Yu. Ya., and Ivanov-Shits, A. K. (1978). *Elektrokhimiya* **14,** 960.
Gurevich, Yu. Ya., and Ivanov-Shits, A. K. (1980). *Elektrokhimiya* **16,** 3.
Gurevich, Yu. Ya., and Ivanov-Shits, A. K. (1982). *Fiz. Tverdogo Tela* **24,** 795.
Gurevich, Yu. Ya., and Kharkats, Yu. I. (1976a). *Elektrokhimiya* **12,** 1768.
Gurevich, Yu. Ya., and Kharkats, Yu. I. (1976b). *Dokl. Akad. Nauk SSSR* **229,** 367.
Gurevich, Yu. Ya., and Kharkats, Yu. I. (1978). *J. Phys. Chem. Solids* **39,** 751.
Gurevich, Yu. Ya., and Kharkats, Yu. I. (1980). *Elektrokhimiya* **16,** 777.
Gurevich, Yu. Ya., and Kharkats, Yu. I. (1985). *Fiz. Tverdogo Tela* **27,** 1977.
Gurevich, Yu. Ya., and Kharkats, Yu. I. (1986a). *Physics Reports* **139,** 201.
Gurevich, Yu. Ya., and Kharkats, Yu. I. (1986b). *J. Electroanal. Chem.* **200,** 3.
Gurevich, Yu. Ya., and Kharkats, Yu. I. (1987). *Fiz. Tverdogo Tela* **29,** 111.
Gurov, K. P. (1978). "Fenomenologicheskaya Termodinamika neobratimykh protsessov." Nauka, Moscow.
Hagenmuller, P., and van Gool, W., eds. (1977). "Solid Electrolytes." Academic Press, New York.
Hartmann, B., Rickert, H., and Schendler, W. (1976). *Electrochim. Acta* **21,** 319.
Hasegawa, A. (1985). *Solid State Ionics* **15,** 81.
Hebb, M. H. (1952). *J. Chem. Phys.* **20,** 185.
Hersh, H. N. (1971). *J. Electrochem. Soc.* **118,** 144C.
Heyne, L. (1973). *In* "Fast Ion Transport in Solids" (W. van Gool, ed.), p. 123. North-Holland, Amsterdam.

Heyne, L. (1977). *In* "Solid Electrolytes" (S. Geller, ed.), p. 169. Springer-Verlag, Berlin.
Honders, A., Young, E. W. H., de Wit, J. H. W., and Broers, G. H. J. (1983a). *Solid State Ionics* **8,** 115.
Honders, A., Young, E. W. H., van Heeren, A. H., de Wit, J. H. W., and Broers, G. H. J. (1983b). *Solid State Ionics.* **9/10,** 375.
Honders, A., der Kinderen, J. M., van Heeren, A. H., de Wit, J. H. W., and Broers, G. H. J. (1984). *Solid State Ionics* **14,** 205.
Honders, A., der Kinderen, J. M., van Heeren, A. H., de Wit, J. H. W., and Broers, G. H. J. (1985). *Solid State Ionics* **15,** 265.
Hooper, A. (1977). *J. Phys. D.* **10,** 1487.
Horvatić, M., and Vučić, Z. (1984). *Solid State Ionics* **13,** 117.
Hull, M. N., and Pilla, A. A. (1971). *J. Electrochem. Soc.* **118,** 72.
Ikeda, H., Tada, K., Narukawa, S., and Ooe, Y. (1976). *Denki Kagaku* **44,** 535.
Ikeda, H., and Tada, K. (1980). *In* "Applications of Solid Electrolytes" (T. Takahashi and A. Kozawa, eds.), p. 40. JEC Press Inc., Cleveland, Ohio.
Ishikawa, T., and Miyatani, S. (1977). *J. Phys. Soc. Japan* **42,** 159.
Ivanov-Shits, A. K. (1979). *Elektrokhimiya* **15,** 688.
Ivanov-Shits, A. K. (1985). *In* "Tochechnye Defekty i Ionnyi Perenos v Tverdykh Telakh" (Abst. Dokl.), p. 16. Krasnoyarsk.
Ivanov-Shits, A. K., D'yakov, V. A., Borovkov, V. S., and Pushkov, B. I. (1976). *Elektrokhimiya* **12,** 612.
Ivanov-Shits, A. K., Tsvetnova, L. A., and Borovkov, V. S. (1978a). *Elektrokhimiya* **14,** 485.
Ivanov-Shits, A. K., Tsvetnova, L. A., and Borovkov, V. S. (1978b). *Elektrokhimiya* **14,** 1689.
Ivanov-Shits, A. K., Borovkov, V. S., Shirokov, Yu. V., Mishtenko, A. V., and Krasnova, T. M. (1980). *Elektrokhimiya* **16,** 985.
Ivanov-Shits, A. K., Borovkov, V. S., and Tsvetnova, L. A. (1983). *Elektrokhimiya* **19,** 267.
Ivanov-Shits, A. K., Tsvetnova, L. A., and Borovkov, V. S. (1985). *Elektrokhimiya* **21,** 1703.
Joo, S.-K., Raistrick, I. D., and Huggins, R. A. (1985). *Mat. Res. Bull.* **20,** 897.
Joshi, A. V. (1973). *In* "Fast Ion Transport in Solids' (W. van Gool, ed.), p. 173. North-Holland, Amsterdam.
Joshi, A. V., and Wagner, J. B., Jr. (1972). *J. Phys. Chem. Solids* **33,** 205.
Joshi, A. V., and Wagner, J. B., Jr. (1975). *J. Electrochem. Soc.* **122,** 1071.
Joshi, A. V., and Liang, C. C. (1975). *J. Phys. Chem. Solids* **36,** 927.
Junod, P., Hediger, H., Kilchör, B., and Wullschleger, J. (1977). *Phyl. Mag.* **36,** 941.
Kennedy, J. H. (1972). *In* "Physics of Electrolytes" (J. Hladik, ed.), p. 931. Academic Press, New York.
Kennedy, J. H. (1977a). *J. Electrochem Soc.* **124,** 865.
Kennedy, J. H. (1977b). *Thin Solid Films* **43,** 41.
Kennedy, J. H., and Chen, F. (1969). *J. Electrochem. Soc.* **116,** 207.
Kennedy, J. H., Chen, F., and Hunter, G. (1973). *J. Electrochem Soc.* **120,** 454.
Kerker, G. (1981). *Phys. Rev. B* **23,** 6312.
Kharkats, Yu. I. (1984). *Elektrohimiya.* **20,** 248.
Kirkpatrick, S. (1973). *Rev. Mod. Phys.* **45,** 574.
Kiukkola, K., and Wagner, C. (1957). *J. Electrochem. Soc.* **104,** 308.
Kittel, Ch. (1969). "Thermal Physics." J. Wiley and Sons, Inc., New York.
Kleitz, M., Sapoval, B., and Chabre, V., eds. (1983). "Solid State Ionics" (Proc. 4th Int. Conf.), *Solid State Ionics* **9/10.**
Konev, V. N., Fomenkov, S. A., and Chebotin, V. N. (1985). *Izv. Akad. Nauk SSSR, Neorgan. Mater.* **21,** 202.

Korn, G. A., and Korn, T. M. (1961). "Mathematical Handbook for Scientists and Engineering." McGraw-Hill Book Co., Inc., New York.
Kristofel, N. N. (1985a). *Fiz. Tverdogo Tela* **27**, 2001.
Kristofel, N. N. (1985b). *Fiz. Tverdogo Tela* **27**, 2095.
Kukoz, F. I., Kolomoets, A. M., and Shvetsov, B. S. *Elektrokhimiya* **13**, 92.
Kröger, F. A. (1974). "The Chemistry of Imperfect Crystals," Second Edition. North-Holland, Amsterdam.
Lampert, M. A., and Mark, P. (1970). "Current Injection in Solids." Academic Press, New York.
Lidiard, A. B. (1957). *In* "Handbuch der Physik" (S. Flügge, ed.), Vol. XX, p. 246. Springer-Verlag, Berlin.
Lingras, A. R., and Simkovich, G. (1978). *J. Phys. Chem. Solids* **39**, 1225.
Liu, C., Sundar, H. G. K., and Angell, A. (1985). *Mat. Res. Bull.* **20**, 525.
Lushtick, Ch. B., Vitol, I. K., and Elango, M. A. (1977). *Uspekhi Fiz. Nauk* **122**, 223.
Lusis, A., ed. (1981). "Oksidnye Elektrokhromnye Materialy." Latviiskii Gos. Univ., Riga.
Mahan, G. D., and Roth, W. L., eds. (1976). "Superionic Conductors." Plenum Press, New York.
Malov, Yu. I., Bukun, N. G., and Ukshe, E. A. (1979). *Elektrokhimiya* **15**, 422.
Mazumdar, D., Govingachuryalu, P. A., and Bose, D. N. (1982). *J. Phys. Chem Solids* **43**, 933.
McDougall, J., and Stoner, E. C. (1938). *Trans Roy. Soc (London)* **A237**, 67.
Mikhailov, M. D., and Özişik, M. N. (1984). "Unified Analysis and Solutions of Heat and Mass Diffusion." J. Wiley and Sons, Inc., New York.
Minami, T. (1985). *J. Non-Crystalline Solids* **73**, 273.
Miyatani, S. (1955). *J. Phys. Soc. Japan* **10**, 786.
Miyatani, S. (1958a). *J. Phys. Soc. Japan* **13**, 317.
Miyatani, S. (1958b). *J. Phys. Soc. Japan* **13**, 341.
Miyatani, S. (1968). *J. Phys. Soc. Japan* **24**, 328.
Miyatani, S. (1973). *J. Phys. Soc. Japan* **34**, 423.
Miyatani, S. (1981a). *J. Phys. Soc. Japan* **50**, 1595.
Miyatani, S. (1981b). *Solid State Commun.* **38**, 257.
Miyatani, S. (1984). *J. Phys. Soc. Japan* **53**, 4284.
Miyatani, S. (1985). *J. Phys. Soc. Japan* **54**, 639.
Miyatani, S., Toyota, Y., Yanagihara, T., and Iida, K. (1967). *J. Phys. Soc. Japan* **23**, 35.
Mizusaki, J., Fueki, K., and Mukaibo, T. (1978). *Bull. Chem. Soc. Japan* **51**, 694.
Mizusaki, J., Fueki, K., and Mukaibo, T. (1979). *Bull. Chem. Soc. Japan* **52**, 1890.
Mizusaki, J., and Fueki, K. (1980). *Revue Chimie Miner.* **17**, 356.
Mizusaki, J., and Fueki, K. (1982). *Solid State Ionics* **6**, 85.
Moizhes, B. Ya. (1983). *Fiz. Tverdogo Tela* **25**, 924.
Mostafa, S. N., Amer, S. M., and Eissa, E. A. M. (1982). *J. Electroanal. Chem.* **133**, 125.
Mostafa, S. N., and Abd-Elneheem, M. A. (1985). *Electrochim. Acta* **30**, 635.
Mott, N. F., and Davis, E. A. (1979). "Electron Processes in Non-Crystalline Materials," Second Edition. Clarendon Press, Oxford.
Murin, I. V., Ivanov-Shits, A. K., Tsvetnova, L. A., Chernov, S. V., and Borovkov, V. S. (1982). *Vestnik LGU* **N 10**, 118.
Nakayama, N. (1968). *J. Phys. Soc. Japan* **25**, 290.
Nemoshkalenko, V. V., Aleshkin, V. G., and Panchenko, M. T. (1976). *Dokl. Akad. Nauk SSSR* **231**, 585.
Nernst, W. (1888). *Z. Phys. Chem.* **2**, 613.
Nernst, W. (1900). *Z. Electrochem.* **6**, 41.
Nigmatullin, R. Sh., Nasyrov, I. K., Karamov, F. A., and Salikhov, I. A. (1984). *In* "Ionika Tverdogo Tela" (Proc. III Seminar), p. 4. Vilnius. Gos. Univ., Vilnius.

Nimon, E. S., L'vov, A. L., and Pridatko, I. A. (1977). *Elektrokhimiya* **13**, 600.
Nimon, E. S., Rotenberg, A. Z., L'vov, A. L., and Pridatko, I. A. (1979). *Elektrokhimiya* **15**, 217.
Nimon, E. S., Rotenberg, Z. A., and L'vov, A. L. (1980). *Elektrokhimiya* **16**, 1437.
Nimon, E. S., L'vov, A. L., and Pridatko, I. A. (1981). *Elektrokhimiya* **17**, 1076.
Obayashi, H., Nagai, R., Gotoh, A., Mochizuki, S., and Kudo, T. (1981). *Mat. Res. Bull.* **16**, 587.
von Oehsen, U., and Schmalzried, H. (1981). *Ber. Bunsenges. Phys. Chem.* **85**, 7.
Ohachi, T. (1974). Thesis. Doshisha University, Kyoto, Japan.
Ohachi, T., and Taniguchi, I. (1977). *Electrochimica Acta* **22**, 747.
Ohachi, T., and Taniguchi, I. (1981). *Solid State Ionics* **3/4**, 89.
Ohtsuka, H., and Yamaji, A. (1983). *Solid State Ionics* **8**, 43.
Owens, B. B., and Argue, G. R. (1967). *Science* **157**, 308.
Owens, B. B., Oxley, J. E., and Sammels, A. F. (1977). *In* "Solid Electrolytes" (S. Geller, ed.), p. 67. Springer-Verlag, Berlin.
Oxley, J. E. (1972). *J. Electrochem. Soc.* **119**, 110C.
Pankove, J. I., ed. (1980). "Display Devices" (Topics in Applied Physics), Vol. 40. Springer-Verlag, Berlin.
Patterson, J. W., Bogren, E. C., and Rapp, R. A. (1967). *J. Electrochem. Soc.* **114**, 752.
Patterson, J. W. (1974). *In* "Electrical Conductivity in Ceramic and Glass" (N. N. Tallan, ed.), Part B, p. 453. M. Dekker Inc., New York.
Perram, J. W., ed. (1983). "The Physics of Superionic Conductors and Electrode Materials." Plenum Press, New York.
Perthuis, H., Velasco, G., and Collomban, Ph. (1984). *Japanese J. Appl. Phys.* **23**, 534.
Perthuis, H., and Colomban, Ph. (1985). *J. Mat. Sci. Letters* **4**, 344.
Pham-Thi, M., Adet, Ph., Velasco, G., and Colomban, Ph. (1986). *Appl. Phys. Lett.* **48**, 1348.
Pleskov, Yu. V., and Gurevich, Yu. Ya. (1986). "Semiconductor Photoelectrochemistry." Plenum Press, New York.
Radhakrishna, S., Hariharan, K., and Jagadeesh, M. S. (1979). *J. Appl. Phys.* **50**, 4883.
Raichenko, A. I. (1981). "Matematicheskaya Teoriya Diffuzii v Prilozheniyakh." Naukova Dumka, Kiev.
Raistrick, I. D., Mark, A. J., and Huggins, R. A. (1981). *Solid State Ionics* **5**, 351.
Raleigh, D. O. (1966). *J. Phys. Chem.* **70**, 689.
Raleigh, D. O. (1967a). *J. Phys. Chem.* **71**, 1785.
Raleigh, D. O. (1967b). *J. Electrochem Soc.* **114**, 493.
Raleigh, D. O. (1969). *Z. Phys. Chemie.* **63**, 319.
Raleigh, D. O., and Crowe, H. R. (1969). *J. Electrochem. Soc.* **116**, 40.
Ramasesha, S. (1982). *J. Solid State Chem.* **41**, 333.
Rhoderick, E. H. (1978). "Metal-Semiconductor Contacts." Clarendon Press, Oxford.
Rice, C. E., and Bridenbaugh, P. M. (1981). *Appl. Phys. Lett.* **38**, 59.
Rickert, H. (1982). "Electrochemistry of Solids." Springer-Verlag, Berlin.
Rickert, H., and Steiner, R. (1966). *Z. Phys. Chem. NF* **49**, 127.
Rickert, H., and El-Miligy, A. A. (1968). *Z. Metallkunde* **59**, 635.
Rickert, H., and Weppner, W. (1974). *Z. Naturforsch.* **29a**, 1849.
Rickert, H., Sattler, V., and Weddle, Ch. (1975). *Z. Phys. Chem. NF* **98**, 339.
Ross, A., and Schoonman, J. (1984). *Solid State Ionics* **13**, 205.
Salamon, M. B., ed. (1979). "Physics of Superionic Conductors." Springer-Verlag, Berlin.
Sasaki, J., Mizusaki, J., Yamauchi, S., and Fueki, K. (1981). *Bull. Chem. Soc. Japan* **54**, 2444.
Schmalzried, H. (1962). *Z. Electrochem.* **66**, 572.
Schmalzried, H. (1980). *Progr. Solid State Chem.* **13**, 119.
Schoonman, J., Wolfert, A., and Untereker, D. F. (1983). *Solid State Ionics* **11**, 187.

Schnell, Ph., Velasco, G., and Colomban, Ph. (1981). *Solid State Ionics* **5,** 281.
Schulz, H. (1982). *In* "Annual Rev. Mat. Sci.," Vol. 12, p. 351. Ann. Rev. Inc., Palo Alto, California.
Scott, R. F. (1972). *Radio-Electronics* **43,** N 3, 53.
Seeger, K. (1973). "Semiconductor Physics." Springer-Verlag, New York.
Seevers, R., De Mizzio, J., and Farrington, G. C. (1983). *J. Solid State Chem.* **50,** 146.
Sekido, S., and Ninomiya, Y. (1981). *Solid State Ionics* **3/4,** 153.
Shahi, K. (1977). *Phys. Stat. Sol.* **A41,** 11.
Shirokov, Yu. V., Pushkov, B. I., Borovkov, V. S., and Lukovtsev, P. D. (1972). *Elektokhimiya* **8,** 579.
Shklovskii, B. I., and Efros, A. L. (1979). "Elektronnye Svoistva Legirovannykh Poluprovodnikov." Nauka, Moscow.
Shukla, A. K., and Schmalzried, H. (1979). *Z. Phys. Chemie NF.* **118,** 59.
Sibley, W. A., and Pooley, D. (1974). *In* "Treatise on Materials Science and Technology" (H. Herman, ed.), Vol. 5, p. 45. Academic Press, New York.
Singh, S., Van Uitert, L. G., and Zydzik, G. I. (1980). US Patent 4 187 530.
Smith, P. V. (1976). *J. Phys. Chem. Solids* **37,** 581.
Smith, R. A. (1978). "Semiconductors," Second Edition. Cambridge Univ. Press, London.
Sohége, J., and Funke, K. (1984). *Ber. Bunsenges. Phys.. Chem.* **88,** 657.
Starostin, N. V., and Ganin, V. A. (1973). *Fiz. Tverdogo Tela* **15,** 3404.
Starostin, N. V., and Shepilov, M. G. (1975). *Fiz. Tverdogo Tela* **17,** 822.
Stoneham, A. M. (1985). *Phyl. Mag.* **B 51,** 161.
Strikha, V. I. (1974). "Teoreticheskie Osnovy Raboty Kontakta Metall-Poluprovodnik." Naukova Dumka, Kiew.
Strock, L. W. (1936). *Z. Phys. Chem.* **31,** 132.
Studenyak, I. P., Kovach, D.Sh., Pan'ko, V. V., Kovach, E. G., and Borets, A. N. (1984). *Fiz. Tverdogo Tela* **26,** 2598.
Sudarikov, S. A. (1973). *Zh. Fiz. Khim.* **47,** 1766.
Takahashi, T., Fueki, K., Owens, B. B., and Vincent, C. A. eds. (1981). "Solid Electrolytes-Solid State Ionics and Galvanic Cells" (Proc. 3rd Int. Conf.), *Solid State Ionics* **3/4**.
Takahashi, T., and Yamamoto, O. (1972). *J. Electrochem. Soc.* **119,** 1735.
Takahashi, T., Nomura, E., and Yamamoto, O. (1973). *Denki Kagaku* **41,** 723.
Takahashi, T., Yamamoto, O., Yamada, S., and Hayashi, S. (1979). *J. Electrochem. Soc.* **125,** 1654.
Tanase, S., and Takahashi, T. (1985). *Rev. Sci. Instr.* **56,** 1964.
Treier, V. V. (1978). "Elektrokhimicheskie Pribory." Sovetskōe Radio, Moscow.
Tubandt, C., and Lorenz, F. (1913). *Z. Phys. Chem.* **87.** 513.
Tubandt, C. (1932). *In* "Handbuch der Experimentalphysik" (W. Wien and F. Harms, eds.), Vol. 21, Part 1, p. 383. Academische Verlagsgesellschaft, Leipzig.
Ukshe, E. A., and Bukun, N. G. (1977). "Tverdye elektrolity." Nauka, Moscow.
Ukshe, E. A., Vershinin, N. N., and Malov, Yu. I. (1982). *Zarubezhnaya elektronika* **N 7,** 53.
Ulstrup, J. (1980). "Charge Transfer Processes in Condensed Media." Springer-Verlag, Berlin.
Valverde, N. (1970). *Z. Phys. Chemie NF.* **70,** 113, **70,** 128.
Valverde, N. (1981). *J. Appl. Electrochem.* **11,** 305.
Vashishta, P., Mundy, J. N., and Shenoy, G. K., eds. (1979). "Fast Ion Transport in Solids," Elsevier, North-Holland, Amsterdam.
Vecher, A. A., and Vecher, D. V. (1968). *Zh. Fiz. Khim.* **42,** 799.
Vershinin, N. N., Dermanchuk, E. P., Bukun, N. G., and Ukshe, E. A. (1981). *Elektrokhimiya* **17,** 383.

Vershinin, N. N., Malov, Yu. I., and Ukshe, E. A. (1982). *Elektrokhimiya* **18,** 255.
Vetter, K. J. (1961). "Elektrochemische Kinetik," Springer-Verlag, Berlin.
Wada, T., and Wagner, J. B., Jr. (1979). *In* "Fast Ion Transport in Solids" (P. Vashishta, J. N. Mundy, and G K. Shenoy, eds.), p. 585. Elsevier, North-Holland, Amsterdam.
Warburg, E. (1884). *Wied. Ann.* **21,** 622.
Wagner, C. (1933). *Z. Phys. Chem.* **B21,** 25.
Wagner, C. (1953). *J. Chem. Phys.* **21,** 1819.
Wagner, C. (1956). *Z. Electrochem.* **60,** 4.
Wagner, C. (1957). *In* "Proc. Seventh Meeting CITCE, Lindau, 1955," p. 361. Butterworth Publishers, London.
Wagner, C. (1966). *In* "Adv. Electrochemistry and Electrochem. Eng." (P. Delahey, ed.), Vol. 4, p. 1. Interscience, New York.
Wagner, C. (1972). *Progr. Solid State Chemistry* **1,** 1.
Wagner, J. B., and Wagner, C. (1957). *J. Chem. Phys.* **26,** 1602.
Wagner, J. B., Jr. (1976). *In* "Electrode Processes in Solid State Ionics" (M. Kleitz and J. Dupuy, eds.), p. 185. D. Reidel Publ. Comp., Dordrecht, Holland.
Weiss, K. (1968). *Z. Phys. Chemie NF.* **59,** 242.
Weiss, K. (1969a). *Ber. Bunsengen. Phys. Chemie* **73,** 338.
Weiss, K. (1969b). *Ber. Bunsengen. Phys. Chemie* **73,** 344.
Weiss, K. (1969c). *Z. Phys. Chemie NF* **67,** 86.
Weiss, K. (1970). *Ber. Bunsenges. Phys. Chemie* **74,** 227, 235.
Weiss, K. (1971). *Electrochim. Acta* **16,** 201.
Weiss, K. (1972). *Ber. Bunsenges. Phys. Chemie* **76,** 379.
Wen, C. J., Boukamp, B. A., Huggins, R. A., and Weppner, W. (1979). *J. Electrochem. Soc.* **126,** 2258.
Weppner, W. (1976). *Z. Naturforsch.* **a31,** 1336.
Weppner, W. (1977a). *Electrochim. Acta* **22,** 721.
Weppner, W. (1977b). *J. Solid State Chem.* **20,** 305.
Weppner, W. (1981). *Solid State Ionics* **3/4,** 1.
Weppner, W., and Huggins, R. A. (1977a). *J. Electrochem. Soc.* **124,** 35.
Weppner, W., and Huggins, R. A. (1977b). *J. Electrochem. Soc.* **124,** 1569.
Weppner, W., and Huggins, R. A. (1977c). *J. Solid State Chem.* **22,** 297.
Weppner, W., and Huggins, R. A. (1978). *J. Electrochem. Soc.* **125,** 7.
Wieger, G. A. (1976). *J. Less-Common Metals* **48,** 269.
Xue, R., and Chen, L. (1986). *Solid State Ionics* **18/19,** 1134.
Yamamoto, T., Kikkawa, S., and Koizumi, M. (1985). *Solid State Ionics* **17,** 63.
Yokota, I. (1953). *J. Phys. Soc. Japan* **8,** 595.
Yokota, I. (1961). *J. Phys. Soc. Japan* **16,** 2213.
Yokota, I., and Miyatani, S. (1962). *Japan J. Appl. Phys.* **1,** 144.
Yokota, I., and Miyatani, S. (1981). *Solid State Ionics* **3/4,** 17.
Yushina, L. D. (1979). Authorship. cert. No. 684628, H01G9/22..
Yushina, L. D., Karpachev, S. V., and Terekhov, V. I. (1979). *In* "Fast Ion Transport in Solids" (P. Vashishta, J. N. Mundy, and G. K. Shenoy, eds.), p. 121. Elsevier, North-Holland, Amsterdam.
Ziman, J. M. (1979). "Models of Disorder." Cambridge Univ. Press, London.

Appendix

TABLE A

ELECTRONIC CONDUCTIVITY, σ_{el}, OF SOME SUPERIONIC MATERIALS*

SIM	σ_{el}, $\Omega^{-1}\,cm^{-1}$	Ref.	Remarks
1	2	3	4
α-AgI	$\sigma_h = 5.2 \times 10^{-10}$ (550 K)	1	$\sigma_h(T) = 3.5 \times 10^3 \exp(-1.4/kT)$
	$\sigma_h = 3.3 \times 10^{-10}$ (550 K)	2	$\sigma_h(T) = 2.76 \times 10^3 \exp(-1.41/kT)$
	$\sigma_h = 1.5 \times 10^{-7}$ (422 K)	3	$\sigma_h(T) = 5.68 \times 10^4 \exp(-0.97/kT)$
AgI + 50 m/o=Fe_2O_3	$\sigma_e = 1.8 \times 10^{-6}$	4	$\sigma_i = 7.4 \times 10^{-4}$
Ag_3SI	$\sigma_e = 1.6 \times 10^{-4}$	5	
	$\sigma_e = 10^{-8}$	6	
	$\sigma_e = 10^{-6}$	7	
	$\sigma_e = 6 \times 10^{-6}$	8	
	$\sigma_e = 10^{-5}$	9	Technique SD**
Ag_4RbI_5	$\sigma_e = 5 \times 10^{-7}$	10	Technique SD
	$\sigma_e = 10^{-9}$	11	Technique SD
	$\sigma_e = 10^{-9}$	9	Technique SD
	$\sigma_e = 10^{-11}$	12	
	$\sigma_e = 10^{-8}$	13	
	$\sigma_e = 8 \times 10^{-11}$	14	
	$\sigma_h = 10^{-12}$	14	
	$\sigma_h = 7 \times 10^{-19}$	15	
AgI–QI, Q=Me_4N, Me_2Et_2N, Et_4N	$\sigma_e < 10^{-10}$	16	$\sigma_i = 0.02$–0.06

TABLE A-*continued*

SIM	σ_{el}, Ω^{-1} cm^{-1}	Ref.	Remarks
1	2	3	4
$Ag_7I_4AsO_4$	$\sigma_e \sim 10^{-8}$	17	$\sigma_i = 0.04$
$Ag_7I_4VO_4$	$\sigma_e \sim 10^{-8}$	17	$\sigma_i = 0.07$
$Ag_7I_4PO_4$	$\sigma_e < 10^{-8}$	18	
$Ag_{19}I_{15}P_2O_7$	$\sigma_e < 10^{-8}$	18	
$Ag_6I_4WO_4$	$\sigma_e < 10^{-8}$	19	
CuX–$C_6H_{12}N_4$RX, X=I, Br, Cl, R=H, CH_3	$\sigma_h < 10^{-11}$ (373 K)	20	
CuBr–$DTDBr_2$	$\sigma_h = 4 \times 10^{-13}$	21	$\sigma_i = 0.02$
CuBr–$C_5H_{11}NCH_3Br$	$\sigma_h = 3.2 \times 10^{-12}$ (383 K)	22	$\sigma_i = 0.1$ (383 K)
$CuPb_3Br_3$	$\sigma_h = 10^{-11}$ (498 K)	23	$\sigma_i = 0.03$ (498 K)
Cu_3RbCl_4	$\sigma_h = 2 \times 10^{-16}$	24	$\sigma_i = 2.25 \times 10^{-3}$
Cu_4KI_5	$\sigma_h = 2 \times 10^{-7}$ (554 K)	25	$\sigma_i = 0.1$ (554 K)
$Cu_4RbI_{1.75}Cl_{3.25}$	$\sigma_e \sim 10^{-13}$ $\sigma_{el} \sim 10^{-10}$	26 27	$\sigma_i = 0.3$
$CuCl_2$	$\sigma_h = 4 \times 10^{-7}$ (655 K)	28	

Li_3N	$\sigma_e < 10^{-12}$	29	$\sigma_i = 1.2 \times 10^{-3}$
	$\sigma_{el} \sim 10^{-4}$	30	Technique SD
	$\sigma_e = 1.7 \times 10^{-4}$	30	$E_a = 0.32$ eV
Li_5NI_2	$\sigma_{el} \sim 10^{-8}$	31	
$LiAlSiO_4$	$\sigma_e = 5 \times 10^{-9}$ (550 K)	32	$\sigma_i = 2 \times 10^{-4}$ (550 K)
Li_3AlN_2	$\sigma_{el} \sim 10^{-11}$	33	$\sigma_i = 10^{-8}$
$Li_{3.6}Ge_{0.8}V_{0.4}O_4$	$\sigma_e = 2.38 \times 10^{-8}$	34	
	$\sigma_h = 6.6 \times 10^{-10}$	34	
$Li_{14}Zn(GeO_4)_4$	$\sigma_e = 3.17 \times 10^{-10}$	35	$E_a = 1.02$ eV
	$\sigma_h = 5.69 \times 10^{-15}$	35	$E_a = 1.39$ eV
$Li_{12}Zn_2(GeO_4)_4$	$\sigma_e = 3.63 \times 10^{-10}$	35	$E_a = 0.89$ eV
	$\sigma_h = 2.49 \times 10^{-14}$	35	$E_a = 1.49$ eV
Ag–β–Al_2O_3	$\sigma_e = 2 \times 10^{-7}$ (833 K)	36	
Na–β–Al_2O_3	$\sigma_e = 4 \times 10^{-8}$ (383 K)	37	$\sigma_e = 2.8 \times 10^5 \exp(-0.8/kT)$
β-PbF_2	$\sigma_e = 3 \times 10^{-6}$ (765 K)	38	(+)Ni, NiF_2|SIM|Au(−)
	$\sigma_h = 10^{-8}$ (765 K)	38	(+)Ni, NiF_2|SIM|Au(−)
	$\sigma_e = 6 \times 10^{-9}$ (765 K)	38	(+)Au|SIM|Cu, CuF_2(−)
	$\sigma_h = 7 \times 10^{-5}$ (765 K)	38	(+)Au|SIM|Cu, CuF_2(−)
	$\sigma_h = 1.7 \times 10^{-20}$ (400 K)	39	$\sigma_i = 2.5 \times 10^{-6}$
	$\sigma_e = 4.7 \times 10^{-9}$ (325 K)	40	
	$\sigma_e = 6.4 \times 10^{-8}$	41	
	$\sigma_e = 6.9 \times 10^{-10}$	42	$\sigma_i = 1.8 \times 10^{-8}$
	$\sigma_e = 3.5 \times 10^{-7}$ (373 K)	42	$\sigma_i = 10^{-5}$ (373 K)

TABLE A-*continued*

SIM	σ_{el}, Ω^{-1} cm^{-1}	Ref.	Remarks
1	2	3	4
PbF_2 + 0.3 m/ORbF	$\sigma_e = 6.2 \times 10^{-9}$ $\sigma_e = 10^{-8}$ (373 K)	42 42	$\sigma_i = 7.5 \times 10^{-6}$ $\sigma_i = 1.8 \times 10^{-5}$ (373 K)
$PbF_2 + YF_3$	$\sigma_e = 3.7 \times 10^{-8}$ $\sigma_e = 1.2 \times 10^{-6}$ (373 K)	42 42	$\sigma_i = 4.1 \times 10^{-7}$ $\sigma_i = 1.4 \times 10^{-5}$ (373 K)
$Pb_{0.85}U_{0.15} \times F_{2.3}$	$\sigma_e = 2.3 \times 10^{-6}$ (512 K) $\sigma_h = 8.1 \times 10^{-12}$ (512 K)	43 43	$\sigma_i = 0.029$ (512 K)
$BaSnF_4$	$\sigma_{el} = 3 \times 10^{-5}$ (500 K)	44	$\sigma_i = 0.01$ (500 K) $E_a = 0.34$ eV
$RbSn_2F_5$	$\sigma_{el} < 10^{-8}$	45	
CaF_2	$\sigma_h = 5.3 \times 10^{-8}$ (873 K) $\sigma_h = 1.1 \times 10^{-8}$ (873 K) $\sigma_e = 2.3 \times 10^{-6}$ (873 K)	46 46 46	Ni, NiF_2|SIM|C Co, CoF_2|SIM|C Y, YF_3|SIM|C
$LaF_3 + SrF_2$	$\sigma_{el} < 10^{-10}$	47	Ni, NiF_2|SIM|Ag
PbI_2	$\sigma_h = 1.4 \times 10^{-16}$ (435 K) $\sigma_h = 1.1 \times 10^{-8}$ (564 K) $\sigma_h = 5 \times 10^{-8}$ $\sigma_h = 9 \times 10^{-11}$	48 48 49 49	Pb|SIM|C Pb|SIM|C (+)I_2|SIM|C(−) (+)C|SIM|I_2(−)
$GaBr_3$	$\sigma_e = 5 \times 10^{-8}$ (413 K)	50	$\sigma_i = 1.5 \times 10^{-5}$ (400 K)

* If not otherwise indicated, the values of σ_{el} were determined by polarisation technique at room temperature.
** Self discharge of galvanic cell.

REFERENCES FOR TABLE A

1. Ilschner, B. (1958). *J. Chem. Phys.* **28,** 1109.
2. Mizusaki, J., Fueki, K., and Mukaibo, T. (1978). *Bull. Chem. Soc. Japan* **51,** 694.
3. Mazumdar, D., Govingachuryalu, P. A., and Bose, D. N. (1982). *J. Phys. Chem. Solids* **43,** 933.
4. Zhao, Z.-Y., Wang, C.-Y., and Chen, L.-Q. (1984). *Acta Physica Sinica* **33,** 1205.
5. Yushina, L. D., Karpachev, S. V., and Ovchinnikov, Yu. M. (1970). *Elektrokhimiya* **6,** 1391.
6. Takahashi, T., and Yamamoto, O. (1966). *Electrochim. Acta* **11,** 779.
7. Hoshino, H., Yanagiva, H., and Shimoji, M. (1974). *J. Chem. Soc. Faraday Trans.* (part I) **70,** 281.
8. Shirokov, Yu. V., Pushkov, B. I., Borovkov, V. S., and Lukovtsev, P. D. (1972). *Elektrokhimiya* **8,** 579.
9. Khachaturyan, N. A., Tyurin, V. S., and Borovkov, V. S. (1975). *Elektrokhimiya* **11,** 666.
10. Kleimenov, A. I., Prokopets, V. E., Plotnikova, O. A., Stroganova, I. Ya., and Tsar'kov, M. S. (1973). *Elektrotekn. Prom. Ser. Khim. Fiz. Istochniki Toka* **N4,** 10.
11. Takahashi, T., Yamamoto, O., and Ikeda, C. (1969). *Denki Kagaku* **37,** 843.
12. Argue, G. R., Groce, I. J., and Owens, B. B. (1970). *In* "Power Sources 2" (D. H. Collins, ed.), p. 241. Pergamon Press, Oxford.
13. Scrosati, B. (1971). *J. Appl. Chem. and Biotechnol.* **21,** 223.
14. Kukoz, F. I., Kolomoets, A. M., and Shvetsov, V. S. (1977). *Elektrokhimiya* **13,** 92.
15. Ivanov-Shits, A. K. (1979). *Elektrokhimiya* **15,** 688.
16. Owens, B. B. (1970). *J. Electrochem. Soc.* **117,** 1536.
17. Scrosati, B., Papaleo, F., Pistoia, G., and Lazzari, M. (1975). *J. Electrochem. Soc.* **122,** 339.
18. Takahashi, T., Ikeda, S., and Yamamoto, O. (1972). *J. Electrochem. Soc.* **119,** 477.
19. Takahashi, T., Ikeda, S., and Yamamoto, O. (1973). *J. Electrochem. Soc.* **120,** 647.
20. Takahashi, T., Yamamoto, O., and Ikeda, S. (1973). *J. Electrochem. Soc.* **120,** 1431.
21. Lazzari, M., Pace, R. C., and Scrosati, B. (1975). *Electrochim. Acta* **20,** 331.
22. Takahashi, T., Wakabayashi, N., and Yamamoto, O. (1976). *J. Electrochem. Soc.* **123,** 129.
23. Takahashi, T., Yamamoto, O., and Takahashi, H. (1973). *J. Solid State Chem.* **21,** 37.
24. Matsui, T., and Wagner, J. B., Jr. (1977). *J. Electrochem. Soc.* **124,** 941.
25. Bonino, F., and Lazzari, M. (1976/1977). *J. Power Sources* **1,** 103.
26. Takahashi, T., Yamamoto, O., Yamada, S., and Hayashi, S. (1979). *J. Electrochem. Soc.* **126,** 1654.
27. Chaney, C., Shriver, D. F., and Whitmore, D. H. (1981). *Solid State Ionics* **5,** 505.
28. Joshi, A. V. (1973). *In* "Fast Ion Transport in Solids" (W. van. Gool, ed.), p. 173. North-Holland, Amsterdam.
29. Alpen von, U., Rabenau, A., and Talat, G. H. (1977). *Appl. Phys. Lett.* **30,** 621.
30. Bittihn, R. (1983). *Solid State Ionics* **8,** 83.
31. Obayashi, H., Nagai, R., Gotoh, A., Mochizuki, S., and Kudo, T. (1981). *Mat. Res. Bull.* **16,** 587.
32. Alpen von, U., Schöherr, E., Schulz, H., and Talat, G. H. (1977). *Electrochim. Acta* **22,** 805.
33. Yamane, H., Kikkawa, S., and Koizumi, M. (1985). *Solid State Ionics* **15,** 51.
34. Bose, M., Basu, A., Mazumdar, D., and Bose, D. N. (1985). *Solid State Ionics* **15,** 101.
35. Mazumdar, D., Bose, D. N., and Mukherjee, M. L. (1984). *Solid State Ionics* **14,** 143.
36. Whittingham, M. S., and Huggins, R. A. (1971). *J. Electrochem. Soc.* **118,** 1.

37. Evtushenko, V. V., Bukun, N. G., and Ukshe, E. A. (1975). *Elektrohimiya* **11**, 1007.
38. Fang, W. C., and Rapp, R. A. (1977). *J. Electrochem. Soc.* **124**, 315 C.
39. Joshi, A. V., and Liang, C. C. (1975). *J. Phys. Chem. Solids* **36**, 927.
40. Schoonman, J., Korteweg, G. A., and Bonne, R. W. (1975). *Solid State Commun.* **16**, 9.
41. Kennedy, J. H., and Miles, R. C. (1976). *J. Electrochem. Soc.* **123**, 47.
42. Murin, I. V., Glumov, A. V., and Glumov, O. V. (1979). *Elektrokhimiya* **15**, 1119.
43. Murin, I. V., Glumov, O. V., and Kozhina, I. I. (1980). *Vestnik LGU* **N 22**, 87.
44. Dĕnés, G., Birchall, T., Sayer, M., and Bell, M. F. (1984). *Solid State Ionics* **13**, 213.
45. Murin, I. V., and Chernov, S. V. (1982). *Vestnik LGU* **N 10**, 105.
46. Reddy, S. N. S., and Rapp, R. A. (1977). *J. Electrochem. Soc.* **124**, 314 C.
47. Murin, I. V., Glumov, O. V., and Amelin, Yu. V. (1980). *Zh. Prikladn. Khimii.* **53**, 1474.
48. Lingras, A. P., and Simkovich, G. (1978). *J. Phys. Chem. Solids* **39**, 1225.
49. Schoonman, J., Wolfert, A., and Untereker, D. F. (1983). *Solid State Ionics* **11**, 187.
50. Hönle, W., Gerlach, G., Weppner, W., and Simon, A. (1986). *J. Solid State Chem.* **61**, 171.

Index

E

H

I

L

Q

R

S

Contents of Previous Volumes

Volume 4 **Physics of III–V Compounds**

Volume 5 **Infrared Detectors**

Volume 6 **Injection Phenomena**

Volume 7 **Application and Devices Part A**

Part B

T. Misawa, IMPATT Diodes
H. C. Okean, Tunnel Diodes
Robert B. Campbell and Hung-Chi Chang, Silicon Carbide Junction Devices
R. E. Enstrom, H. Kressel, and L. Krassner, High-Temperature Power Rectifiers of $GaAs_{1-x}P_x$

Volume 8 **Transport and Optical Phenomena**

Richard J. Stirn, Band Structure and Galvanomagnetic Effects in III–V Compounds with Indirect Band Gaps
Roland W. Ure, Jr., Thermoelectric Effects in III–V Compounds
Herbert Piller, Faraday Rotation
H. Barry Bebb and E. W. Williams, Photoluminescence I: Theory
E. W. Williams and H. Barry Bebb, Photoluminescence II: Gallium Arsenide

Volume 9 **Modulation Techniques**

B. O. Seraphin, Electroreflectance
R. L. Aggarwal, Modulated Interband Magnetooptics
Daniel F. Blossey and Paul Handler, Electroabsorption
Bruno Batz, Thermal and Wavelength Modulation Spectroscopy
Ivar Balslev, Piezooptical Effects
D. E. Aspnes and N. Bottka, Electric-Field Effects on the Dielectric Function of Semiconductors and Insulators

Volume 10 **Transport Phenomena**

R. L. Rode, Low-Field Electron Transport
J. D. Wiley, Mobility of Holes in III–V Compounds
C. M. Wolfe and G. E. Stillman, Apparent Mobility Enhancement in Inhomogeneous Crystals
Robert L. Peterson, The Magnetophonon Effect

Volume 11 **Solar Cells**

Harold J. Hovel, Introduction; Carrier Collection, Spectral Response, and Photocurrent; Solar Cell Electrical Characteristics; Efficiency; Thickness; Other Solar Cell Devices; Radiation Effects; Temperature and Intensity; Solar Cell Technology

Volume 12 **Infrared Detectors (II)**

W. L. Eiseman, J. D. Merriam, and R. F. Potter, Operational Characteristics of Infrared Photodetectors
Peter R. Bratt, Impurity Germanium and Silicon Infrared Detectors
E. H. Putley, InSb Submillimeter Photoconductive Detectors
G. E. Stillman, C. M. Wolfe, and J. O. Dimmock, Far-Infrared Photoconductivity in High Purity GaAs
G. E. Stillman and C. M. Wolfe, Avalanche Photodiodes
P. L. Richards, The Josephson Junction as a Detector of Microwave and Far-Infrared Radiation
E. H. Putley, The Pyroelectric Detector—An Update

Volume 13 **Cadmium Telluride**

Volume 14 **Lasers, Junctions, Transport**

Volume 15 **Contacts, Junctions, Emitters**

Volume 16 **Defects, (HgCd)Se, (HgCd)Te**

Volume 17 **CW Processing of Silicon and Other Semiconductors**

Volume 18 **Mercury Cadmium Telluride**

Volume 19 **Deep Levels, GaAs, Alloys, Photochemistry**

Volume 20 **Semi-Insulating GaAs**

Volume 21 **Hydrogenated Amorphous Silicon Part A**

Part B

Part C

Part D

Volume 22 **Lightwave Communications Technology**
Part A

Part B

Part C

Part D

Part E

Volume 23 **Pulsed Laser Processing of Semiconductors**

Volume 24 **Applications of Multiquantum Wells, Selective Doping, and Superlattices**

C. Weisbuch, Fundamental Properties of III–V Semiconductor Two-Dimensional Quantized Structures: The Basis for Optical and Electronic Device Applications
H. Morkoç and H. Unlu, Factors Affecting the Performance of (Al, Ga)As/GaAs and (Al, Ga)As/InGaAs Modulation-Doped Field-Effect Transistors: Microwave and Digital Applications
N. T. Linh, Two-Dimensional Electron Gas FETs: Microwave Applications
M. Abe et al., Ultra-High-Speed HEMT Integrated Circuits
D. S. Chemla, D. A. B. Miller, and P. W. Smith, Nonlinear Optical Properties of Multiple Quantum Well Structures for Optical Signal Processing
F. Capasso, Graded-Gap and Superlattice Devices by Band-gap Engineering
W. T. Tsang, Quantum Confinement Heterostructure Semiconductor Lasers
G. C. Osbourn et al., Principles and Applications of Semiconductor Strained-Layer Superlattices

Volume 25 **Diluted Magnetic Semiconductors**

W. Giriat and J. K. Furdyna, Crystal Structure, Composition, and Materials Preparation of Diluted Magnetic Semiconductors
W. M. Becker, Band Structure and Optical Properties of Wide-Gap $A^{II}_{1-x}Mn_xB^{VI}$ Alloys at Zero Magnetic Field
Saul Oseroff and Pieter H. Keesom, Magnetic Properties: Macroscopic Studies
T. Giebultowicz and T. M. Holden, Neutron Scattering Studies of the Magnetic Structure and Dynamics of Diluted Magnetic Semiconductors
J. Kossut, Band Structure and Quantum Transport Phenomena in Narrow-Gap Diluted Magnetic Semiconductors
C. Riqaux, Magnetooptics in Narrow Gap Diluted Magnetic Semiconductors
J. A. Gaj, Magnetooptical Properties of Large-Gap Diluted Magnetic Semiconductors
J. Mycielski, Shallow Acceptors in Diluted Magnetic Semiconductors: Splitting, Boil-off, Giant Negative Magnetoresistance
A. K. Ramas and S. Rodriquez, Raman Scattering in Diluted Magnetic Semiconductors
P. A. Wolff, Theory of Bound Magnetic Polarons in Semimagnetic Semiconductors